Excel

Get the Results You Want!

YEAR 11 PHYSICS

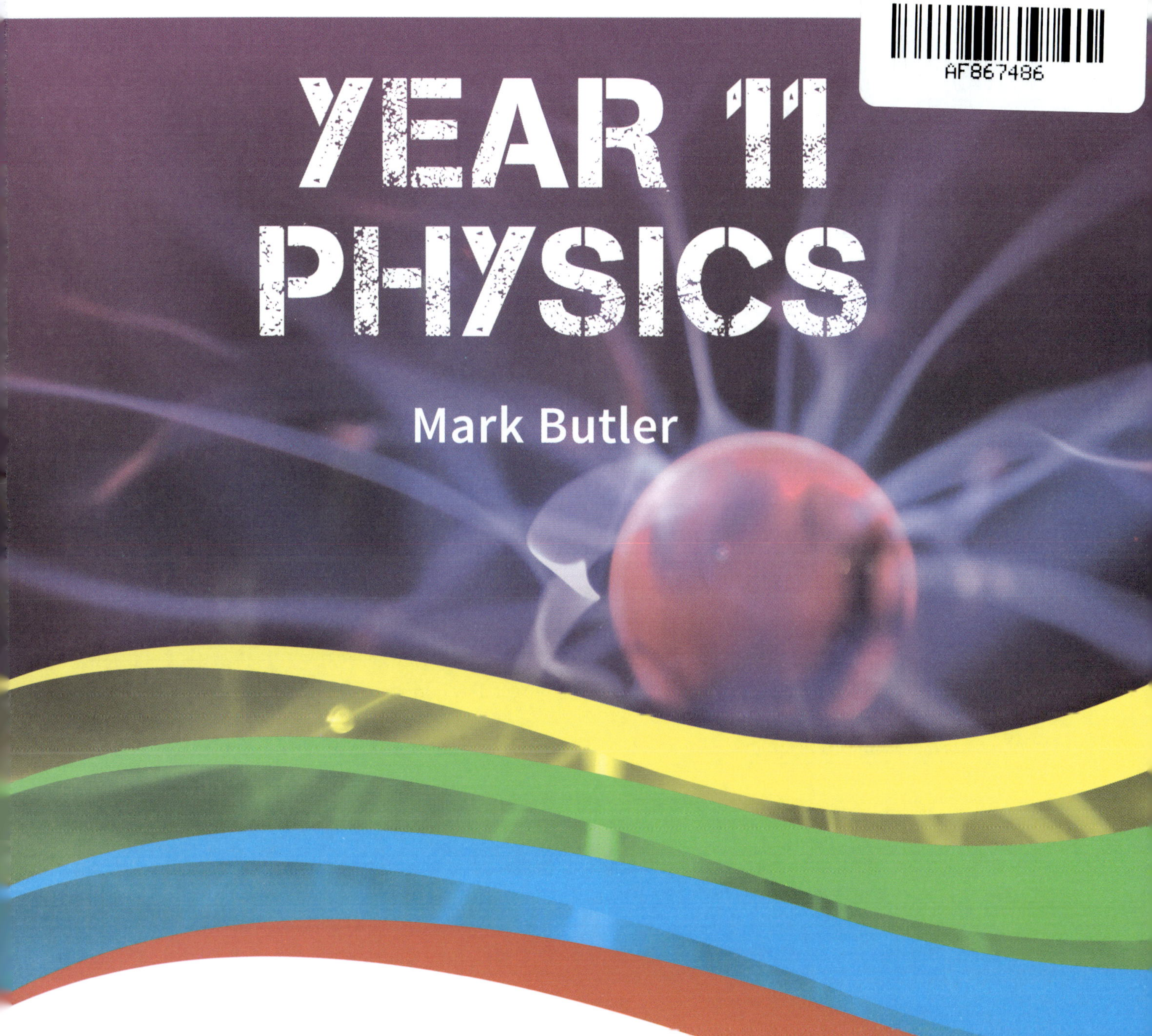

Mark Butler

PASCAL PRESS

CONTENTS

CHAPTER 1 MOTION IN A STRAIGHT LINE

MODULE 1
KINEMATICS

INQUIRY QUESTION:

How is the motion of an object moving in a straight line described and predicted?

Galileo and Newton showed that linear (straight line) motion could be described and predicted using graphs and by using algebraic relationships.

1 Introduction to measurement in physics

Before we can understand motion and other topics in physics we must be familiar with the science of measurement. A summary of the fundamental principles of scientific measurement follows.

Significant figures

- When writing quantities, the *number of significant figures* is used to specify the accuracy to which the quantity is known.
- Unless explicitly stated the **uncertainty** in a given quantity is assumed to be half of the smallest factor of 10 specified in the answer. Hence 10.23 m implies that the quantity lies in the range 10.23 +/− 0.005 m, and 45.2 m implies an answer in the range 45.2 +/− 0.05 m.

uncertainty: the range of values within which the true value is believed to lie

- The following rules are used to determine the number of significant figures in a quantity:
  - All non-zero numbers are significant.
  - For numbers less than 1, zeros between the decimal place and the first number are not significant; for example, 0.0056 has 2 significant figures.
  - Zeros that are placed between numbers are significant; for example, 40.002 m has 5 significant figures.
  - Zeros at the end of numbers are only significant if they are indicated to be so by scientific notation or if they are after a number behind the decimal place.
  - For example, 7800 m has 2 significant figures, 7.800×10^3 has 4 significant figures, and 6.40 has 3 significant figures.
- When solving problems your answer should generally not have more significant figures than the least-accurate quantity used in the calculation. That is, your answer should have the same number of significant figures as the quantity with the fewest significant figures used in your calculation.

EXAMPLE 1

You are given the following data: x = 04.500 m and y = 16 200 m

a Find the number of significant figures in x and y.

b Find y/x and xy.

Recall the rule for the significance of zeros on each side of the decimal place

Answer:

a Zeros to the left of numbers are not significant, but numbers to right after the decimal point are significant and hence x has 4 significant figures. Zeros at the end of numbers are not significant unless they are after the decimal place or scientific notation is used, hence y has 3 significant figures.

b As the least number of significant figures in the data values was for y, which has 3, the answer should also have 3 significant figures.

Thus $y/x = \dfrac{16200 \text{ m}}{4.500 \text{ m}} = 3.60 \times 10^3$,

as we need to specify 3 significant figures (note the units cancel out).

$xy = 16\,200 \text{ m} \times 4.500 \text{ m} = 7.29 \times 10^5 \text{ m}^2$, again written to specify 3 significant figures of accuracy.

Scientific notation

- Very large and very small numbers may be expressed using scientific notation.
- When using scientific notation, place the decimal place after the first number in the answer.
- Recall from mathematics that $10^{-x} = 1/10^x$, and this allows us to express small numbers using scientific notation.
- Note also that $10^x 10^y = 10^{x+y}$ and that $10^x/10^y = 10^{x-y}$.

EXAMPLE 2

Express the following numbers in scientific notation with 4 significant figures.

a **$6.674\,08 \times 10^{-11}$**

b **0.16 027**

c **299 792 458**

Remember to place the decimal place after the first number when changing to scientific notation

Answer:

a 6.674×10^{-11}

b 1.603×10^{-1}

Note that this must be rounded up to 4 significant figures.

c 2.998×10^{8}

Note that it was also necessary to round up this quantity.

Order of magnitude

- Sometimes approximate values in science are given in terms of *order of magnitude*, where the order of magnitude refers to the closest multiple of ten. For example, 2200 has an order of magnitude of 10^3, while 79 has an order of magnitude of 10^2.
- Because $10^{\frac{1}{2}} = 3.16$, numbers below 3.16×10^b have an order of magnitude of 10^b, while numbers above 3.16×10^b have an order of magnitude of 10^{b+1}.

EXAMPLE 3

There are 6.02×10^{23} atoms in 1 gram of hydrogen. What order of magnitude is this?

Remember that the order of magnitude refers to the closest multiple of ten

Answer:

As 6.02×10^{23} is closer to 10^{24} than 10^{23}, the order of magnitude for the number of atoms in 1 gram of hydrogen is 24.

Système Internationale (SI) units

- Physics uses **SI units**. Motion uses the **fundamental SI units** of the metre (m), kilogram (kg) and second (s), as well as the **derived SI units** made up from these fundamental units. For example, acceleration has the derived SI units of ms^{-2}.

SI unit: a system of units (*Système Internationale*) used by scientists based on fundamental standards for length, mass and time, named the metre, kilogram and second, respectively

fundamental SI unit: one of a number of unrelated units that form the basis of a system of units. Fundamental units cannot be broken into simpler units but can be combined together to form more complex, derived units

derived SI unit: a measure of a physical property that can be made up of a combination of fundamental units

- The SI system uses the following prefixes:
  - giga = G = $\times 10^{9}$
  - mega = M = $\times 10^{6}$
  - kilo = k = $\times 10^{3}$
  - milli = m = $\times 10^{-3}$
  - micro = μ = $\times 10^{-6}$
  - nano = n = $\times 10^{-9}$

EXAMPLE 4

Convert the following quantities to fundamental SI units.

a **3.45 Mm** b **27.8 ms**

c **467 ns** d **25 kmh^{-1}**

You should commit the most common SI prefixes to memory

Answer:

Replacing the prefix with the appropriate order of 10 gives:

a 3.45×10^{6} m

b 27.8×10^{-3} s = 2.78×10^{-2} seconds

c 467×10^{-9} s = 4.67×10^{-7} seconds

d There are 60×60 seconds in 1 hour and hence:

$$25\ kmh^{-1} = \frac{25 \times 10^{3}}{60 \times 60}\ ms^{-1} = 6.9\ ms^{-1}$$

KEY QUESTIONS

1 **How is the uncertainty in a measurement expressed?**

2 **Where should the decimal point be placed when scientific notation is used?**

Answers p. 17

Dimensions and units

- Physics quantities can be expressed in terms of fundamental dimension such as mass (m), length (l) or time (t).
- The SI system uses agreed standards to define the kilogram, metre and second as the fundamental units of mass, length and time.
- Physics constants are all written in terms of SI units. To ensure physics equations give the correct answers, SI units should always be used.
- If you are given data to use in a problem that uses SI prefixes or other units, you must convert these to the fundamental SI units before substituting them into equations.
- Answers to problems should be given in SI units.
- SI prefixes and scientific notation may be used to simplify answers.

EXAMPLE 5

A car travels with a constant speed of 120 kmh^{-1} for 25 minutes. Find the distance travelled by the car in metres. Note that distance = speed × time.

Answer:

Convert all given data to SI units first

Convert to SI units:

$$120\ kmh^{-1} = 120 \times \frac{1000}{60 \times 60}$$
$$= 33.33\ ms^{-1}$$

25 minutes = 25 × 60
= 1500 seconds

Distance = speed × time = 33.33 × 1500
= 50 000 m
= 5.00×10^4 m

Unit analysis

- When we write equations in physics we are saying one side of the equation is exactly equal to the other side. This means that not only is the *magnitude* of both sides the same but the *units* of both sides and the *direction* of the quantities (if they are vectors) must be the same.
- If the units of one side of a given equation are not the same as the units on the other side, the equation must be incorrect. If the units on both sides are the same, the equation may or may not be correct. Thus having the same units on each side is a necessary condition of a correct equation, but it does not prove an equation is correct.
- We can use **unit analysis** to check equations or to find the units of an unknown quantity or constant in an equation. For example, $a = v - u/t$ and, as the right-hand side has the units of velocity/time = $ms^{-1}/s = ms^{-2}$, the left-hand side (i.e. the acceleration) must also have the units ms^{-2}.

unit analysis: equating the units on both sides of an equation to find the units of one of the quantities or a constant in the equation

EXAMPLE 6

Newton showed that two objects attract one another with a force of gravity, which is related to the masses (m_1 and m_2) of the objects and their distance apart (r) by the equation:

$$F = Gm_1m_2/r^2$$

where F is the force, measured in $kgms^{-2}$
G is a constant, called the *universal constant of gravitation*
m is the mass and
r is the distance apart

Figure 1.1 shows the variables that determine the force of gravity between objects.

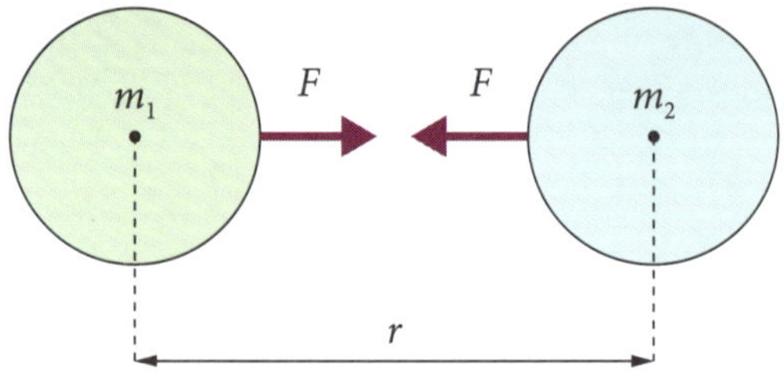

Figure 1.1 Force of gravity between two objects

Use unit analysis to find the SI units of the universal constant of gravitation.

Answer:

Recall that the units of the quantity on one side of the equation must be the same as the units of the quantity on the other side of the equation

First rearrange the standard equation to make G the subject:

$$G = Fr^2/m_1m_2$$

The units on the right-hand side must be the same as the units on the left-hand side. As the units of mass are kilograms and the units of distance are metres, substituting in the units gives:

Units of $G = (kg\ ms^{-2})(m^2)/kg \times kg$
$= m^3\ kg^{-1}\ s^{-2}$

Check this equation by substituting the units of G into Newton's equation to see if it produces the units of force:

Units of $Gm_1m_2/r^2 = (m^3\ kg^{-1}\ s^{-2})\ kg \times kg/m^2$
$= kg\ ms^{-2}$, as expected.

Vectors and scalar quantities

- All measured quantities are either **vectors** or **scalars**. Vectors have a magnitude (size) and direction, while scalar quantities have magnitude but no direction in space. For example, force is a vector because it operates in a specific direction but energy is a scalar as it is not a directed quantity and only has magnitude.

vector: a measurable quantity that has magnitude (size) and a direction in space
scalar: a measurable quantity that has only magnitude

- Vector quantities in equations are sometimes differentiated from scalar quantities by placing a small arrow above or below the vectors or by using bold type for the vector quantities.
- We can represent vectors diagrammatically as arrows in which the direction operates. The length of the arrow represents the magnitude of the vector. For example,

$4 ms^{-1}$ west could be represented by the arrow vector shown in Figure 1.2.

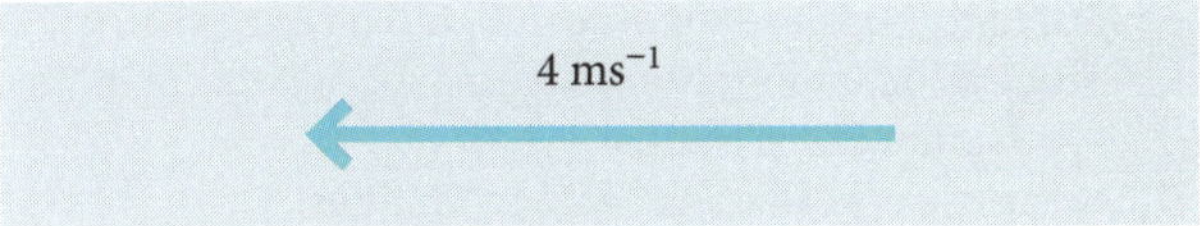

Figure 1.2 Vector representing a velocity of 4 ms^{-1} west

Adding and subtracting vector quantities

➔ To add two vectors diagrammatically, we place the arrows that represent the vectors 'head to tail'. The vector drawn from the 'free tail' to the 'free head' is the *resultant* (sum) of the vectors. For example, Figure 1.3 shows the vector diagram to add 40 m west to 20 m west.

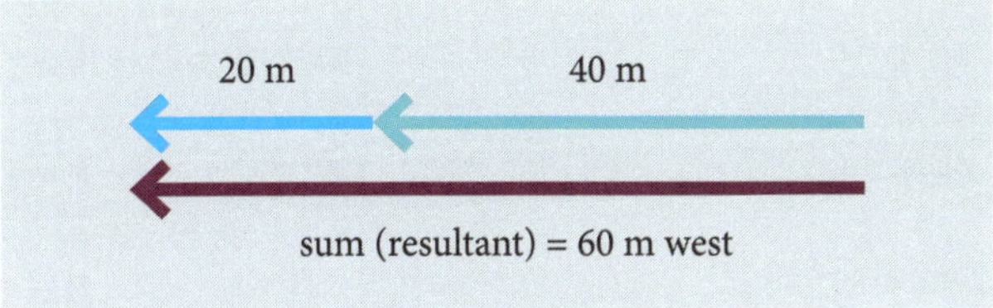

Figure 1.3 Vector addition diagram

➔ A *negative vector* is a vector in the opposite direction. For example:

−40 m north = 40 south

➔ Thus, to subtract one vector from another we reverse the direction of the negative vector and then add them. For example:

40 m west − 20 m east = 40 m west + 20 m west = 60 west

➔ In one dimension we can also add and subtract vectors algebraically. To do this we simply call vectors in one direction *positive*, and vectors in the other direction *negative*.

EXAMPLE 7

Determine the resultant of:
25 m west − 20 m east − 10 m west + 40 m east

Find the resultant vector:

a using vector diagrams and

b by calling one direction positive.

A negative vector can be replaced by a positive vector in the opposite direction

Answer:

a To solve with a vector diagram, change the direction of negative vectors and then add all the vectors by placing them head to tail. The resultant is the vector that goes from the free tail to the free head:

25 m west − 20 m east − 10 m west + 40 m east
= 25 m west + 20 m west +10 m east + 40 m east

Adding these vectors using a vector diagram in Figure 1.4 shows that the resultant vector is 5 m east.

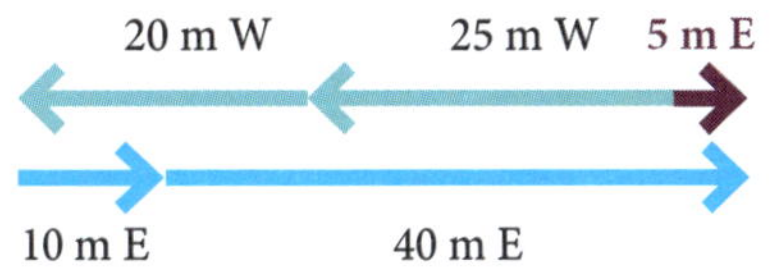

Figure 1.4 Adding vectors

b To solve algebraically, call west positive and east negative. The vector equation then becomes:

25 m west − 20 m east − 10 m west + 40 m east
= 25 − (−20) − 10 − 40 = −5 m = 5 m east (as we said east is negative)

➔ KEY QUESTIONS

3 **Why are units important in physics?**

4 **What is the difference between a dimension and a unit?**

5 **What is the difference between a fundamental unit and a derived unit?**

Answers ➲ p. 17

Measuring change and rate of change

➔ The change (Δ) in a quantity is given by subtracting the initial amount from the final amount.

Change = Δ = final amount − initial amount

➔ The rate of change (Δ/t) of a quantity is a measure of the amount of change per second. We can determine the rate of change by dividing the change by the time over which the change occurred.

Rate of change = change/time

change: the difference between the initial amount from the final amount

rate of change: the change in a quantity per unit time; in SI units this would be the change per second

EXAMPLE 8

A new strand of fast-growing bean plant grows from a height of 4.0 cm to 8.5 cm in 24 hours.

a Determine the change of height of the plant in metres.

b Calculate the rate of change of height in ms^{-1} during the 24 hours.

Answer:

a Change = final amount – initial amount
$= 8.5 - 4.0 = 4.5$ cm
$= 0.045$ m

b Rate of change = change/time

$$= \frac{0.045}{24 \times 60 \times 60}$$
$$= 5.2 \times 10^{-7}\ \text{ms}^{-1}$$

Change is final amount minus initial amount, but the rate of change is the change divided by time

Rate of change of motion quantities

➔ Velocity (v) is the rate of change of displacement, which can be expressed as:

$$v = \Delta s/t$$

where Δs is the change in velocity and
t is the elapsed time

This is why velocity has the units ms^{-1}.

➔ Acceleration (a) is the rate of change of velocity, which can be expressed as:

$$a = \Delta v/t = (v - u)/t$$

where u is the initial velocity and
v is the final velocity

This is why acceleration has the units $\text{ms}^{-1}/\text{s} = \text{ms}^{-2}$.

EXAMPLE 9

A racing car travelling at 80 ms^{-1} applies the brakes and comes to rest in 5 seconds.

a Calculate the change in velocity.

b Calculate the average rate of change of velocity.

If the car comes to rest, the final velocity will be zero

Answer:

Call the direction the car is travelling in positive.

The initial speed was 80 ms^{-1} and the final speed was 0 ms^{-1}.

a Change = final amount – initial amount
$= 0 - 80$
$= -80\ \text{ms}^{-1}$

The negative sign indicates the change is in the opposite direction of the motion.

b Rate of change = change/time

$$= \frac{-80}{5}$$
$$= -16\ \text{ms}^{-2}$$

Hence the car accelerates in the opposite direction of the motion at 16 ms^{-2} for 5 seconds.

2 Motion quantities

» Students describe uniform straight-line (rectilinear) motion and uniformly accelerated motion through:
- qualitative descriptions
- the use of scalar and vector quantities.

➔ Table 1.1 lists and defines terms (variables) that can be used to qualitatively describe different types of motion.

➔ We will see later (p. 9) that by relating these variables algebraically to one another we can derive expressions that enable motion to be described quantitatively.

Table 1.1 Variables used to describe motion

Quantity	Symbol	SI units	Description	Vector/ scalar
Distance		m	Total length of journey	scalar
Speed		ms^{-1}	Rate of change of distance	scalar
Displacement	s	m	Directed distance	vector
Initial velocity	u	ms^{-1}	Rate of change of displacement at time = 0	vector
Final velocity	v	ms^{-1}	Rate of change of displacement at time = t	vector
Acceleration	a	ms^{-2}	Rate of change of velocity	vector
Time	t	s	Elapsed time	scalar

➔ The average velocity is given by:

$$v_{average} = \text{total displacement/total time}$$

➔ If the acceleration is constant, the average velocity is also given by:

$$v_{average} = (v + u)/2$$

➔ The instantaneous velocity is the velocity over a very (vanishingly) small time interval at the time of interest, which can be represented mathematically as:

$$v_{instantaneous} = \Delta s/\Delta t, \text{ as } \Delta t \rightarrow 0$$

➔ The acceleration refers to how much the velocity changes each second. As velocity is a vector, acceleration may refer to a change in the magnitude or a change in the direction of the velocity:

$$a = \Delta v/\Delta t = (v - u)/t$$

EXAMPLE 10

A car travels a total distance of 0.625 km in a straight line by travelling at constant velocity of 15 ms^{-1} for 30 seconds and then accelerating at 0.5 ms^{-2} for the next 10 seconds. Find:

a **the instantaneous velocity after 34 seconds**

b **the average velocity of the car for the entire journey**

c **the average velocity of the car during the time it was accelerating.**

Remember that the journey has two parts: one at constant velocity and the other with constant acceleration

Answer:

a The car's velocity increases by 0.50 ms^{-2} each second after 30 seconds. Thus at 34 seconds its velocity would have increased from 15 ms^{-1} to 17 ms^{-1}.

b $v_{average}$ = total displacement/total time

$$= \frac{625 \text{ m}}{40 \text{ s}}$$

$$= 15.6 \text{ ms}^{-1}$$

c After 10 seconds accelerating at 0.5 ms^{-2} the velocity would have increased to 15 + (10 × 0.5) = 20 ms^{-1} and, as the acceleration was constant during the last 10 seconds, we can use:

$$v_{average} = (v + u)/2$$

$$= \frac{15+20}{2}$$

$$= 17.5 \text{ ms}^{-1}$$

3 Relative motion in a straight line

» Students calculate the relative velocity of two objects moving along the same line using vector analysis.

➔ We can calculate the velocity of one object relative to another by subtracting the velocity of the observer from that of the object:

$$v_{relative} = v_{object} - v_{observer}$$

➔ Note that this equation involves the subtraction of one vector from another vector so the direction of the vectors needs to be taken into account when solving the equation. This can be done using vector diagrams or, in the case of one dimensional motion, by specifying the direction of each velocity as either positive or negative.

➔ To take the direction of vectors into account in one dimension, call one direction positive and the other direction negative. Then specify the direction of each vector quantity by making it positive or negative. To find the relative velocity simply substitute the velocities, with the correct signs, into the above equation. The sign of the relative velocity that results will determine its direction.

EXAMPLE 11

A red car is moving east at 40 ms^{-1} and a blue car is travelling west at 30 ms^{-1}. Find the velocity of the blue car with respect to the red car.

Answer:

If we call east positive and west negative:

$v_{observer}$ = 40 ms^{-1}

v_{object} = −30 ms^{-1}

$v_{relative} = v_{object} - v_{observer}$

= −30 ms^{-1} – 40 ms^{-1}

= −70 ms^{-1}

Call one direction positive and the other negative

As the west direction is called negative, the observer sees the object moving with a velocity of 70 ms^{-1} west. That is, from the observer's point of view (from the observer's frame of reference), the object is travelling at 70 ms^{-1} in a westerly direction.

➔ KEY QUESTIONS

6 **What is difference between the change in a measured quantity and the rate of change of the quantity?**

7 **Why is it important to know which measured quantities are vectors and which are scalars?**

Answers ➲ p. 17

4 The equations of motion

» Students use mathematical modelling and graphs, selected from a range of technologies, to analyse and derive relationships between time, distance, displacement, speed, velocity and acceleration in rectilinear motion, including:

- $s = ut + \frac{1}{2}at^2$
- $v = u + at$
- $v^2 = u^2 + 2as$.

Graphing motion

➔ Key points to remember when graphing motion are:
- The independent variable (the one you change) is generally placed on the horizontal axis, and the dependent variable (the one you measure) is placed on the vertical axis.
- The axes should be labelled and include units in brackets.
- Data points should be visible and a smooth 'line of best fit' should always be drawn through the data points.

- The instantaneous velocity is the gradient of the tangent to the curve of the displacement against time graph at the time of interest.
- The instantaneous acceleration is the gradient of the tangent to the curve of the velocity against time graph at the time of interest.
- The displacement after a time t is the area between the line of best fit and the time axis of the velocity against time graph between $t = 0$ and the time t.
- It is good practice to draw graphs in pencil. Choose a scale that enables all the data to be plotted and uses most of the grid paper.
- Recall from mathematics that the **gradient of a graph** is the rise divided by the run (i.e. $m = \Delta y/\Delta x = (y_2 - y_1)/(x_2 - x_1)$).

gradient of a graph: the rise of the graph over the run or $\Delta y/\Delta x$

EXAMPLE 12

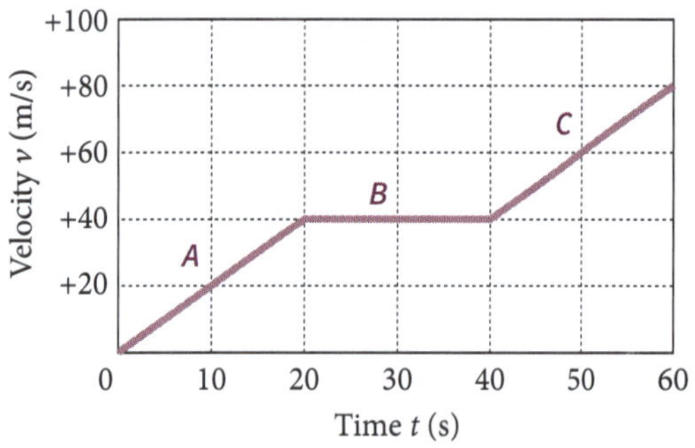

Figure 1.5 Velocity versus time graph

For the motion shown in Figure 1.5, find:

a **the acceleration during the times marked A, B and C**

b **the distance travelled after 40 seconds.**

Recall that the gradient of the v versus t graph is the *acceleration*, and the area underneath gives the *displacement*

Answer:

a The gradient of the graph gives the acceleration:

For section A, $a = \dfrac{40}{20} = 2\ \text{ms}^{-2}$

For section B, $a = \dfrac{0}{40} = 0\ \text{ms}^{-2}$

For section C, $a = \dfrac{40}{20} = 2\ \text{ms}^{-2}$

b Displacement is equal to the area under the graph:

$$s = 40 \times 20 \times \frac{1}{2} + 40 \times 20 = 1200\ \text{m}$$

Deriving the equation of motion

→ Equations of motion that relate motion quantities for objects moving in a straight line with constant velocity can be derived as follows:

- Acceleration is the rate of change of velocity:

  $a = (v - u)/t$

  Thus:

  $v = u + at$

- The average velocity is defined as:

  $v_{av} = s/t$

  For an object accelerating at a constant rate, the average velocity is also given by:

  $v_{av} = (v + u)/2$

- Equating these expressions for the average velocity and substituting $v = at + u$ to remove velocity yields:

  $s = ut + \frac{1}{2}at^2$

→ Rearranging $v = u + at$ as $t = (v - u)/a$ and substituting this into $s = ut + \frac{1}{2}at^2$ to remove time yields:

$v^2 = u^2 + 2as$

→ Motion with constant acceleration in a straight line can be described algebraically using these equations of motion:

$$s = ut + \frac{1}{2}at^2$$
$$v = u + at$$
$$v^2 = u^2 + 2as$$

Note that the equations of motion involve vector quantities (v, u, s and a) and, for straight line motion, the direction can be taken into account by calling all the vectors in one direction positive, and all those in the opposite direction negative.

EXAMPLE 13

A car travelling at 20 ms^{-1} accelerates at 2.0 ms^{-2} for 3.0 seconds and then moves at constant velocity.

a **Determine the final velocity of the car.**

b **How far did the car travel during the 3.0 seconds it was accelerating?**

Answer:

Extract the raw data from the question first and then choose the equation of motion that will allow you to find the answer from the given data

a $v = u + at$
$= 2 \times 3 + 20$
$= 26\ \text{ms}^{-1}$

b $s = ut + \frac{1}{2}at^2$
$= 20 \times 3 + \frac{1}{2}(2 \times 3)^2$
$= 69\ \text{m}$

Using the equations of motion to solve problems

➔ To solve problems using these equations:

- Read the question carefully and determine what motion quantities are given and what you need to find.
- Change all given values into SI units and replace any prefixes with the appropriate multiple of 10.
- Call all the vectors in one direction positive and all the vectors in the opposite direction negative.
- Check to see if the initial or final velocity in the situation is zero.
- Look at what quantities are given and what needs to be calculated. Then choose the equation of motion that has all these quantities in it.
- Substitute the values into the equation and solve it for the unknown quantity.
- Write your answer with units and a direction (if it is a vector quantity).

EXAMPLE 14

A jet travelling at 280 ms^{-1} encounters some turbulence and reduces its velocity to 245 ms^{-1}. If the jet travelled a distance of 2 km while it was decelerating, find:

a the change in velocity of the jet

b the magnitude and direction of the jet's acceleration

c the time over which the jet changed its velocity.

Change to SI units and recall that change is final minus initial

Answer:

Extract given motion quantities and convert to SI units.

Raw data: $u = 280\text{ms}^{-1}$, $v = 245\text{ ms}^{-1}$, $s = 2000\text{ m}$

Call the direction of the initial velocity positive.

a $\Delta v = v - u$

$= 245 - 280$

$= 35\text{ ms}^{-1}$ or 35 ms^{-1}

in the direction opposite to the initial velocity.

b As we do not have time we need to use:

$v^2 = u^2 + 2as$

Making the acceleration the subject of the equation gives:

$a = (v^2 - u^2)/2s$

$$= \frac{245^2 - 280^2}{2 \times 2000}$$

$= -4.6\text{ ms}^{-2}$

Hence $a = 4.6\text{ ms}^{-2}$ in the opposite direction to the initial velocity.

c $v = u + at$

Rearranging for time gives $t = (v - u)/a$

$$\text{Hence } t = \frac{245 - 280}{-4.6}$$

$= 7.6\text{ s}$

Acceleration due to gravity

➔ Near the Earth's surface, if we neglect air friction, all objects fall with a constant **acceleration due to gravity** (g) of 9.8 ms^{-2}, irrespective of the mass of the object.

acceleration due to gravity: the rate at which, in the absence of other forces, an object accelerates downwards due to the force of gravity; near the Earth's surface, $g = 9.8\text{ ms}^{-2}$

➔ Generally upwards quantities are positive and downward quantities are negative to accommodate the vector nature of motion quantities when calculating the motion of objects being thrown in the air or falling.

➔ It is helpful when calculating the motion under the influence of gravity to remember that, for an object thrown upwards, provided air friction is neglected, the initial velocity when it leaves the ground will be equal to the final velocity when it returns to the initial position, but the velocities will be in opposite directions. Calling upwards positive, then:

$u = -v$

➔ Note also that, at the top of the flight of an object thrown upwards, the final velocity is zero. That is, final vertical velocity $v = 0$ at the top of the motion.

EXAMPLE 15

A ball is thrown upwards from ground level with an initial velocity of 12 ms^{-1}. Assuming that air friction is negligible, find:

a the maximum height reached by the ball

b the velocity of the ball just before it reaches the ground

c the time for the ball to return to the ground.

Answer:

Call upwards positive. Extract raw data:

$u = 12\text{ ms}^{-1}$, $a = g = -9.8\text{ ms}^{-2}$

Note that the velocity at the top of the flight is zero and gravity is downwards (negative)

a As time is not given, we use:

$v^2 = u^2 + 2as$

But at the top of the flight, $v = 0$. Thus:

$0 = u^2 + 2as$

Making the displacement the subject of the equation:

$s = -u^2/2a$

$$= \frac{-12^2}{2 \times -9.8}$$

$= 7.4\text{ m}$

b If we neglect air friction the motion is symmetrical and the final velocity will have the same magnitude as the initial velocity. Hence $v = 12\text{ ms}^{-1}$

$= 12\text{ ms}^{-1}$ downwards.

c Using $v = at + u$ and substituting $v = -u$ gives $-u = at + u$.

Hence $t = -2u/a$

$$= \frac{-(2\times12)}{-9.8}$$

$= 2.5$ seconds

➔ KEY QUESTIONS

8 How can we find the gradient at a specific time if the graph is a curve?

9 What does the gradient of a velocity against time graph represent?

10 What does the area under a velocity against time graph represent?

11 How can we take the vector nature of quantities into account when doing calculations of motion in one dimension?

Answers ➲ p. 17

5 Measuring and graphing motion

» Students conduct a practical investigation to gather data to facilitate the analysis of instantaneous and average velocity through:
- quantitative, firsthand measurements
- the graphical representation and interpretation of data.

FIRSTHAND INVESTIGATION 1

Using a ticker-tape timer to investigate motion

In the experiment illustrated in Figure 1.6, a ticker-tape timer that uses carbon paper to mark a paper tape with a dot for each 0.02 seconds is used to measure motion.

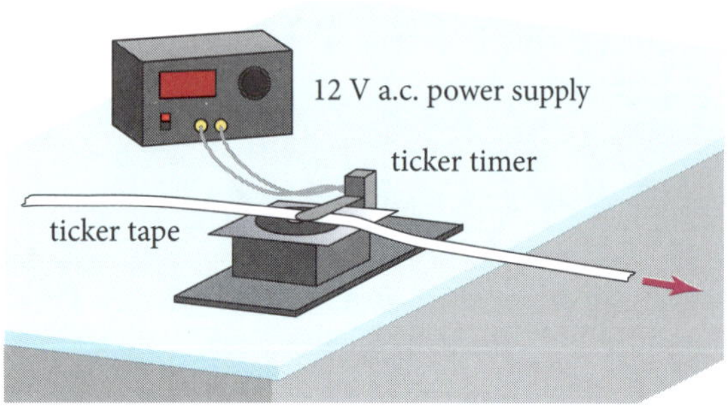

Figure 1.6 A simple experiment to investigate motion

By pulling a tape at constant slow velocity, constant fast velocity, and with acceleration, we obtain tapes similar to those shown in Figure 1.7.

Figure 1.7 Ticker tapes showing slow uniform, fast uniform and accelerated motion

Displacement

Because the time between dots in the ticker tape timer is 0.02 seconds it is possible to measure the total distance travelled (displacement = s) after each 0.02 seconds. These results can be placed in a table and graphed, as shown in the resulting displacement against time graph in Figure 1.8.

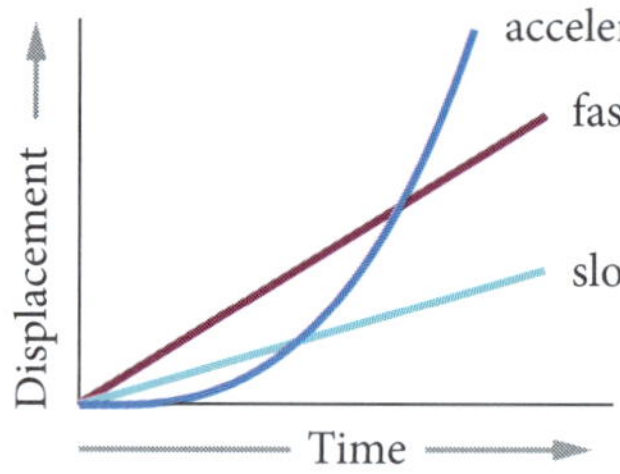

Figure 1.8 Displacement against time graph

Velocity

The instantaneous velocity at any time can be determined by dividing the distance between adjacent dots at that time by 0.02 ($v = s/t$). This enables the velocity to be measured for each time and these results can be placed in a table and graphed, as shown in the velocity against time graph in Figure 1.9.

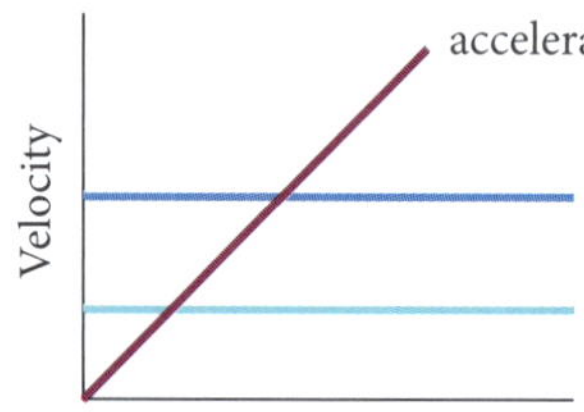

Figure 1.9 Velocity against time graph

Acceleration

The actual acceleration of the third tape can be determined by measuring the gradient of the velocity versus time graph for the accelerating tape.

6 Measuring specific motion quantities

» Students conduct practical investigations, selecting from a range of technologies, to record and analyse the motion of objects in a variety of situations in one dimension in order to measure or calculate:

- time
- distance
- displacement
- speed
- velocity
- acceleration.

FIRSTHAND INVESTIGATION 2

Using a sonic ranger to analyse the motion of a body accelerating at a constant rate

A sonic ranger and data logger can be used to measure and analyse the motion of an object accelerating at a constant rate, as shown in Figure 1.10.

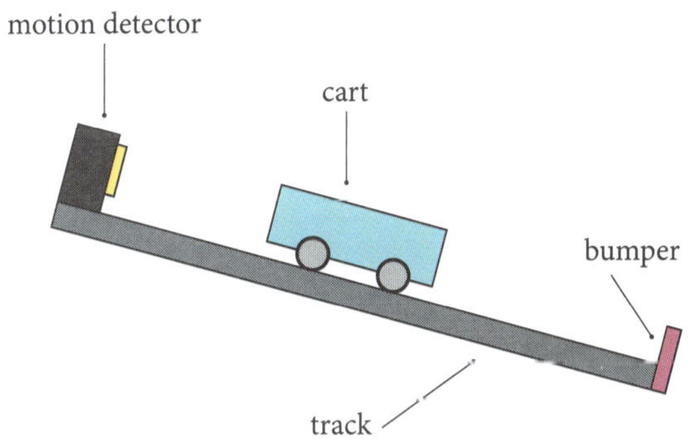

Figure 1.10 Measuring the position of a cart rolling down an inclined plane

After measuring the position as a function of time, the data-logger software can be used to display displacement versus time and velocity versus time graphs, as shown in Figure 1.11.

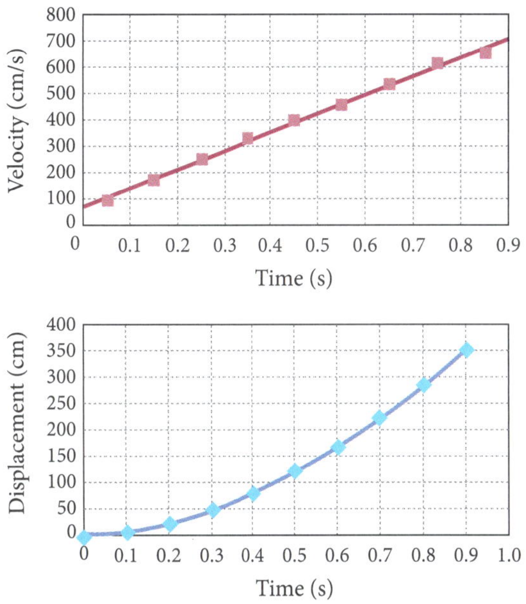

Figure 1.11 Velocity versus time and displacement versus time graphs for an object moving with constant acceleration

The velocity versus time graph is a straight line, indicating the acceleration is constant. The acceleration of the object is equal to the gradient of the line of best fit on the velocity versus time graph. For the graph shown above:

$$\begin{aligned} \text{Gradient} &= a \\ &= \text{rise/run} \\ &= \frac{500 - 200}{0.6 - 0.2} \\ &= 750\ \text{cms}^{-2} \\ &= 7.5\ \text{ms}^{-2} \end{aligned}$$

The displacement at $t = 0.9$ seconds, $s = 3.5$ m, can be read directly from the displacement versus time graph, or can be estimated from the area under the velocity versus time graph:

$$\begin{aligned} \text{Area} &= \text{displacement} \\ &= s \\ &= 650 \times 0.9 \times 0.5 + 50 \times 0.9 \\ &= 338\ \text{cm} \\ &= 3.4\ \text{m} \end{aligned}$$

CHAPTER SYLLABUS CHECKLIST

Are you able to answer these questions from the syllabus for this chapter? Tick each question as you go through the checklist if you are able to answer it. If you cannot answer a question, turn to the relevant page in the study guide to find the answer. For NESA key word meanings, go to www.educationstandards.nsw.edu.au and search 'key words'.

	FOR A COMPLETE UNDERSTANDING OF THIS TOPIC:	PAGE NO.	✓
1	Can I write quantities with the correct number of significant figures to reflect the precision of the measurement?	4	
2	Can I use scientific notation and SI prefixes to write quantities?	4–5	
3	Can I use unit analysis to find the units of an unknown quantity in an equation?	6	
4	Can I differentiate between a vector quantity and a scalar quantity?	6	
5	Can I give examples of vector and scalar quantities?	6	
6	Can I represent a vector quantity as an arrow to show its direction and magnitude?	7	
7	Can I recall the symbols that are used to represent them and the SI units of the key motion quantities such as displacement, initial and final velocity, and acceleration?	8	
8	Can I differentiate between the independent and dependent variable in an experiment?	9	
9	Can I draw displacement versus time graphs and velocity versus time graphs with labelled axes with units, clear data points and the line of best fit drawn?	10	
10	Can I find the gradient of a graph and its units?	10	
11	Can I use the gradient of a displacement versus time graph to find the velocity at a specified time?	10	
12	Can I use the gradient of a velocity versus time graph to find the velocity at a specified time?	10	
13	Can I find the displacement of an object at any time by measuring the area under the velocity against time graph?	10	
14	Can I rearrange each equation of motion to make a different quantity the subject of the equation?	11	
15	Can I use the equations of motion to solve a wide range of problems involving motion in one dimension?	11	

A QUICK NOTE! All the **dark green panels** with information (like the one below) have the **syllabus inquiry questions** and all the **light green panels** (like the one below) have the **syllabus dot points** for each topic. Note the checklist questions in the table above are based on these.

INQUIRY QUESTION:

How is the motion of an object moving in a straight line described and predicted?

Galileo and Newton showed that linear (straight line) motion could be described and predicted using graphs and by using algebraic relationships.

2 Motion quantities

» Students describe uniform straight-line (rectilinear) motion and uniformly accelerated motion through:
- qualitative descriptions
- the use of scalar and vector quantities.

Physics Stage 6 Syllabus

Motion in a Straight Line

Inquiry question: How is the motion of an object moving in a straight line described and predicted?

Students:
- describe uniform straight-line (rectilinear) motion and uniformly accelerated motion through:
  - qualitative descriptions
  - the use of scalar and vector quantities (ACSPH060)
- conduct a practical investigation to gather data to facilitate the analysis of instantaneous and average velocity through:
  - quantitative, first-hand measurements
  - the graphical representation and interpretation of data (ACSPH061)

YEAR 11 EXAM-TYPE QUESTIONS

Now for the real thing! The following questions are modelled on the types of questions you will face in the yearly examination in Year 11. Think about it: if you get extensive practice at answering these sorts of questions, you will be more confident in answering them in the yearly examination. It makes sense, doesn't it?

Another advantage for your exam preparation is the format of the answers: they are deliberately structured to give you strategies on how to answer examination questions. This will help you aim for full marks!

- For each objective-response question you will have the correct answer and an explanation, and reasons why all the other answers are incorrect.
- For each extended-response question you will have a detailed answer marked with ticks to indicate what part of the question gains which marks and also, when needed, an examiner's plan (Examiner Maximiser / EM) to help you get full marks.

When you mark your work, highlight any questions you found difficult and earmark these areas for extra study.

Objective-response questions (1 mark each)

1 Several trials of a solar-powered car are sketched on the velocity versus time graph shown in Figure 1.12.

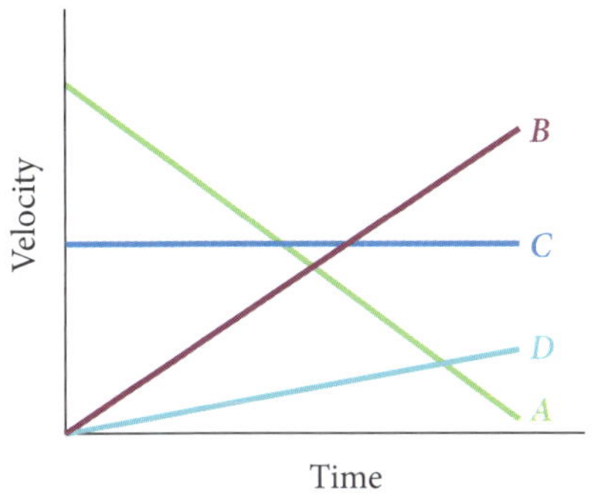

Figure 1.12 Velocity against time graph

Which of the following is a correct statement about the trials?

A The car travelled a greater distance in trial *B* than in trial *A*.

B At the end of the trial, the car was moving fastest in trial *C*.

C The acceleration was negative in trial *A*.

D The acceleration was positive in trial *C*.

2 Which of the following statements about acceleration is correct?

A Acceleration is equal to the change in speed.

B Acceleration is equal to the rate of change of velocity.

C Acceleration is equal to the rate of change of speed of a body.

D Acceleration can only be zero or a constant value.

3 A car travels north at a constant speed of 20 ms^{-1} for 16 minutes and then returns to its starting point. The entire journey takes 40 minutes. Which of the following observations about the journey is correct?

	Average speed (ms^{-1})	Distance travelled (m)	Average velocity (ms^{-1})
A	16	19 200	0
B	15	19 200	15
C	16	38 400	0
D	20	38 400	20

4 Which of the following statements about the motion graph shown in Figure 1.13 is correct?

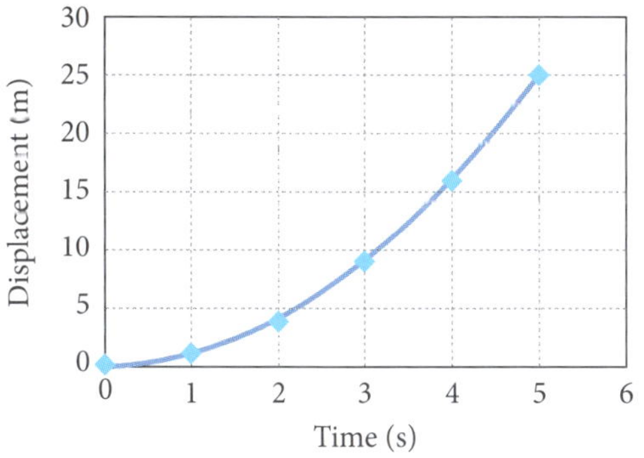

Figure 1.13 Displacement versus time graph

A The instantaneous velocity at 4 seconds is greater than the average velocity for the entire journey.

B The instantaneous velocity at 4 seconds is less than the average velocity for the entire journey.

C The rate of change of displacement is constant throughout the journey.

D The instantaneous velocity is greater at 3 seconds than at 3.5 seconds.

5 The graph shown in Figure 1.14 represents the motion of a bouncing ball.

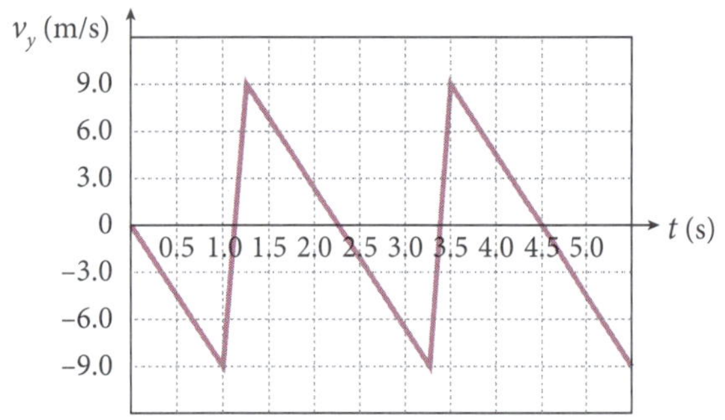

Figure 1.14 Velocity versus time graph for a bouncing ball

Which of the following statements about the graph is true?

A The ball is moving upwards between 1 and 1.25 seconds.

B The ball is moving upwards between 1.25 and 2.25 seconds.

C The ball's acceleration is always positive.

D The graph shows the ball hitting the ground four times.

Extended-response questions

6 A girl rides a bike at 15 ms^{-1} east trying to escape a dog that was running at 17.5 ms^{-1} east. Determine the velocity of the dog relative to the girl. Include a labelled vector diagram in your answer. (3 marks)

7 A group of students conducts an experiment to measure the displacement of a toy car as a function of time. Their results are shown in Figure 1.15.

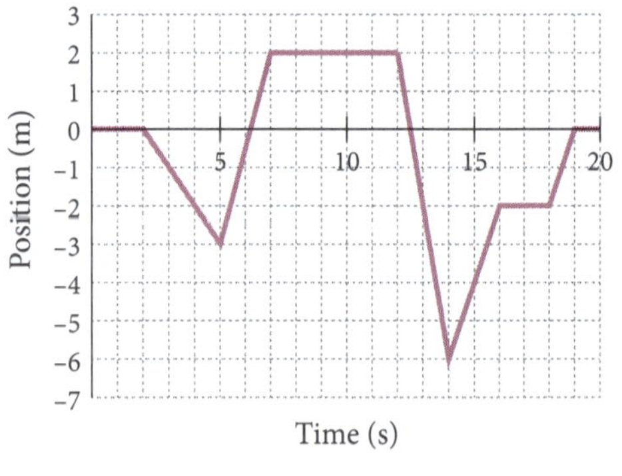

Figure 1.15 Displacement versus time graph for a toy car

a During which time intervals is the car not moving? (2 marks)

b Use the graph to determine the velocity between:

i 2 and 5 seconds

ii 5 and 7 seconds. (2 marks)

c Determine the total distance travelled after 20 seconds and the final displacement. (2 marks)

8 A motorcycle travelling at 20 ms^{-1} accelerates at 12.5 ms^{-2} for 3 seconds.

a Calculate the final velocity of the motorcycle in ms^{-1} and kmh^{-1}. (2 marks)

b Determine the distance travelled during the acceleration. (1 mark)

9 A car initially travelling at 45 ms^{-1} applies the brakes and comes to rest after travelling 150 m.

a Find the direction and magnitude of the acceleration of the car while it was braking. (2 marks)

b Find the time over which the brakes were applied. (1 mark)

10 On the Earth objects fall down with an acceleration of 9.8 ms^{-2} if air resistance is neglected. Consider a golf ball dropped from the top of a 70 m high building. You may consider the initial velocity of the ball to be zero.

a Find the final velocity just before it hits the ground. (1 mark)

b Calculate the time for the golf ball to reach the ground. (1 mark)

c Sketch a velocity versus time graph, an acceleration versus time graph and a displacement versus time graph for the motion of the ball. (3 marks)

ANSWERS

KEY QUESTIONS

Key questions ➲ p. 5

1 Uncertainty can be expressed explicitly (for example, 34.2 ± 0.2 m) or implicitly as half the smallest digit stated (for example, 45.4 implies 45.4 ± 0.05).

2 Scientific notation always uses numbers between 1 and 10 multiplied by a power of ten. Hence the decimal place is always placed after the first number when numbers are represented in scientific notation.

Key questions ➲ p. 7

3 To investigate nature and test theories in physics, physical quantities must be measured. Measurement is therefore of fundamental importance in physics and to ensure all scientists can compare their measurements with one another they must agree on the units they are using to make their measurements.

4 A dimension is the type of quantity being measured (e.g. length, mass or time), while a unit is the agreed standard we use to measure a specific dimension (e.g. metres, kilograms or seconds).

5 A fundamental unit is the most basic set of units used in a system of units, while derived units are made up from more than one fundamental unit. For example, metres and seconds are fundamental units but the units of acceleration ms^{-2} make a derived unit.

Key questions ➲ p. 9

6 The change in a quantity is the difference between the initial and final amount of the quantity, while the rate of change is how much change happens per unit time. Thus the rate of change is a measure of how fast the change occurred.

7 It is important to know which quantities are vectors and which are scalars because scalars add and subtract like numbers, but the direction must be taken into account when adding or subtracting vectors.

Key questions ➲ p. 12

8 The instantaneous gradient at a point on the curve can be found by finding the gradient of a tangent drawn to the line of best fit at the point of interest.

9 The gradient of the velocity against time graph measures the acceleration $= \Delta v/\Delta t$.

10 The area under the velocity against time graph is the displacement.

11 We can take the vector nature of quantities into account in one dimension motion by calling all the quantities in one direction positive and all the quantities in the other direction negative.

YEAR 11 EXAM-TYPE QUESTIONS

Objective-response questions

1 EM This question requires students to demonstrate a good understanding of how the gradient of a velocity against time graph relates to acceleration.

C. Because the gradient = the velocity, which is negative. **A** is incorrect because displacement = area under curve and the area under curve *A* is greater than the area under curve *B*. **B** is wrong because the fastest-moving car is *B* at the end of the time shown, not car *C*. **D** is wrong because gradient = acceleration and the acceleration of curve $C = 0$.

2 EM This question is a simple definition but also requires students to differentiate between speed and velocity.

B. This is the definition of acceleration. Answers **A**, **C** and **D** are all incorrect definitions.

3 EM This question requires students to convert to SI units, solve a two-part calculation and understand the difference between average velocity and average speed.

C. We have to calculate the total distance travelled. For the trip out $s = vt = 20 \times (16 \times 60) = 19\,200$ m and, as we also travel back to the start, the total displacement $= 2 \times 19\,200 = 38\,400$ m.

Hence average speed = total distance/total time

$$= \frac{38\,400}{40 \times 60}$$

$$= 16 \text{ ms}^{-1}$$

As it returns to where it started the total displacement is zero and hence the average velocity must also be zero. **A**, **B** and **D** all have incorrect values.

4 EM This question requires students to have an understanding of the instantaneous velocity and how it relates to the gradient of the displacement against time graph.

A. The average velocity is $\frac{25}{5} = 5 \text{ ms}^{-1}$ but the instantaneous velocity is the gradient of the tangent to the curve at $t = 4$ seconds, which is greater than 5 ms^{-1}. **B** is incorrect because the gradient at 4 s is greater than the average gradient. **C** is wrong because rate of change of position is the velocity = gradient, which changes continuously. **D** is wrong because the gradient increases as time passes and hence the instantaneous velocity increases throughout the journey.

5 EM In this question students must relate a velocity against time graph to a complex motion and determine which sections of the graph are related to which parts of the motion.

B. The area under the graph = displacement, which is only positive when the ball is moving upwards. **A** is incorrect because the velocity is negative (downwards) during this time. **C** is wrong because the gradient = acceleration, which is positive and negative on the graph. **D** is wrong because the graph shows only two bounces when the velocity changes from negative (downwards) to positive (upwards).

Extended-response questions

6 EM This question requires students to understand relative motion and be able to draw and use a vector subtraction diagram.

Vector subtraction: $v_{rel} = v_{object} - v_{observer} = v_{dog} - v_{girl}$

The correct vector diagram is shown in Figure A1.1.

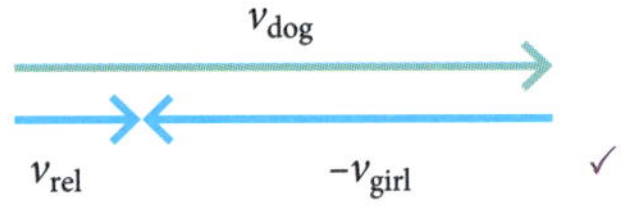

Figure A1.1 Vector diagram

Therefore $v_{rel} = 2.5\ ms^{-1}$ ✓ east ✓

7 EM This question requires students to relate motion to a displacement against time graph, understand that the gradient of such a graph is equal to the velocity and differentiate between distance and velocity.

a $t = 0$ to 2 s, $t = 7$ to 12 s, $t = 16$ to 18 s, $t = 19$ to 20 s ✓✓

b i $v = \text{gradient} = \frac{-3}{3} = -1.0\ ms^{-1}$ ✓

ii $v = \text{gradient} = \frac{+5}{2} = +2.5\ ms^{-1}$ ✓

c $d = (3 \times 2) + (2 \times 2) + (6 \times 2)$
$= 22$ m ✓
$s = 0$ m ✓

8 EM This question requires rearranging and substitution into two of the equations of motion.

a i $v = u + at = 20 + (12.5 \times 3) = 57.5\ ms^{-1}$ ✓

ii $\frac{57.5}{1000} \times (60 \times 60) = 207\ kmh^{-1}$ ✓

b $s = ut + \frac{1}{2}at^2$
$= 20 \times 3 + \frac{1}{2}(12.5)(3)^2$
$= 116$ m ✓

9 EM This question requires students to demonstrate an understanding of the equations of motion and the application of boundary conditions to solve a problem (i.e. final velocity = 0).

a $v^2 = u^2 + 2as$ and hence $a = (v^2 - u^2)/2s = \frac{0 + 45^2}{2 \times 150}$
$= 6.75\ ms^{-2}$ ✓ in the opposite direction to the initial velocity ✓

b $v = u + at$ and hence $t = (v - u)/a = \frac{0 - 45}{-6.75} = 6.7$ s ✓

10 EM This question requires students to demonstrate an ability to solve problems with the equations of motion using the acceleration due to gravity. The problem also requires students to demonstrate an understanding of the relationship between motion and motion graphs.

Call all downwards vector quantities negative and all upwards vector quantities positive.

a $v^2 = u^2 + 2as$, hence $v = \sqrt{2 \times -9.8 \times -70}$
$= 37\ ms^{-1}$ downwards ✓

b $s = ut + \frac{1}{2}at^2$ and, as $u = 0$, we have:

$t = \sqrt{2s/a} = \sqrt{\frac{2 \times -70}{-9.8}} = 3.8$ s ✓

c The correct graphs are shown in Figure A1.2. Note the gradient of the velocity against time graph should be $-9.8\ ms^{-2}$.

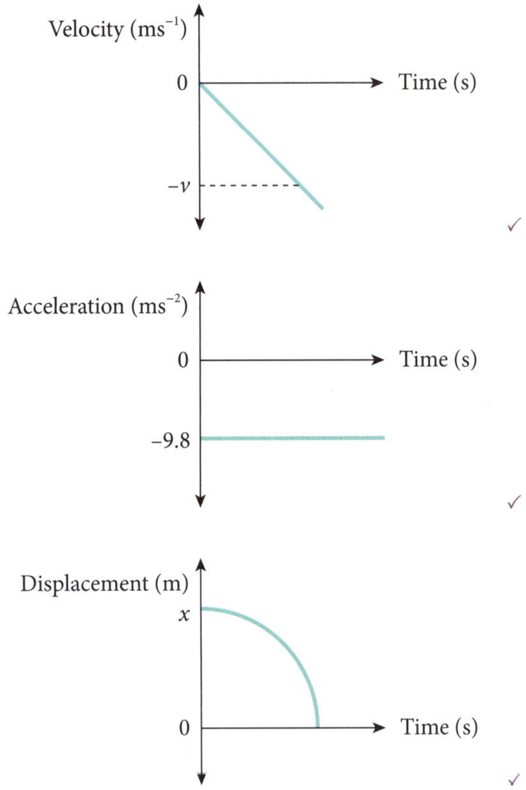

Figure A1.2 Motion graphs

CHAPTER 2 MOTION ON A PLANE

INQUIRY QUESTION:

How is the motion of an object that changes its direction of movement on a plane described?

Changes in direction and relative motion of moving objects on a plane (i.e. in two dimensions) can be analysed using vectors.

1 Vector analysis

» Students analyse vectors in one and two dimensions to:
- resolve a vector into two perpendicular components
- add two perpendicular vector components to obtain a single vector.

Vectors in two dimensions

➔ A vector quantity can be represented by an arrow drawn in the direction the vector points. The length of the arrow represents the magnitude of the vector.

For example, if a car was travelling at 40 ms^{-1} 30° towards the east from the north we could represent its velocity by the arrow vector shown in Figure 2.1.

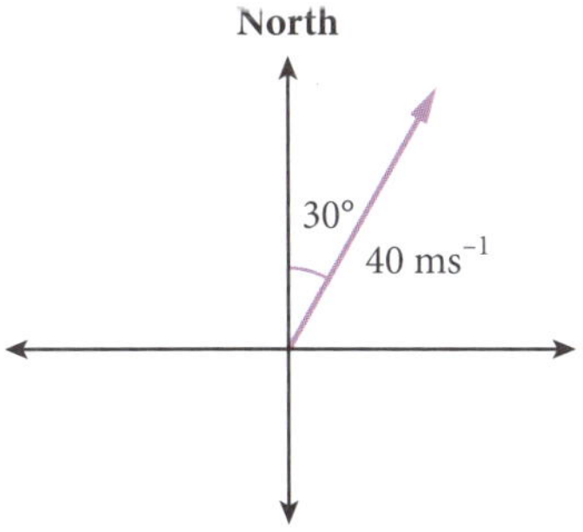

Figure 2.1 Representing a vector diagrammatically

We can state the direction of this vector as N30°E or simply as a **bearing**, N30°, which implies it is 30° towards the east (clockwise) from north.

bearing: the direction of a vector towards the east from north; hence a vector pointing SE has a bearing of N135°

EXAMPLE 1

Write the direction of the following vectors in term of a bearing from north:

a **20 m NE**

b **8 m SW**

c **14 m at S30°E**

Bearing is the angle clockwise from north

Answer:

Working out the angle clockwise from north gives:

a 20 m N45°

b 8 m N225°

c 14 m N150°

Adding and subtracting two-dimensional vectors

➔ To add two vectors, we place the arrows that represent them 'head to tail'. The vector drawn from the 'free tail' to the 'free head' is the **resultant** (sum) of the vectors.

For example, Figure 2.2 shows how we can add the vectors 50 north and 90 east by placing them head to tail and then drawing the resultant vector from the free tail to the free head. The magnitude of the resultant vector is obtained from Pythagoras' theorem and its direction can be determined from trigonometry.

resultant: the single vector that results when two or more vectors are added together

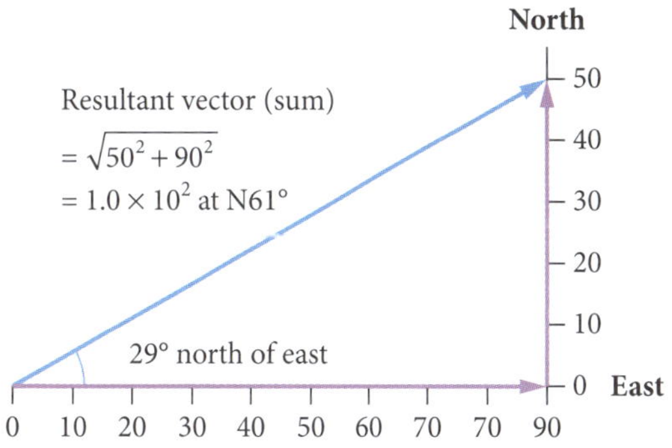

Figure 2.2 Vector diagram required to add vectors

- A negative vector is a vector in the opposite direction as the stated direction. For example, $-5\ \text{ms}^{-1}$ north = $5\ \text{ms}^{-1}$ south.
- To subtract vectors, simply reverse the direction of the negative vector and then add them. For example, $14\ \text{ms}^{-1}$ east – $10\ \text{ms}^{-1}$ west = $14\ \text{ms}^{-1}$ east + $10\ \text{ms}^{-1}$ east = $24\ \text{ms}^{-1}$ east.

EXAMPLE 2

Add the displacement vectors 25 m east and 65 m south.

Draw the vectors head to tail first

Answer:

Draw the vectors head to tail. The resultant is then drawn from the free vector tail to the free vector head. Figure 2.3 shows the vector diagram required to show that the resultant vector is 70 m N159°.

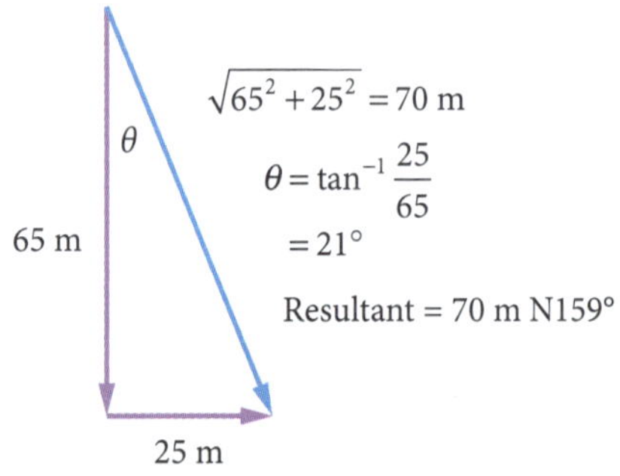

Figure 2.3 Vector diagram

Components of a vector

- Because vectors add in this way, we can also break any vector into two mutually perpendicular component vectors. For example, the vector v shown in Figure 2.4 can be thought of as being made of a **component vector** in the y direction and a component vector in the x direction.

component vectors: mutually perpendicular vectors that can be thought of as being equivalent to the vector that is formed when they are added together; any vector can be broken into two mutually perpendicular component vectors

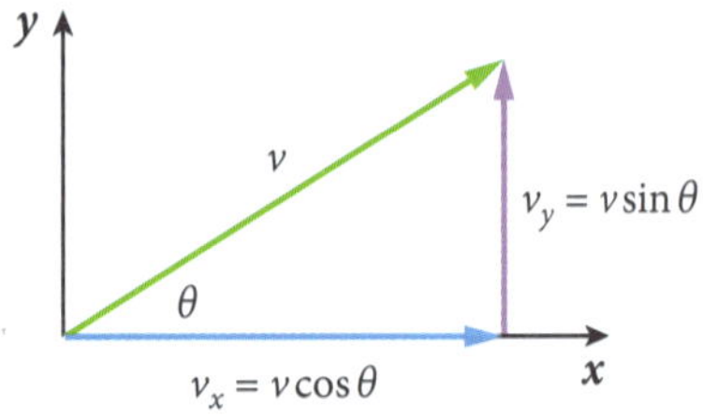

Figure 2.4 Breaking a vector into its component vectors

- The magnitude of the component vectors can be determined using trigonometry:

  $v_x = v\cos\theta$ and

  $v_y = v\sin\theta$

- Vectors can be added to one another or subtracted from one another by breaking the vectors into components and adding or subtracting the relevant components. The resultant vector is then determined by combining the final two components.

EXAMPLE 3

A ball is thrown at $12\ \text{ms}^{-1}$ at an angle of 30° above the horizontal, as shown in Figure 2.5. Find the horizontal and the vertical components of the initial velocity.

Draw a vector diagram with a perpendicular and vertical vector added together to give the resultant vector

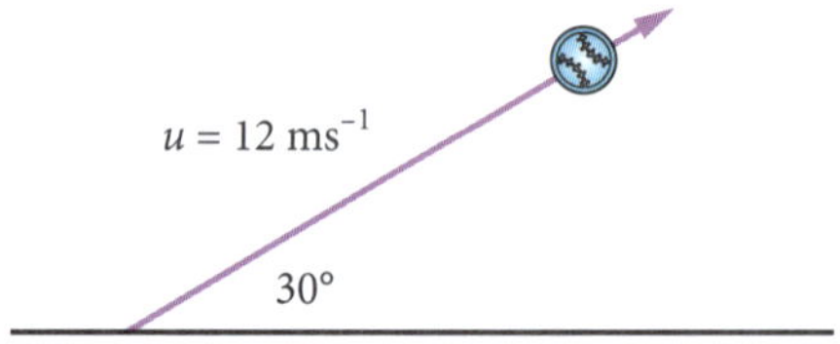

Figure 2.5 Ball being thrown in the air

Answer:

Breaking the vector into components and using trigonometry, as shown in Figure 2.6, we obtain:

Vertical component = 12 sin 30° = $6\ \text{ms}^{-1}$ upwards

Horizontal component = 12 cos 30° = $10\ \text{ms}^{-1}$ to the right

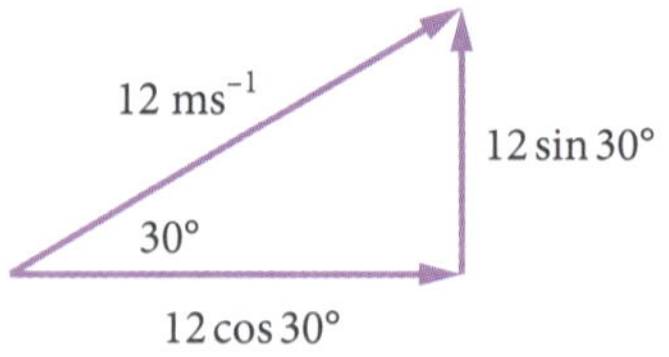

Figure 2.6 Breaking a vector into its component vectors

KEY QUESTIONS

1. How can two vector quantities be added together?
2. How can one vector quantity be subtracted from another vector quantity?
3. How can we break a vector into two perpendicular component vectors?

Answers p. 27

2 Displacement as a vector quantity

» Students represent the distance and displacement of objects moving on a horizontal plane using:
- vector addition
- resolution of components of vectors.

➔ Distance is the total length of a journey, while displacement is the directed length. The following discussion illustrates the concept.

Consider a car that travels 4 km north and then 4.0 km east, as shown in Figure 2.7. The distance travelled would be 8 km, but the displacement would be 5.66 km at N45°.

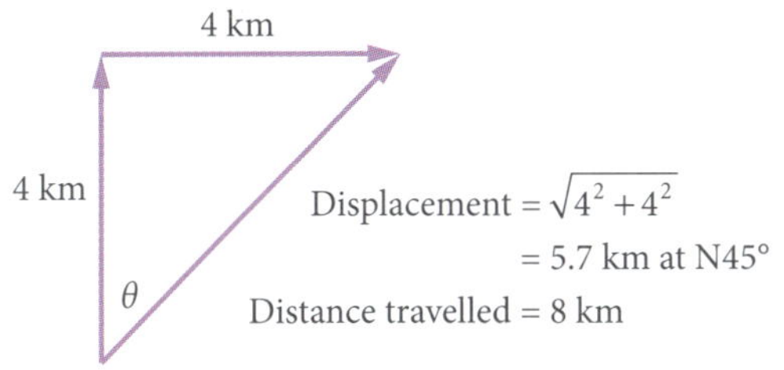

Figure 2.7 Comparing distance travelled with displacement

➔ Alternatively, we could use components to describe the displacement by saying the displacement was made up of the two components, 4 km north plus 4 km east.

➔ We can use vector components to represent the velocity of an object moving on a plane by saying the velocity is made up of a north–south component and an east–west component.

EXAMPLE 4

Find the total velocity of a car that is moving at 40 ms^{-1} north and 20 ms^{-1} east.

Answer:

The resultant is the sum of the two perpendicular component vectors

Simply add the component vectors by drawing the vectors head to tail, and then find the resultant vector by drawing from the free tail to the free head, as shown in Figure 2.8.

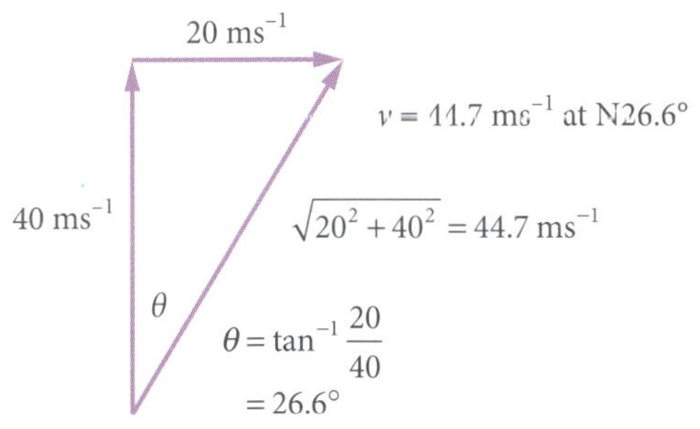

Figure 2.8 Adding component vectors

3 Analysing change in motion quantities

» Students describe and analyse algebraically, graphically and with vector diagrams, the ways in which the motion of objects changes, including:
- velocity
- displacement.

Change of vectors and scalars

➔ The change (Δ) of a quantity can be expressed as:

Change = Δ = final quantity – initial quantity

➔ Change of a scalar quantity occurs when the size of the scalar quantity changes.

➔ Change of a vector quantity may represent a change in the size and/or direction of the vector.

Change in motion quantities

➔ The change in displacement (s) can be expressed as:

Δs = final displacement – initial displacement

➔ Note that the rate of change of displacement (how much the displacement changes each second) is the velocity (v):

$v = \Delta s/t$

➔ The change in velocity can be expressed as:

$\Delta v = v - u$

➔ The rate of change of velocity (how much the velocity changes each second) is the acceleration (a):

$a = \Delta v/t$

➔ As both displacement and velocity are vectors, calculating the change in these quantities requires the relevant vectors to be subtracted from one another.

➔ For example, consider the change in velocity of an object moving around a circle at constant speed. As shown in Figure 2.9, the change in velocity is towards the centre of the circle.

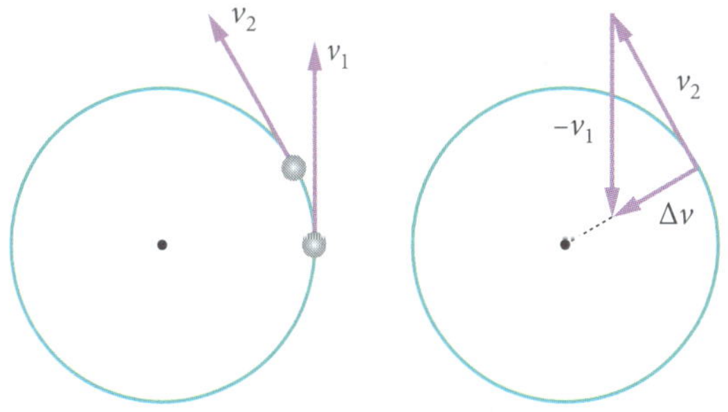

Figure 2.9 Change in velocity (Δv) for an object undergoing uniform circular motion

EXAMPLE 5

A car travels at 25 ms^{-1} north, turns left, and continues in a westerly direction at 30 ms^{-1}. Find the change in the velocity of the car.

Answer:

Subtract the initial velocity vector from the final velocity vector to find the change

Apply the relationship change = final quantity – initial quantity.

Thus $\Delta v = v - u$

To solve this vector equation, reverse the direction of the initial velocity (the negative vector) and add it to the final velocity, as shown in Figure 2.10. This shows that the change in velocity of the car is 39 ms^{-1} N230.2°.

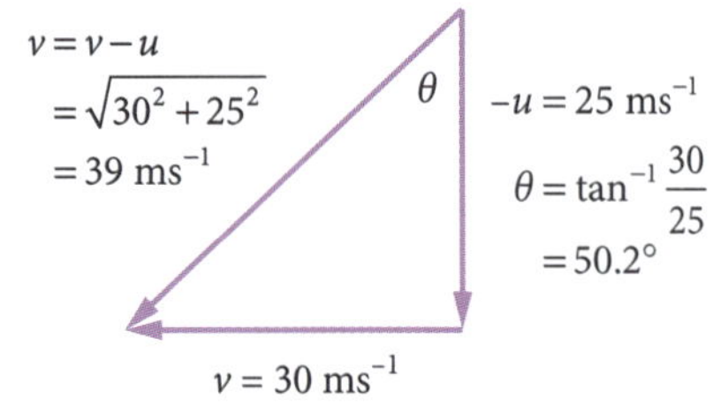

Figure 2.10 Vector diagram to find change in velocity

4 Relative motion on a plane

» Students describe and analyse, using vector analysis, the relative positions and motions of one object relative to another object on a plane.

- The term **relative motion** is used to describe the motion of one object with respect to another object.
- The motion of one object relative to another moving object on a plane can be determined by making measurements using a set of axes fixed to the moving observer. We call the set of axes that we use to make measurements a **frame of reference** and in this case we are using the observer's frame of reference to make our measurements.

relative motion: the motion of one object as observed from the point of view of another object

frame of reference: the set of axes used to measure the position or motion of an object; when an observer is moving, objects measured from their point of view are measured from a frame of reference centred on and moving with the observer

- The *position* of one object relative to another is determined by subtracting the position of the observer:

$$s_{relative} = s_{object} - s_{observer}$$

- The *velocity* of one object relative to another can be determined by subtracting the velocity of the observer from that of the object:

$$v_{relative} = v_{object} - v_{observer}$$

EXAMPLE 6

Car *A* is travelling at 20 ms^{-1} west, while car *B* travels at 35 ms^{-1} north. Find the velocity of car *A* with respect to an observer in car *B*.

Answer:

We subtract the velocity of the observer to find the relative velocity

Drawing a vector diagram to represent the equation $v_{relative} = v_{object} - v_{observer}$ will allow the relative velocity to be determined. To do this we reverse the direction of the negative vector and then add the vectors, as shown in Figure 2.11. This yields:

$$v_{relative} = \sqrt{20^2 + 35^2}$$
$$= 40.3 \text{ ms}^{-1} \text{ at N209.7°}$$

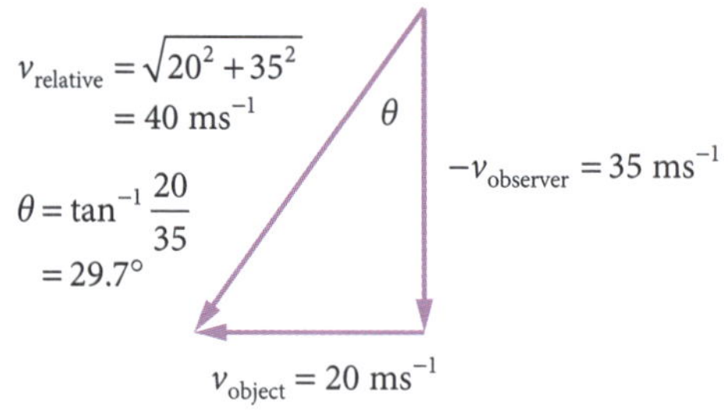

Figure 2.11 Vector diagram to find relative velocity

→ KEY QUESTIONS

4 How can we find the velocity of one moving object relative to another moving object?

5 How can we measure the absolute velocity of an object in a moving medium?

6 How can the change in a vector quantity be determined?

Answers ⊃ p. 27

5 Motion in a moving medium

» Students analyse the relative motion of objects in two dimensions in a variety of situations, for example:
- a boat on a flowing river relative to the bank
- two moving cars
- an aeroplane in a crosswind relative to the ground.

→ The absolute motion of an object moving with respect to a moving medium is given by adding the velocity of the object and the velocity of the medium:

$$v_{\text{absolute}} = v_{\text{object}} + v_{\text{medium}}$$

For example, to find the velocity of a jet aircraft flying in a crosswind we add the velocity vector of the wind to the velocity vector of the plane. To find the velocity of a boat moving in a river with respect to the bank of the river we add the velocity of the river to the velocity of the boat.

EXAMPLE 7

A yacht sails at 5ms^{-1} in a northerly direction with respect to the water. If an easterly current of 3 ms^{-1} is running, what is the boat's absolute velocity?

Answer:

We can use a vector diagram like the one shown in Figure 2.12 to add the velocity of the boat to the velocity of the current. This yields a resultant velocity of:

$$v = \sqrt{3^2 + 5^2}$$
$$= 5.83 \text{ ms}^{-1} \text{ N}31°$$

We must add the velocity of the medium (the water) to find the absolute velocity

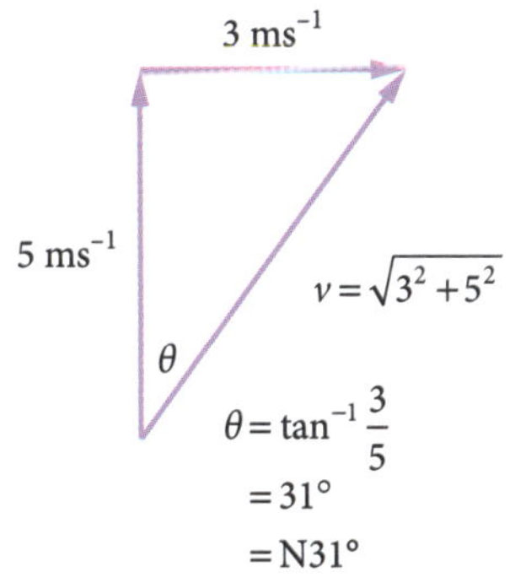

Figure 2.12 Vector diagram to find total velocity

CHAPTER SYLLABUS CHECKLIST

Are you able to answer these questions from the syllabus for this chapter? Tick each question as you go through the checklist if you are able to answer it. If you cannot answer a question, turn to the relevant page in the study guide to find the answer. For NESA key word meanings, go to www.educationstandards.nsw.edu.au and search 'key words'.

FOR A COMPLETE UNDERSTANDING OF THIS TOPIC:		PAGE NO.	✓
1	Can I represent a vector diagrammatically?	19	
2	Can I use the bearing from north to represent the direction of a vector?	19	
3	Can I draw vector diagrams and use them to add vectors together to obtain the magnitude and direction of the resultant vector?	19	
4	Can I draw and use vector diagrams to subtract one vector from another?	20	
5	Can I break a vector into two mutually perpendicular components?	20	
6	Can I add the components of a vector to determine the magnitude and direction of the resultant vector?	21	
7	Can I draw and use a vector diagram to determine the magnitude and direction of the change in a vector quantity?	21	
8	Can I use vector subtraction to determine the magnitude and direction of the velocity of one moving object relative to another moving object?	22	
9	Can I use vector addition to find the absolute velocity of an object moving with respect to a moving medium (e.g. the velocity of a plane moving in a crosswind)?	23	

YEAR 11 EXAM-TYPE QUESTIONS

Objective-response questions

(1 mark each)

1 A helicopter traveling north at 40 ms^{-1} encounters a crosswind blowing at 30 ms^{-1} from west to east. What is the velocity of the helicopter relative to the ground?

A 70 ms^{-1} NE
B 50 ms^{-1} NE
C 70 ms^{-1} N37°E
D 50 ms^{-1} N37°E

2 A car drives at 40 ms^{-1} in the direction N120°. Which of the following statements about the car's components of motion is correct?

A The car is travelling at 20 ms^{-1} towards the north and 34.6 ms^{-1} towards the east.
B The car is travelling at 34.6 ms^{-1} towards the north and 20 ms^{-1} towards the east.
C The car is travelling at 20 ms^{-1} towards the south and 34.6 ms^{-1} towards the east.
D The car is travelling at 34.6 ms^{-1} towards the north and 34.6 ms^{-1} towards the east.

3 Given that the vertical component of a ball's velocity is 16 ms^{-1} downwards and the horizontal component of the ball's velocity is 12 ms^{-1}, what is the velocity of the ball?

A 20 ms^{-1} at 53° below the horizontal
B 28 ms^{-1} at 53° below the horizontal
C 20 ms^{-1} at 39° below the horizontal
D 28 ms^{-1} at 39° below the horizontal

4 A bird flies north with a velocity of 5 ms^{-1} near a helicopter flying west with a velocity of 40 ms^{-1}. Which of the following vector diagrams in Figure 2.13 correctly shows the relative velocity of the bird as measured by an observer in the helicopter?

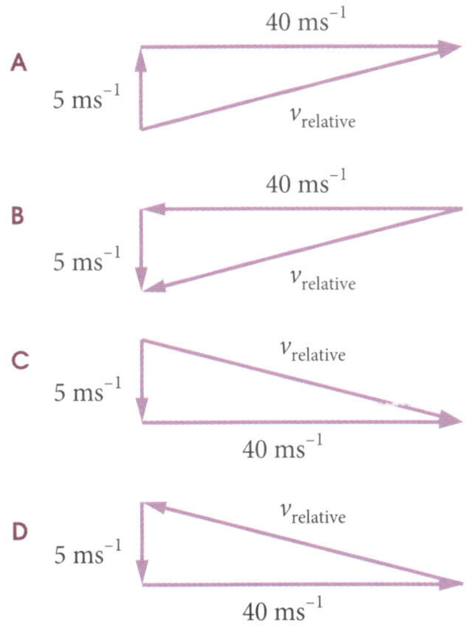

Figure 2.13 Vector diagrams

5 If a ball moving north with constant velocity begins to accelerate at a constant rate towards the east, what path would it describe?

A It would continue to travel in a straight line but in a different direction.
B It would start to move in a circle.
C It would move in a parabolic path as its displacement changes at a constant rate in the northerly direction but increases at an increasing rate in the easterly direction.
D It is impossible to determine the path it will take.

Extended-response questions

6 A ball is projected with an initial vertical velocity component of 5 ms^{-1} and an initial horizontal velocity component of 3 ms^{-1} as shown in Figure 2.14. Find the magnitude and direction of the initial velocity. (2 marks)

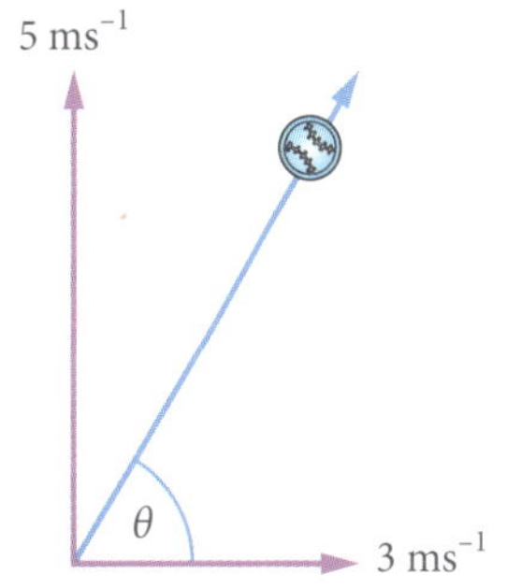

Figure 2.14 Velocity components of a ball thrown in the air

7 A hot air balloon is drifting east at 14 ms^{-1} while a car drives north at 30 ms^{-1}.

a Find the velocity of the car relative to an observer in the balloon. (2 marks)
b Find the velocity of the balloon relative to an observer in the car. (2 marks)

8 A certain species of butterfly can fly with a constant velocity of 1.2 ms^{-1} through still air. The butterfly flies to a flower 10 m north of its initial position while a light wind of 1 ms^{-1} blows from east to west.

a How long would it take the butterfly to reach the flower if no wind was blowing? (1 mark)
b What direction must the butterfly fly in if it is to travel directly north? (1 mark)
c What will the butterfly's velocity be as it heads towards the flower? (1 mark)
d How long would it take the butterfly to reach the flower in the wind? (1 mark)

9 A drone was flying east at 30 ms^{-1} when a wind started to blow at 25 ms^{-1} from south to north.

a Calculate the direction of the drone's flight path when the wind is blowing. (1 mark)

b Calculate the velocity of the drone with respect to the ground after the wind started blowing. (2 marks)

c Calculate the displacement of the drone 10 minutes after the wind started to blow from the position of the drone when the wind started. (1 mark)

10 A boy on a skateboard travelling at 5.0 ms^{-1} south sees a girl on a bicycle travelling at 13 ms^{-1} to the north-west with respect to him.

a Draw a vector diagram to represent this situation. (2 marks)

b Use the vector diagram to calculate the girl's velocity relative to the ground. (2 marks)

ANSWERS

KEY QUESTIONS

Key questions ➲ p. 20

1 We add vectors by drawing them as arrows in the appropriate direction and then placing the vectors to be added 'head' to 'tail'. The vector drawn from the free 'tail' to the free 'head' is the sum of the vectors.

2 We subtract vectors by reversing the direction of the negative vector and then adding them.

3 As shown in Figure A2.1, we can break a vector down into two mutually perpendicular components by drawing a vector diagram to make the vector the sum (hypotenuse) of the two mutually perpendicular component vectors.

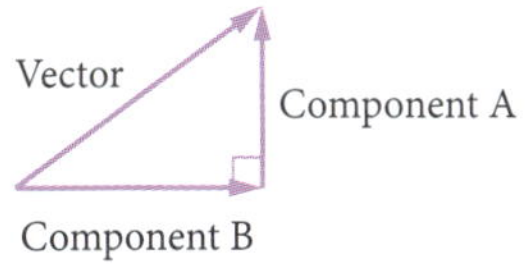

Figure A2.1 Mutually perpendicular components of a vector

Key questions ➲ p. 22

4 To find the velocity of one object relative to another we perform a vector subtraction taking the observer's velocity from the object's velocity.

5 We add the velocity of the object to the velocity of the medium.

6 The change in a vector quantity can be determined by subtracting the initial vector from the final vector.

YEAR 11 EXAM-TYPE QUESTIONS

Objective-response questions

1 **D.** The vectors for velocity of helicopter and wind are added, as shown in Figure A2.2. **A**, **B** and **C** have the wrong magnitude or direction.

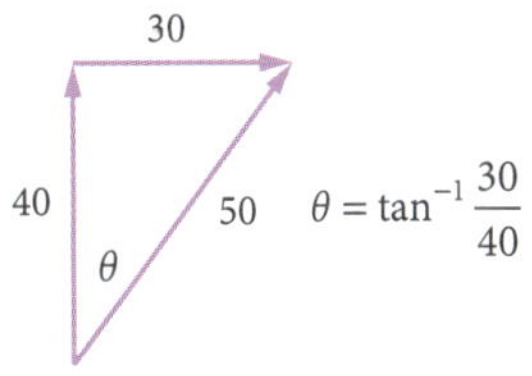

Figure A2.2 Vector diagram

2 **C.** Velocity is resolved into perpendicular components, as shown in Figure A2.3. **A**, **B** and **D** are wrong as they do not have a southerly component of 20 ms^{-1}.

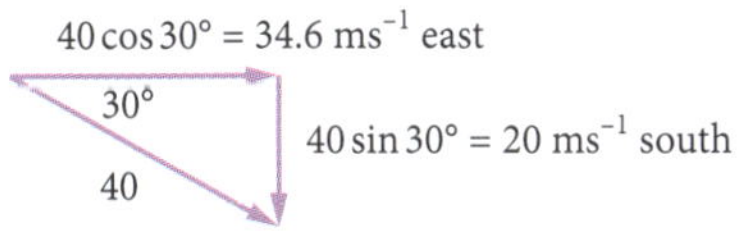

Figure A2.3 Resolving a vector into its components

3 **A.** The two component vectors are added, as shown in Figure A2.4. **B**, **C** and **D** are incorrect as they have either the wrong magnitude or direction.

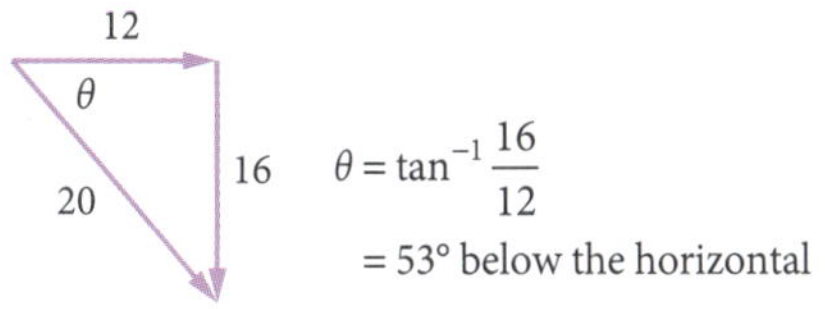

Figure A2.4 Adding component vectors

4 **A.** Subtract the velocity of the observer (the helicopter) from the object (the bird). **B**, **C** and **D** are wrong as they have the vector diagram drawn incorrectly.

5 **C.** Just like a projectile fired horizontally, it will describe a parabolic path. **A** is wrong because its velocity is changing continually towards the east so its direction must also be continually changing. **B** is wrong because the acceleration is only towards the east and moving in a circle would require it to move east then west. **D** is wrong because it is possible to determine the path as it is just like a projectile thrown horizontally near the Earth's surface.

Extended-response questions

6 EM This question requires the addition of two perpendicular vectors and for students to find the magnitude and direction of the resultant vector.

Adding the component vectors, as shown in Figure A2.5, gives 5.8 ms^{-1} ✓ at 59° above the horizontal. ✓

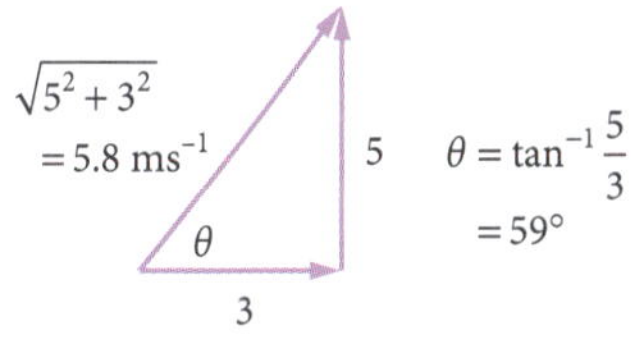

Figure A2.5 Vector addition

7 EM This question requires a sound understanding of relative velocity and vector subtraction.

In both cases we subtract the velocity of the observer.

a Subtract the velocity of the observer as shown in Figure A2.6 by reversing the direction of the negative vector and adding. This yields: $v_{relative} = 33.1\ ms^{-1}$ ✓ N335° ✓

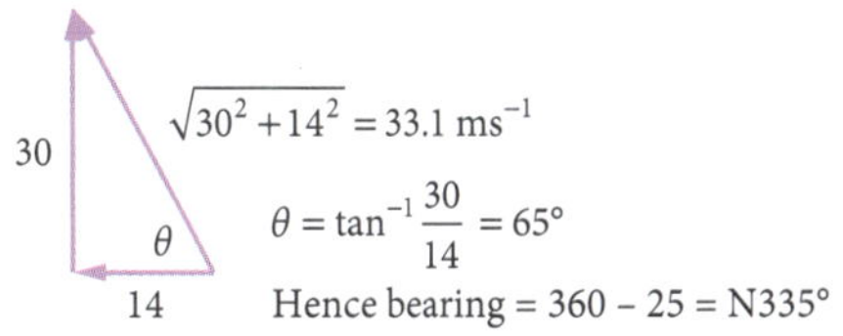

Figure A2.6 Vector diagram required to find relative velocity

b Subtract the velocity of the observer, as shown in Figure A2.7 by reversing the direction of the negative vector and adding. This yields: $v_{relative} = 33.1\ ms^{-1}$ ✓ N155° ✓

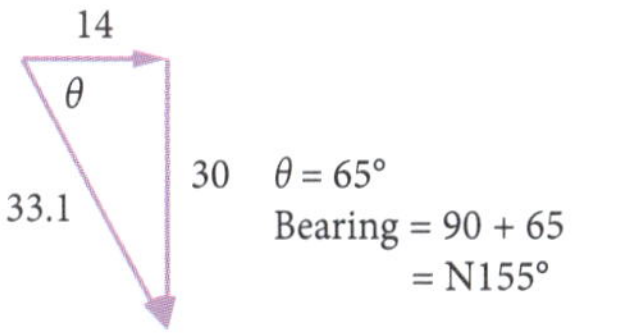

Figure A2.7 Vector diagram required to find relative velocity

8 **EM** This is a relative motion question that requires the student to understand the effect of a moving medium on velocity and direction of travel to travel in a straight line to a specific destination. Students also need to use their knowledge of motion at constant velocity to solve the problem.

a $v = s/t$ and hence $t = s/v = \frac{10}{1.2} = 8.3$ s ✓

b The butterfly must fly in the direction that will have an easterly component to cancel the wind speed if it is to fly directly north. Thus the butterfly must fly at an angle θ as shown in Figure A2.8. Solving the vector diagram gives:

$\theta = \tan^{-1}\frac{1}{1.2}$

$= \text{N}39.8°$ ✓

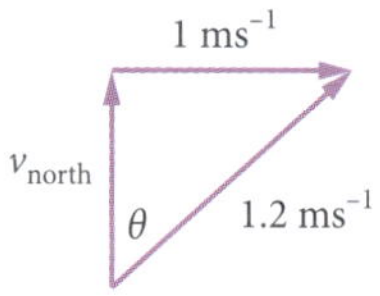

Figure A2.8 Vector diagram

c Solving the triangle in Figure A2.8 for v_{north} gives:

$v_{north} = \sqrt{1.2^2 - 1^2} = 0.66 \text{ ms}^{-1}$ north ✓

d $t = s/v = \frac{10}{0.66} = 15$ s ✓

9 **EM** This question requires students to demonstrate an understanding of motion in a moving medium, relative motion and motion at constant velocity.

a Adding the vectors as shown in Figure A2.9, yields $\theta = \text{N}50.2°$ ✓

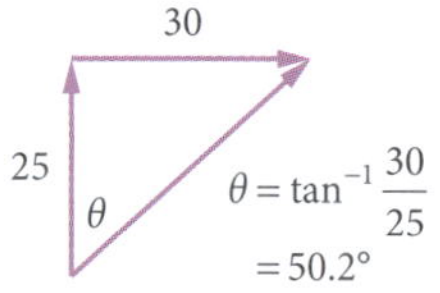

Figure A2.9 Vector diagram

b Solving for v_{drone} in the vector diagram in Figure 2.23 gives:

$v_{drone} = \sqrt{30^2 + 25^2} = 39.1 \text{ ms}^{-1}$ ✓ N50.2° ✓

c $s = vt = 39.1 \times (10 \times 60) = 23.4$ km N50.2° from the point where the wind started to blow. ✓

10 **EM** This question requires students to demonstrate an understanding of relative motion in two dimensions and to perform a vector subtraction. The equation is:

$v_{relative} = v_{object} - v_{observer}$

The observer is the boy and relative velocity is 13 ms^{-1} towards the north-west.

a The correct vector diagram is shown in Figure A2.10. Here 1 mark is allocated for correct values and 1 mark for arrows drawn at the correct end of vectors.

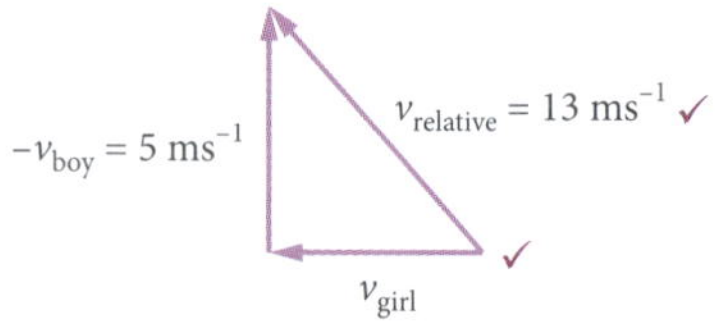

Figure A2.10 Vector diagram

b $v_{object} = \sqrt{13^2 - 5^2} = 12$

$v_{object} = v_{girl} = 12 \text{ ms}^{-1}$ ✓ west ✓

CHAPTER 3 FORCES

INQUIRY QUESTION:

How are forces produced between objects and what effects do forces produce?

Forces between objects can be exerted through direct contact between the objects or at a distance by the action of fields. In either case an equal and opposite force is exerted on each object. If there is a net force exerted on an object it will accelerate in the direction of the net force.

1 Newton's laws of motion

» Students, using Newton's laws of motion, describe static and dynamic interactions between two or more objects and the changes that result from:
- a contact force
- a force mediated by fields.

Force and momentum

- Isaac Newton introduced the concept of *force* (F). We can think of a force as a push or pull.
- The SI unit of force is the newton:

  newton (N) = $kgms^{-2}$
- Forces can act through direct contact or at a distance through the action of fields such as gravitational fields or magnetic fields.
- Newton also introduced the concept of **momentum** (p). Momentum is vector quantity equal to the product of a moving object's mass and velocity:

$$p = mv$$

where m is the mass
v is the velocity

We can see from this equation that the SI units for momentum are $kgms^{-1}$.

momentum: the product of an object's mass and velocity

- Newton then combined the concepts of force and momentum to propose three laws to explain motion.

Newton's three laws of motion still form the bedrock of *dynamics*, the study of why things move.

First law

A body remains at rest or in motion in a straight line with constant velocity unless acted upon by a net unbalanced force.

- This first law is sometimes called the law of **inertia**.

inertia: the resistance of a body to having its state of motion changed (i.e. its resistance to being accelerated)

Second law

The time rate of change of the momentum is equal to the applied force.

- This law can be expressed in two ways. We can express the second law in terms of *momentum* (p):

$$F = \Delta p / \Delta t$$

where F is the applied force
Δp is the change in momentum
Δt is the time over which the force is applied

Or in terms of *acceleration* (a):

$$F = ma$$

Third law

For every action there is an equal and opposite reaction.

- This means that forces always act in equal and opposite pairs on different objects. We can express this idea mathematically as:

$$F_{AB} = -F_{BA}$$

where F_{AB} refers to the action or applied force to object B
$-F_{BA}$ refers to the **reaction force** on object A

reaction force: the force operating in the opposite direction to the applied force

Weight and mass

- The force that gravity exerts on an object near the Earth's surface is called the **weight** and as $F = ma$ then:

$$\text{Weight} = mg$$

where m is the mass
g is the acceleration due to gravity at the point

weight: the force that gravity exerts on an object

- Note that g is the local acceleration due to gravity and, because weight is a force, it is measured in newtons.
- Another important difference between mass and weight is that mass is a measure of inertia and is the same everywhere, but weight depends on the strength of the local gravitational field and hence would be different on different planets (see Figure 3.1).

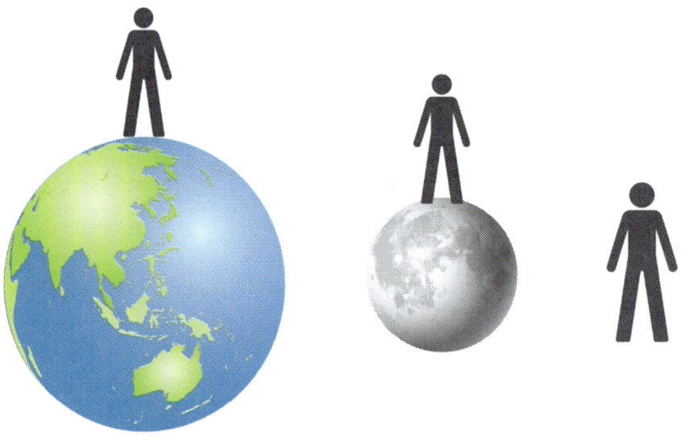

Figure 3.1 Difference between mass and weight

Applying Newton's laws of motion

- When a rope, wire, chain or other similar device is stretched between two objects it will apply an equal and opposite force to each object. We call this force the **tension** in the rope or wire.

tension: the action/reaction forces that a stretched rope, wire, chain, etc., exerts on each end of the rope, chain or wire

- Newton's laws act on entire systems and on each part of the system. For example, consider the following explanations of the simple system shown in Figure 3.2, which we assume is frictionless.

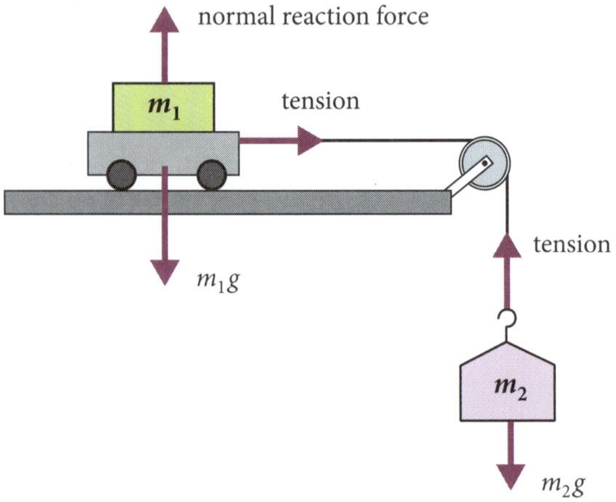

Figure 3.2 A multi-body problem involving forces

For the whole system

- The net force on the system would be the weight of m_2 only and this force would be acting on the total mass of the system $(m_1 + m_2 + m_{trolley})$. Applying Newton's second law to the whole system we get:

$F = ma$

$m_2 g = (m_1 + m_2 + m_{trolley})a$

For mass m_2

- Only two forces are acting on mass m_2: gravity $(m_2 g)$ down, and the tension in the rope (T) upwards, so we can write:

$F = ma$

$m_2 g - T = m_2 a$

For mass m_1 and the trolley

- The force of gravity is balanced by the **normal reaction force** from the table and hence the net force acting on the trolley is the tension in the string, so we can write:

$F = ma$

$T = (m_1 + m_{trolley})a$

- When determining the net force on an object it is often useful to draw a **free body diagram** to represent the size and direction of all the forces acting on the object. For example, there are four forces acing on a plane in flight: the weight force, the lift from the wings, the thrust from the engine and the drag force from air friction. If all these forces are balanced the plane will move with constant velocity. Figure 3.3 shows a free body diagram of the forces acting on an aircraft travelling at constant velocity.

normal reaction force: the reaction force that operates at 90° to the surface on which an object is resting
free body diagram: a diagram showing the size and direction of all the forces acting on an object

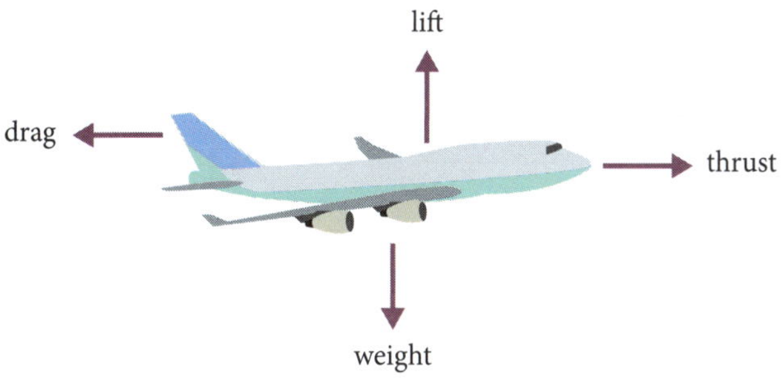

Figure 3.3 Free body diagrams for a plane travelling horizontally at constant velocity

→ KEY QUESTIONS

1 How can we predict the motion of an object when multiple forces are applied to the object?

2 Can a force ever act in isolation; that is, can a force operate without a reaction force?

Answers ➲ p. 39

EXAMPLE 1

A car of mass 1500 kg applies a horizontal force of 2200 N to the road when travelling at a constant velocity of 25 ms^{-1} (Figure 3.4). Assuming that the frictional force acting on the car is constant:

a **How much friction must be opposing the motion when the car is travelling at 25 ms^{-1}?**

b **If the engine applied twice this force, how fast would the car accelerate?**

c **If the engine stopped applying a force to the car when it was initially moving at 25 ms^{-1}, how far would it travel before coming to rest?**

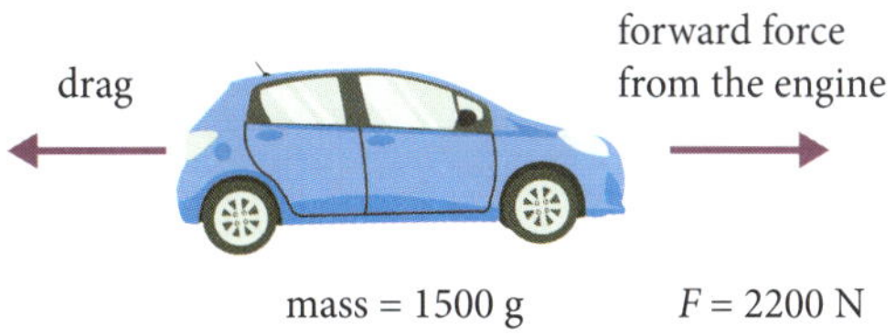

Figure 3.4 Horizontal forces acting on a car

Answer:

Recall that the net force on an object is the vector sum of the applied forces

a As the car moves at constant velocity, the net force on the car must be zero by Newton's first law. Thus the friction force must be equal and opposite to the applied force:

$F_{friction}$ = 2200 N in the opposite direction to the motion

b If the applied force was doubled, the net force on the car would be:

$F_{net} = F_{applied} - F_{friction} = 2 \times 2200 - 2200 = 2200$ N

As $F = ma$, $a = F/m = \frac{2200}{1500} = 1.47$ ms^{-2} in the direction of the applied force.

c If the motor force was removed the car would be slowed by a frictional force of 2200 N from part (a).

As $F = ma$, $a = F/m = \frac{2200}{1500} = 1.47$ ms^{-2} in the opposite direction to the motion. If we call the direction of the car positive, this acceleration can be written as $a = -1.47$ ms^{-2}.

Now $v^2 = 2as + u^2$ and as the car comes to rest $v = 0$ so we can write:

$s = -u^2/2a = \frac{-25^2}{2 \times -1.47} = 212.6$ m

EXAMPLE 2

For the simple multibody system shown in Figure 3.5, find the acceleration of the 1.8 kg mass and the tension in the string. (You may assume the pulley has no mass.)

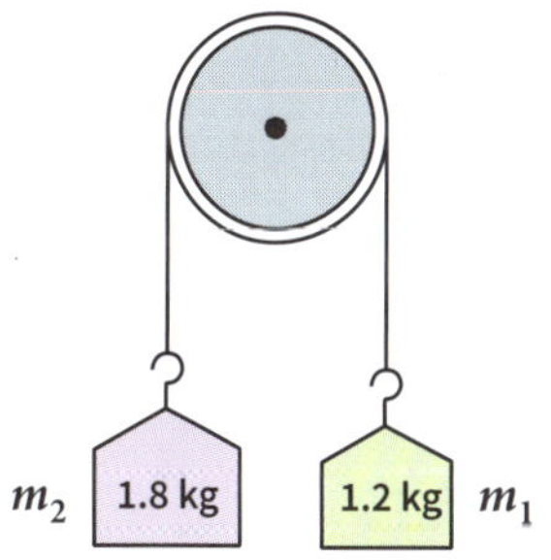

Figure 3.5 Two body system

Answer:

Newton's laws apply to the whole system and each part of the system

Consider the entire system first and assume motion that causes the pulley to rotate anticlockwise is positive. As the weight of each mass acts in opposite directions on the system, the net force on the system would be given by:

$F_{net} = 1.8g - 1.2g = 0.6g$

Now as $F = ma$ for the system, $F_{net} = 0.6g = (1.8 + 1.2)a$

$a = \frac{0.6 \times 9.8}{1.8 + 1.2} = 1.96$ ms^{-2} anticlockwise

To find the tension (T) we write $F_{net} = ma$ for the 1.8 kg block, noting that the net force on the block is given by $mg - T$, which gives:

$F_{net} = ma$

$m_2 g - T = m_2 a$

Hence $T = m_2 g - m_2 a = 1.8 \times 9.8 - 1.8 \times 1.96 = 14.1$ N

2 Forces in equilibrium

» Students explore the concept of net force and equilibrium in one-dimensional and simple two-dimensional contexts using:
- algebraic addition
- vector addition
- vector addition by resolution into components.

→ The vector sum of the forces acting on an object is called the **net force**.

→ If the net force on an object is zero we say the forces are in **equilibrium** and the object must be at rest or moving with constant velocity in a straight line.

net or resultant force: the vector sum of the forces acting on an object

equilibrium: an object or system with no net force; that is, a system in which all the forces cancel out (sum to zero)

→ If there is a non-zero net force acting on an object it will accelerate in the direction of the net force with an acceleration given by:

$$a = F_{net}/m$$

→ Note that acceleration will change the magnitude of the velocity or the direction of the velocity or both, depending on the direction of the acceleration vector and the initial velocity vector.

EXAMPLE 3

A tightrope walker of mass 75 kg balances in the middle of the wire, as shown in Figure 3.6. Find the tension in the tight rope.

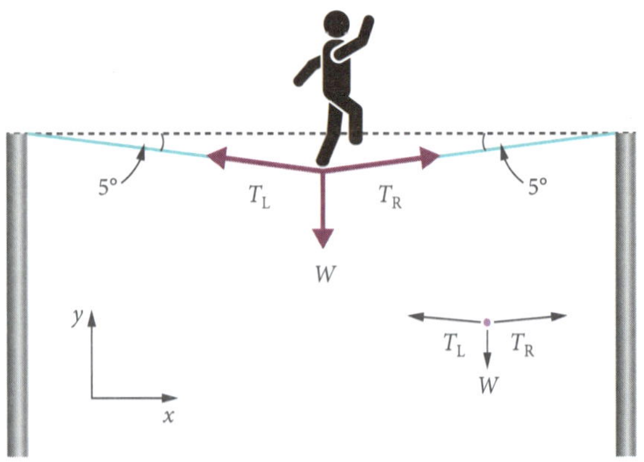

Figure 3.6 Forces acting on a tightrope

Answer:

The vector sum of the forces on a body at rest must sum to zero

At the point of contact of the wire, three forces are acting: the tension pulling right (T_R) and left (T_L) in the tightrope, and the weight of the man. As the man is in the centre of the wire, $T_R = T_L$ and as the tightrope walker is as rest, the net force on the walker must be zero. Hence the vector sum of the tensions and the weight force (W_R) must sum to zero. We can represent this as a vector sum, as shown in Figure 3.7.

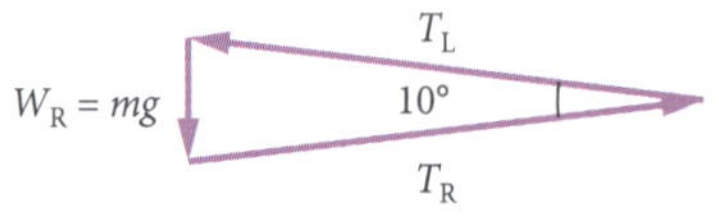

Figure 3.7 The sum of forces acting on the tightrope walker sum to zero

As $T_R = T_L$, this triangle can be broken into two right-angle triangles by drawing a horizontal line to bisect the triangle. The lower triangle that results, shown in Figure 3.8, can be used to calculate the tension in the wire.

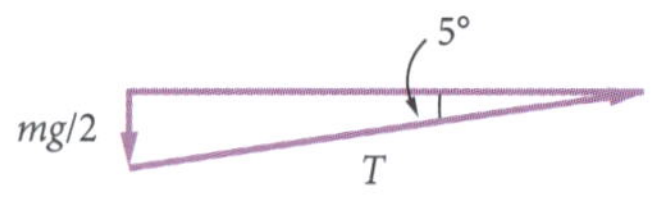

Figure 3.8 Vector diagram used to calculate tension in tight rope

$\sin\theta = mg/2T$ and hence the tension = $T = mg/2\sin\theta$

$$= \frac{75 \times 9.8}{2\sin 5°} = 4217 \text{ N}$$

→ KEY QUESTIONS

3 **What do Newton's laws of motion tell us about the vector sum of the forces applied to an object at rest and one moving with constant velocity?**

4 **If a net force is applied to an object, does the magnitude of the object's velocity always have to change?**

Answers ⊃ p. 39

EXAMPLE 4

Figure 3.9 shows two boys pushing a boat of mass 80 kg. One boy pushes the boat north with a force of 80 N, while the other boy pushes east with a force of 150 N. If the water did not exert any friction on the boat, how fast would it accelerate as the boys pushed it?

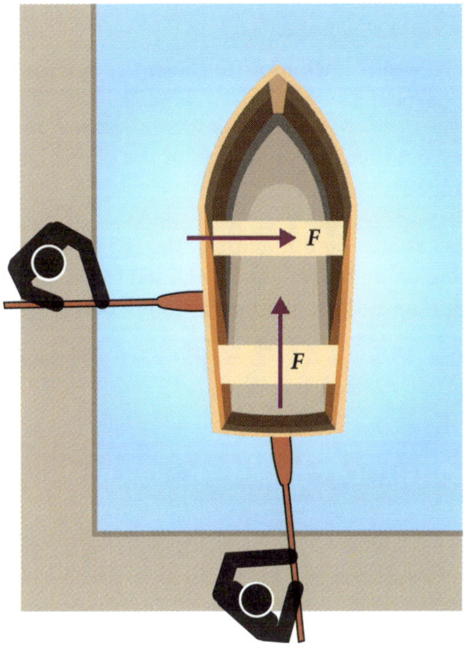

Figure 3.9 Forces acting on a boat

Answer:

To find the acceleration we must first use a vector diagram (Figure 3.10) to find the net force on the boat.

The net force on an object is the vector sum of the applied forces

$$\theta = \tan^{-1}\frac{80}{150} = 28^\circ$$

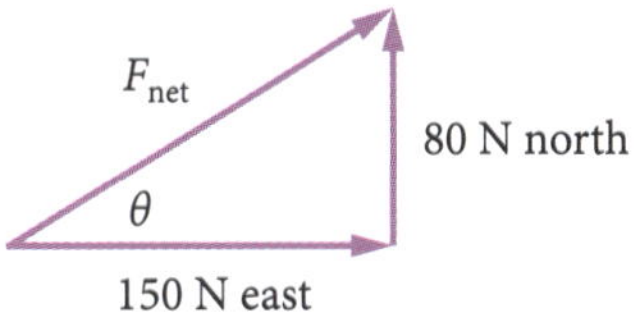

Figure 3.10 Vector diagram required to find resultant force acting on the boat

From Figure 3.10, $F_{net} = \sqrt{80^2 + 150^2} = 170$ N at N62°

As $F = ma$, $a = F/m = \frac{170}{80} = 2.1$ ms^{-2} and N62°

3 Resultant and component forces

» Students solve problems or make quantitative predictions about resultant and component forces by applying the following relationships:

$\vec{F}_{AB} = -\vec{F}_{BA}$

$F_x = F\cos\theta$, $F_y = F\sin\theta$.

- Any vector can be broken into two mutually perpendicular component vectors.
- The **components of a force** vector can be calculated, as shown in Figure 3.11.

components of a force: two mutually perpendicular component-force vectors that when added together would produce the initial-force vector

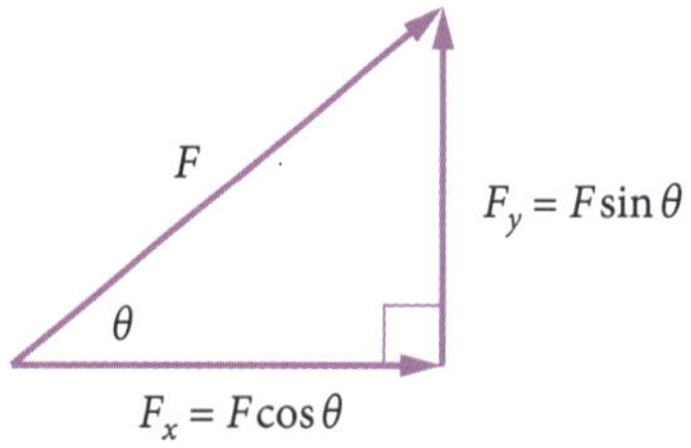

Figure 3.11 Mutually perpendicular component vectors of a force

- Force vectors can be added using vector diagrams or they can be added by breaking each vector into components and adding the components in each direction.
- Components of vectors and many aspects of Newton's laws can be demonstrated by investigating motion on an inclined plane. The forces and components of the weight of an object on an inclined plane are shown in Figure 3.12.

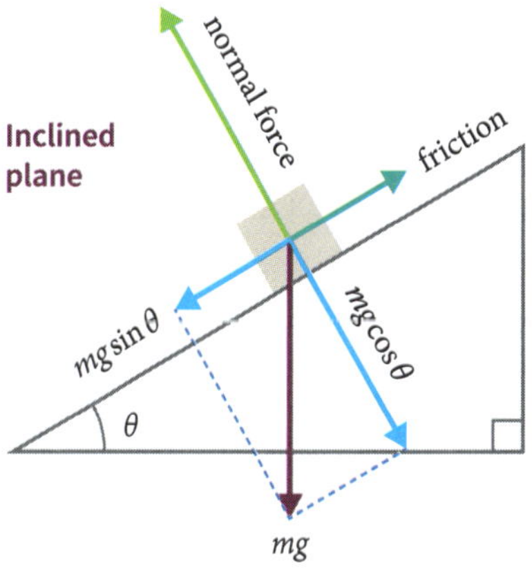

Figure 3.12 Forces acting on a body on an inclined plane

→ Note the *normal reaction force* on the object is the result of Newton's third law of motion, $F_{AB} = -F_{BA}$. The object applies a force into the plane ($mg\cos\theta$ into the plane), and the plane applies an equal and opposite force to the object (normal reaction force = $mg\cos\theta$ out of the plane).

EXAMPLE 5

Two forces are applied to a 30 kg block on a frictionless plane, as shown in Figure 3.13.

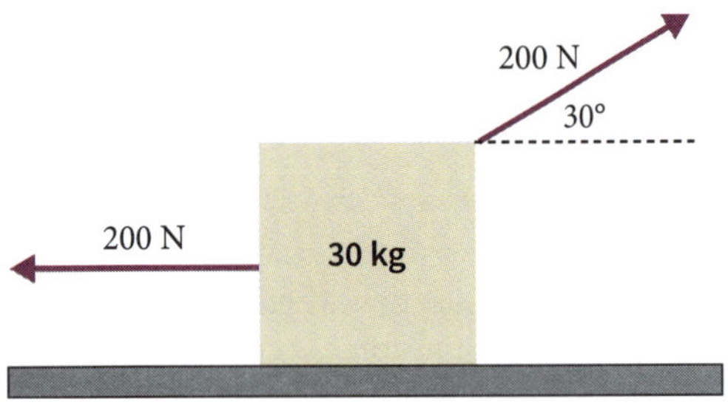

Figure 3.13 Forces acting on a box on a smooth horizontal plane

a **Find the net force on the block.**

b **Calculate the force the block applies to the ground.**

c **Determine the acceleration of the block.**

d **With what sized force would the 200 N force to the right have to be replaced to ensure the block remained at rest?**

Answer:

An upward force will affect the normal reaction force and the force of friction

a The vertical forces on the block will sum to zero, as the upwards component of the force on the right is less than the weight. If we call forces to the left positive, the horizontal component of the force acting to the right is given by:

$F_{horizontal} = -200\cos 30°$

The net horizontal force is therefore:

$F_{net} = 200 - 200\cos 30° = 26.8$ N to the left

b The weight of the block will be reduced by the vertical component of the force on the right and hence the downwards force will be:

$F_{down} = mg - 200\sin 30° = 30 \times 9.8 - 200\sin 30°$
$= 194$ N downwards

c Because the downwards force is balanced by the upwards reaction force, the net force on the block is the horizontal force = 26.8 N.

Now as $F = ma$, $a = F/m = \dfrac{26.8}{30}$

$= 0.89$ ms^{-2} to the left.

d The horizontal component of the force to the right would have to balance the 200 N force to the left.

As $F_{left} = F\cos\theta$ we have:

$$F = F_{left}/\cos\theta = \frac{200}{\cos 30°} = 231 \text{ N}$$

→ KEY QUESTIONS

5 **Do Newton's laws apply to the entire system, or to the parts of a system?**

6 **How can the state of motion of the object be changed?**

7 **What are the components of the weight forces acting on an object on an inclined plane, parallel to the plane and perpendicular to the plane?**

Answers ⊃ p. 39

4 Motion on inclined planes

» Students conduct a practical investigation to explain and predict the motion of objects on inclined planes.

FIRSTHAND INVESTIGATION

Forces on a body on an inclined plane

In this experiment a trolley is placed on an inclined plane and the force parallel to the plane on the trolley is measured for different angles of the plane. The apparatus is illustrated in Figure 3.14.

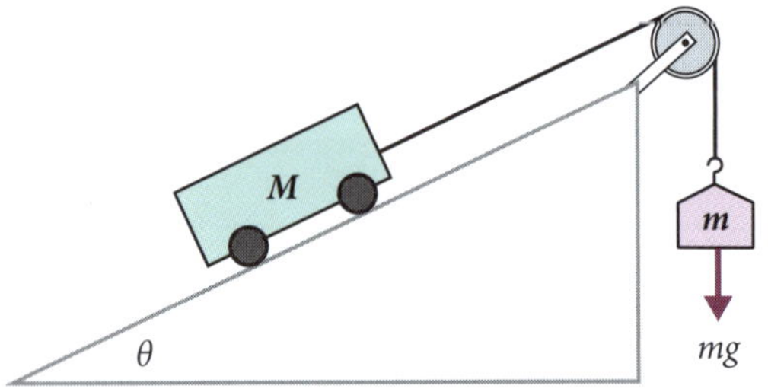

Figure 3.14 Trolley at rest on an inclined plane with all forces in equilibrium

Masses are added to the pan on the right until the forces on the trolley are in equilibrium. When the trolley is at rest, the force up the plane must balance the force down the plane:

$$mg = Mg\sin\theta$$
$$m = M\sin\theta$$

If the theory of components is correct, a plot of the mass m against $\sin\theta$ should give a straight line with a gradient equal to the mass of the trolley M.

Motion of a body on an inclined plane

This experiment (shown in Figure 3.15) is used to test the predictions about motion on an inclined plane. By measuring the length of the inclined plane (s) and the time (t) for the trolley to roll down the plane from rest, we can calculate the acceleration down the plane for different angles of the plane (θ).

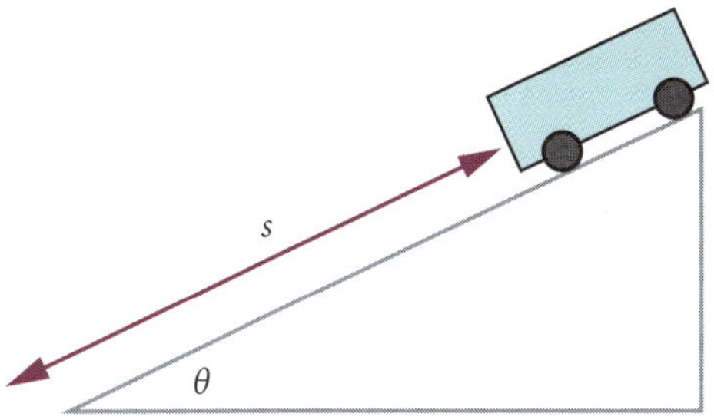

Figure 3.15 Experiment to measure acceleration of a trolley on an inclined plane

The acceleration can be calculated from the equations of motion:

$s = ut + \frac{1}{2}at^2$ and, as $u = 0$ then

$a = 2s/t^2$

Theory suggests the acceleration of the trolley down the plane should be equal to the component of the acceleration due to gravity parallel to the plane:

$a = g\sin\theta$

If the theory is correct, a plot of the measured acceleration against $\sin\theta$ should give a straight line with the gradient equal to the acceleration due to gravity (g). Hence to verify the theory we must plot the measured values of acceleration as a function of $\sin\theta$ and confirm that the graph is a straight line with a gradient = 9.8 ms^{-2}.

CHAPTER SYLLABUS CHECKLIST

Are you able to answer these questions from the syllabus for this chapter? Tick each question as you go through the checklist if you are able to answer it. If you cannot answer a question, turn to the relevant page in the study guide to find the answer. For NESA key word meanings, go to www.educationstandards.nsw.edu.au and search 'key words'.

FOR A COMPLETE UNDERSTANDING OF THIS TOPIC:		PAGE NO.	✓
1	Can I define the term *force* and differentiate between contact forces and forces produced by fields?	29	
2	Can I use Newton's laws of motion and the concept of forces to explain why objects: • remain at rest? • move at constant velocity? • accelerate or decelerate? • change their direction of motion?	29	
3	Can I explain the difference between mass and weight?	30	
4	Can I apply Newton's laws of motion to explain the motion of simple systems involving more than one object and the forces acting between the objects?	30–31	
5	Can I add forces in one and two dimensions to determine the net force on an object or to show that the forces acting on the object are in equilibrium?	32	
6	Can I resolve vectors into perpendicular components?	33	
7	Can I determine the forces acting on an object on an inclined plane, and can I resolve the weight force into components down the plane and perpendicular to the plane?	33	
8	Can I find the net force acting on an object on an inclined plane and use this to predict how the object will move on the plane?	34–35	

YEAR 11 EXAM-TYPE QUESTIONS

Objective-response questions

(1 mark each)

1 **An ice skater is observed to move at constant velocity. What can we say about the forces acting on the skater?**

A There are no forces acting on the skater.
B There is no friction in action.
C There is a small net force acting on the skater in the direction of her motion.
D The vector sum of all the forces acting on the skater must sum to zero.

2 **Which statement best describes the forces that enable a person to jump upwards from the ground?**

A The person applies an upwards force to the ground.
B The person applies less force to the ground than the ground applies to the person.
C The ground applies more force to the person than the person applies to the ground.
D The magnitude of the force applied by the person to the ground is equal to the force applied by the ground to the person.

3 **If we neglect air friction, which of the statements below gives the best description of the acceleration of an object thrown into the air?**

A The acceleration is only downwards after it reaches the top of the flight.
B The acceleration is downwards throughout the entire flight.
C The acceleration changes direction when the ball reaches the top of its flight.
D The acceleration is upwards throughout the entire flight.

4 **A set of traffic lights hangs from two wires, as shown in Figure 3.16.**

Figure 3.16 A traffic light suspended by two wires

How could we best describe the forces acting in this situation?

A The vector sum of the forces applied by the wires to the light must be greater than the force of gravity acting on the light.
B The vector sum of the forces applied by the wires to the light must be less than the force of gravity acting on the light.
C The net force on the light must be downwards.
D The vector sum of the forces applied by the wires to the light must be equal to the force of gravity acting on the light.

5 **Consider a mass on an inclined plane, as shown in Figure 3.17.**

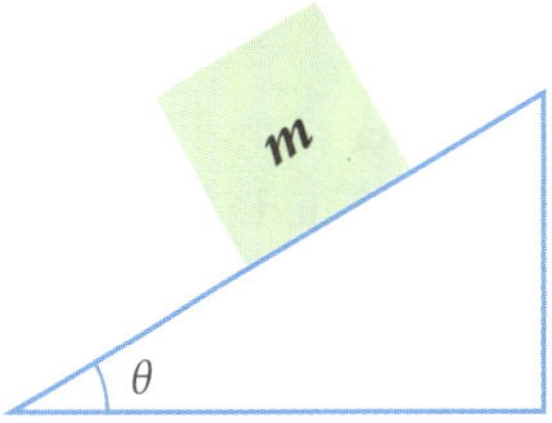

Figure 3.17 A mass on an inclined plane

If the plane was perfectly smooth, on what would the acceleration of the mass down the plane depend?

A The local acceleration due to gravity g and the angle θ.
B The angle θ and the mass M.
C The local acceleration due to gravity g, the mass M and the angle θ.
D The local acceleration due to only gravity g

Extended-response questions

6 **Four puppies pull a bone in different directions with the forces shown in Figure 3.18.**

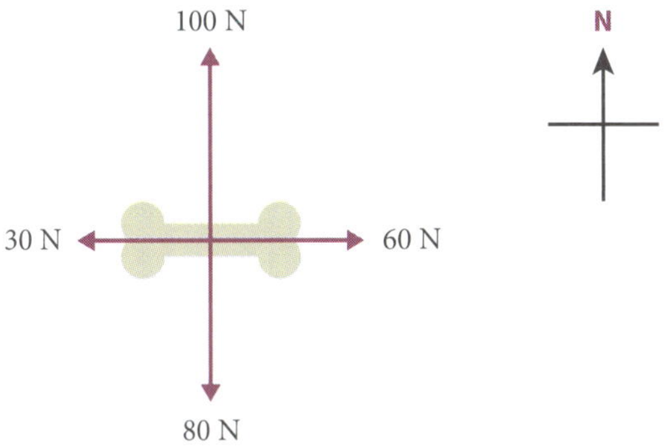

Figure 3.18 Forces applied to a bone by four puppies

a **Find the component of the net force on the bone in the north–south direction.** (1 mark)

b Find the component of the net force on the bone in the east–west direction. (1 mark)

c Find the magnitude and direction of the net force on the bone. (2 marks)

d Find the magnitude and direction of the force that would have to be applied to the bone to stop it from accelerating. (1 mark)

7 A force of 30 N acts on three blocks on a smooth, frictionless plane, as shown in Figure 3.19.

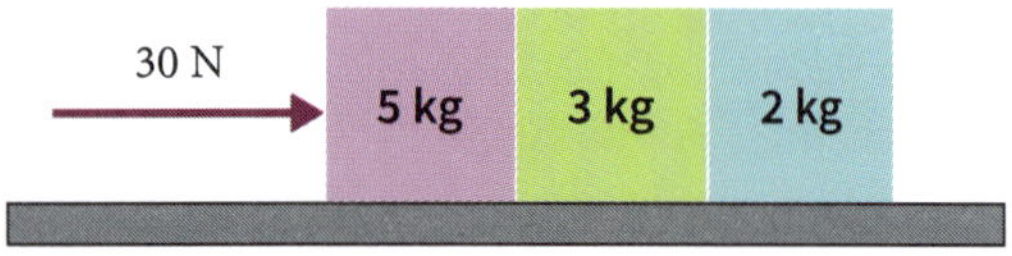

Figure 3.19 Total force acting on three blocks on a smooth, horizontal plane

a Find the acceleration of the system of blocks. (1 mark)

b Find the force applied to the 2 kg block by the 3 kg block. (1 mark)

c Find the net force on the 3 kg block. (1 mark)

d Find the force applied to the 3 kg block by the 2 kg block. (1 mark)

e Find the force applied to the 3 kg block by the 5 kg block. (1 mark)

8 A car of mass 1500 kg travelling at 15 ms^{-1} runs through a 20-metre-long puddle of water that slows the car to 10 ms^{-1}.

a Find the rate at which the car accelerates in the puddle. (1 mark)

b Find the force exerted on the car by the puddle. (1 mark)

c Assuming the force exerted by the puddle is constant, how long would the puddle have to be to bring the car to rest? (1 mark)

9 The experimental apparatus (Attwood's machine) used by a student to measure the acceleration due to gravity is shown in Figure 3.20. You may assume the pulley is weightless, the system has no friction and $m_1 = 4$ kg and $m_2 = 6$ kg.

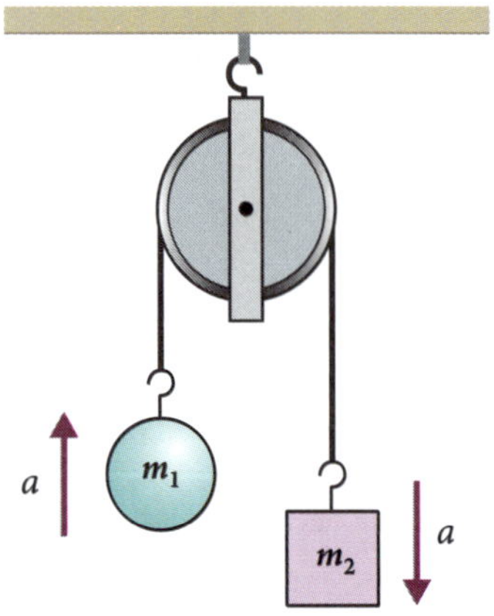

Figure 3.20 Attwood's machine used to measure acceleration due to gravity

a Find the force of gravity on:

i m_1 (1 mark)

ii m_2 (1 mark)

b Determine the acceleration of the system of masses. (2 marks)

c Determine the tension of the string joining the masses. (1 mark)

d Find the net force on m_1. (1 mark)

e Find the tension in the string holding the pulley. (1 mark)

10 A 10 kg block of wood rests on a smooth (frictionless) inclined plane that is inclined at 30° to the horizontal, as shown in Figure 3.21.

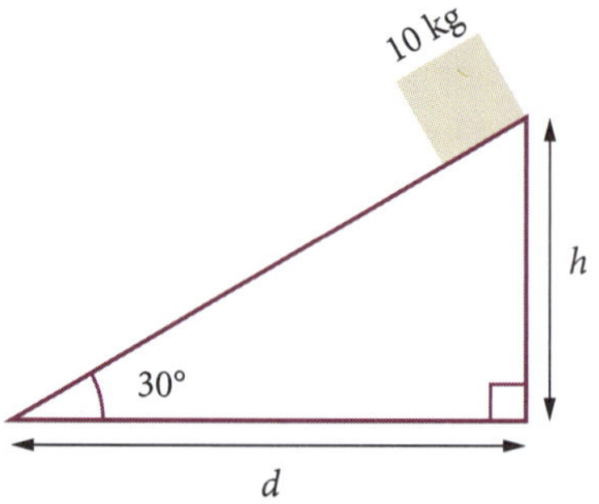

Figure 3.21 Ten-kilogram block on a smooth inclined plane

a Find the weight of the block. (1 mark)

b Find the component of the weight down the plane. (1 mark)

c Find the acceleration of the block down the plane. (1 mark)

d Find the time for the block to reach the bottom of the plane given that the box started from rest at a height of 15 m above the ground (i.e. $h = 15$ m). (1 mark)

e Find the speed of the block when it reaches the bottom of the plane. (1 mark)

ANSWERS

KEY QUESTIONS

Key questions ➲ p. 31

1 The motion will be determined by the net force on the object. The net or resultant force is the vector sum of the forces acting on an object. If there is a net force on the object it will accelerate in the direction of the force with an acceleration given by $a = F_{net}/m$.

2 No, forces always act in equal and opposite pairs.

Key questions ➲ p. 32

3 In both cases the net force acting must be zero and hence the vector sum of the forces acting on a body at rest or one moving with constant velocity must be zero.

4 No, the direction of velocity may also change. If the force was always applied perpendicular to the velocity, the direction of the velocity would continually change but the magnitude of the velocity would remain constant. This is what happens when planets and satellites orbit large central bodies.

Key questions ➲ p. 34

5 Newton's laws can be applied to entire systems or to any part of the system.

6 The state of motion of an object can only change if a net force is applied to the object.

7 The component of the weight of a body parallel to an inclined plane is $mg\sin\theta$ and the component perpendicular to the plane is $mg\cos\theta$.

YEAR 11 EXAM-TYPE QUESTIONS

Objective-response questions

1 **D**. Newton's first law states an object will only move with constant velocity if the net force on the body is zero. **A** is incorrect because there could be many forces on the skater; as long as they sum to zero she will continue to move at constant velocity. **B** is incorrect because friction is a force and another force could be applied to balance it. **C** is incorrect because if there was any net force on the skater she would accelerate.

2 **D**. Newton's third law states that forces always act in equal and opposite pairs. The person moves much further than the Earth ($a = F/m$) because the person has much less mass than the Earth. **A** is incorrect because the ground would apply a downwards force to the person if the person pulled the ground up. **B** and **C** are incorrect because force pairs are always equal by Newton's third law.

3 **B**. The acceleration is produced by the force acting on the body and the only force is gravity downwards hence the acceleration will be downwards. All the other answers are incorrect because they state that the acceleration is not downwards at some time of the flight.

4 **D**. The sum of the forces of a body at rest must be zero by Newton's first law. The other answers are all incorrect because they all imply that a net force would operate on the light. If a net force was on the light it would accelerate in the direction of the force rather than remain at rest.

5 **A**. The component of the acceleration down the plane would be given by $a = g\sin\theta$. Note that the acceleration would only depend on the mass if surface friction was present. **B** and **C** are wrong because acceleration due to gravity is independent of mass. **D** is wrong because a mass will obviously accelerate faster down a steeply inclined plane than a plane with a small incline.

Extended-response questions

6 **EM** This question requires students to add multiple vectors in two dimensions and to find the magnitude and direction of the resultant force. In addition, students have to apply Newton's second law.

- a Calling north positive, the north–south component = 100 – 80 = 20 N north. ✓
- b Calling east positive, the east–west component would be = 60 – 30 = 30 N east. ✓
- c Adding these two vectors, as shown in Figure A3.1, we get net force = 36 N ✓ at N56.3°. ✓

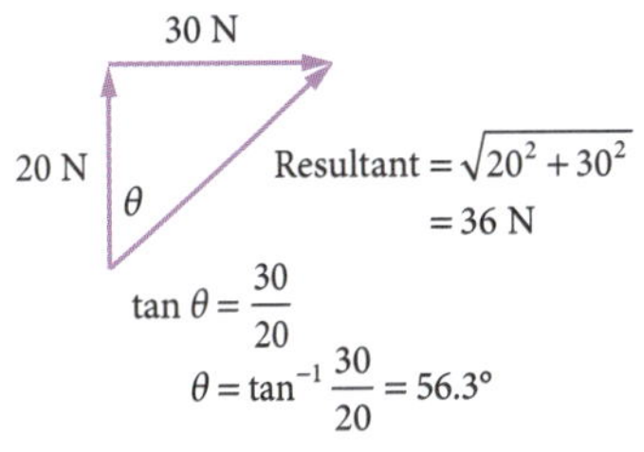

Figure A3.1 Vector addition

- d To balance the applied force, a force of equal magnitude in the opposite direction would have to be applied; that is, F = 36 N at N = 236.3°. ✓

7 **EM** This question requires students to demonstrate a knowledge of Newton's three laws and the concept of applying the laws to a whole system and part of a system.

Calling all vectors to the right positive:

- a $F = ma$ for the system and hence $a = F/m = \frac{30}{10} = 3\ \text{ms}^{-2}$ ✓
- b The only force applied to the 2 kg block is from the 3 kg block. Applying $F = ma$ to the 2 kg block gives $F = ma = 2 \times 3 = 6$ N. ✓
- c Applying $F = ma$ to the 3 kg block, $F = 3 \times 3 = 9$ N. ✓
- d This is the reaction force to the force that the 3 kg block applies to the 2 kg block, hence $F = -6$ N. ✓
- e The net force on the 3 kg block is $F = 9$ N = force applied by the 5 kg block – force applied by the 2 kg block. Hence $9 = F - 6$ or $F = 15$ N. ✓ Alternatively, the force must be great enough to accelerate the 3 kg and 2 kg blocks at 3 ms^{-1}, or $F = ma = (2 + 3)3 = 15$ N. ✓

8 **EM** This question requires students to combine the laws of motion and the equations of motion to solve a problem with resistive force (friction).

- a $v^2 = u^2 + 2as$ and hence $a = (v^2 - u^2)/2s = \frac{10^2 - 15^2}{2 \times 20} = -3.1\ \text{ms}^{-2}$ ✓
- b $F = ma = 1500 \times 3.1 = 4.7 \times 10^3$ N in the opposite direction to the initial velocity. ✓
- c $v^2 = u^2 + 2as$ and the final velocity would be zero ($v = 0$). Thus $s = (-u^2)/2a = = 36$ m. ✓

9 **EM** This question requires students to combine their understanding of Newton's laws, weight and tension. It requires students to apply Newton's laws to a whole system and the parts of a system in a multibody problem.

Calling clockwise motion positive:

a **i** $w = m_2g = 6 \times 9.8 = 58.8$ N ✓ and

ii $w = m_1g = 4 \times 9.8 = 39.2$ N ✓

b $F = ma$ for the system and hence $m_2g - m_1g = (m_2 + m_1)a$ ✓

Or $a = (m_2 - m_1)g/(m_2 + m_1) = \dfrac{(6-4)9.8}{(6+4)}$

$= 1.96$ ms^{-2} clockwise ✓

c Consider $F = ma$ for the mass m_2. The net force on the mass is $m_2g - T$ and, as $F_{net} = ma$ for this block, we can write $m_2g - T = m_2a$ or $T = m_2(g - a) = 6(9.8 - 1.96) = 47.04$ N. ✓

d The net force on block m_1 is given by $T - m_1g = 47.04 - 4 \times 9.8 = 7.8$ N. Alternatively we can use the relationship $F_{net} = ma$ for this block and hence $F_{net} = 4 \times 1.96 = 7.84$ N. ✓

e As the tension is 47.04 N on both sides of the pulley, the total force on the pulley and the tension in the rope holding it would be $2 \times 47.04 = 94.08$ N. ✓

10 **EM** This question requires students to apply Newton's laws, the concept of weight, and the equations of motion to the motion of an object on an inclined plane.

a $w = mg = 10 \times 9.8 = 98$ N ✓

b The component down the plane $= mg\sin 30°$

$= 10 \times 9.8 \times \sin 30°$

$= 49$ N. ✓

c $F = ma$ and hence $a = F/m = \dfrac{49}{10} = 4.9$ ms^{-2} ✓

d Length of the plane $= h/\sin 30° = 30$ m. Now $s = \frac{1}{2}at^2 + ut$ and,

as $u = 0$, $t = \sqrt{\dfrac{2s}{a}} = \sqrt{\dfrac{2\times 30}{4.9}} = 3.5$ s. ✓

e $v = u + at$ and, as $u = 0$, $v = at = 4.9 \times 3.5 = 17.15$ ms^{-1} ✓

CHAPTER 4 FORCES, ACCELERATION AND ENERGY

INQUIRY QUESTION:

How can the motion of objects be explained and analysed?

Newton's laws of motion can be combined with the concepts of work and energy to explain and analyse motion.

1 Newton's second law and friction

» Students apply Newton's first two laws of motion to a variety of everyday situations, including both static and dynamic examples, and include the role played by friction ($f_{\text{friction}} = \mu \vec{F}_N$).

- Newton's laws use the concept of force to explain why objects move and why they remain at rest (**static**).

static: state of being at rest

- Newton's first law tells us that if an object is at rest or moving at constant velocity in a straight line (i.e. undergoing **rectilinear motion**), the vector sum of the forces acting on an object must be zero. Conversely, if the sum of the forces acting on an object is zero then the object will remain at rest or moving with constant velocity. For example, the force of gravity and friction must be equal when a falling body reaches terminal velocity because the body is no longer accelerating. Figure 4.1 shows how the drag forces increase with velocity until the drag forces equal the force of gravity on the body.

rectilinear motion: motion in a straight line

- When you give an object a push on Earth it does not continue to move with constant velocity in a straight line because gravity exerts a downwards force on it and because frictional forces oppose the motion and will eventually bring the moving object to rest.

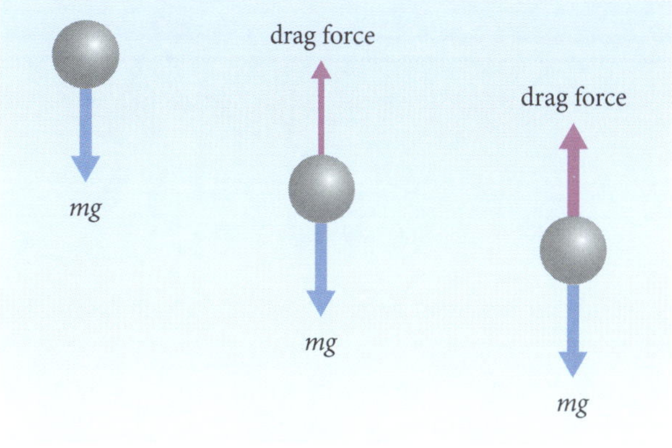

Figure 4.1 Forces on a falling object

- However, Newton's first law is easily demonstrated by astronauts in the weightless environment of orbiting spacecraft. When an astronaut releases a ball, the ball moves in a straight line at constant velocity with respect to the astronaut.
- When calculating motion on Earth we often have to take the force of friction into account. This is difficult for air resistance as this drag force is dependent on the velocity and the shape of the moving object. Surface friction on the other hand is largely independent of the velocity of the moving object and can be subtracted from the applied force to determine the net force on a moving object.
- Surface friction opposes motion and is related to the normal reaction force (F_N) by:

$$f_{\text{friction}} = \mu F_N$$

where μ is the **coefficient of friction**, a measure of the roughness of the surfaces in contact
F_N is the **normal reaction force**

coefficient of friction: a dimensionless quantity that reflects the roughness of two surfaces in contact; the force of friction (f) is related to the coefficient of friction by $f_{\text{friction}} = \mu F_N$
normal reaction force: the component of the reaction force perpendicular to the surface of contact

EXAMPLE 1

Figure 4.2 shows a woman pushing a 40 kg sofa. The sofa was initially at rest and the woman applied a constant horizontal force of 220 N for 2 seconds. If the coefficient of friction between the sofa and the floor was 0.2, find how far the sofa moved during the time it was being pushed.

Figure 4.2 Woman pushing a sofa on a rough floor

Answer:

The net force on a body is the vector sum of the applied forces

The force of friction that opposes the motion is given by:

$f_{friction} = \mu F_N = \mu mg = 0.2 \times 40 \times 9.8 = 78.4$ N

As the friction opposes the applied force, the net force would be:

$F_{net} = F_{applied} - f_{friction} = 220 - 78.4 = 141.6$ N

The acceleration of the sofa would be:

$F = m$ and hence $a = F/m = \dfrac{141.6}{40} = 3.54$ ms^{-2}

The displacement would be given by:

$s = ut + \frac{1}{2}at^2$

As the initial velocity was zero, the displacement of the sofa would be:

$s = \frac{1}{2}at^2 = \frac{1}{2}(3.54)(2)^2 = 7.08$ m

EXAMPLE 2

Figure 4.3 shows a 15 kg block of wood accelerating down an inclined plane at 2 ms^{-2}. If the angle of the plane is $\theta = 30°$, find the coefficient of friction between the box and the plane.

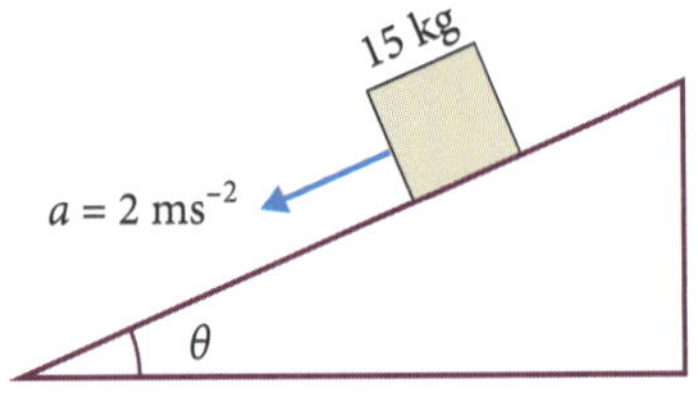

Figure 4.3 A block sliding on a rough inclined plane

Answer:

Friction and the applied force will determine the net force on an accelerating body

As the rate of acceleration is known, the net force on the block can be calculated using Newton's second law:

$F_{net} = ma = 15 \times 2 = 30$ N

Breaking the weight into components, the force down the plane is given by:

$mg\sin\theta = 15 \times 9.8 \sin 30° = 73.5$ N

The force into the plane is $mg\cos\theta$ and hence the normal reaction force is:

$mg\cos\theta = 15 \times 9.8 \cos 30° = 127.3$ N

The net force on the block would be:

F_{net} = force down the plane – force of friction

$F_{net} = mg\sin\theta - \mu mg\cos\theta$

Hence $\mu = (mg\sin\theta - F_{net})/mg\cos\theta = \dfrac{73.5 - 30}{127.3} = 0.34$

→ KEY QUESTIONS

1 **What condition must be met for the centre of mass of an object to remain at rest or move at constant velocity in a straight line?**

2 **How do we take surface friction into account when calculating motion?**

Answers ➲ p. 52

2 Work and energy

» Students apply the special case of conservation of mechanical energy to the quantitative analysis of motion involving:
- work done and change in the kinetic energy of an object undergoing accelerated rectilinear motion in one dimension $W = F_{||}s = Fs\cos\theta$
- changes in gravitational potential energy of an object in a uniform field $\Delta U = mg\Delta h$.

→ **Work** (W) is done when an object moves in response to a force being applied to the object.

work: the energy transferred to an object when a force moves the object

→ The quantity of work performed on the object is given by:

$$W = F_{||}s = Fs\cos\theta$$

where F is the applied force
s is the displacement produced by the force

- Work and energy are measured in joules, where:

  Joule = Nm = kgm^2s^{-2}

- When the force is not in the same direction as the displacement (s), the component of the force in the direction of the displacement is used to calculate the work done.
- For example, the work done by the force (F) on the object shown in Figure 4.4 would be given by work = $(F\cos\theta)d$ as the component of the force in the direction of the displacement is $F\cos\theta$.

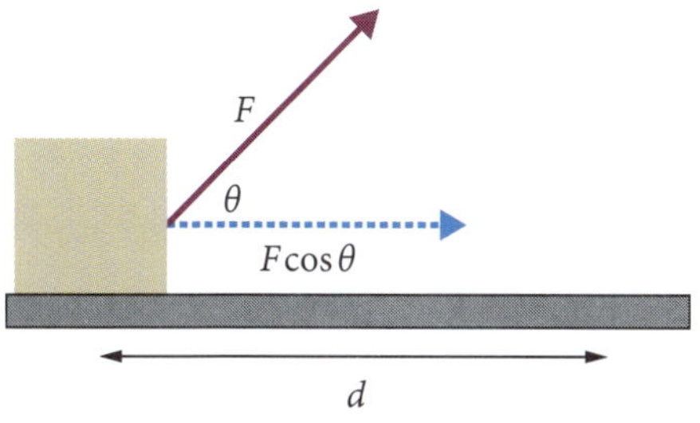

Figure 4.4 Component of the force parallel to the displacement

- We can think of work as energy being transferred from one object to another. Hence to do work an object must have energy. Another way to think of this is to say a body possesses energy if it can do work.
- If the applied force is not constant, the area under a graph of force against displacement is used to determine the work done by the force. For example, the force required to stretch a spring increases as the spring stretches, and the work done to stretch the spring can be found by measuring the area below the force versus displacement graph, as shown in Figure 4.5.

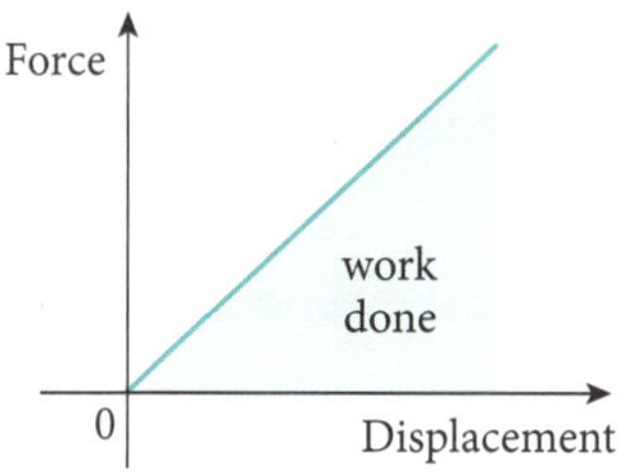

Figure 4.5 Force versus displacement graph

- In a closed system the total energy of the system always remains constant and, while energy may change from one form to another, it can never be created or destroyed. We can use the law of the conservation of energy to determine the energy gained by a body when work is done on the body, as outlined in the next two points.
- The work to lift an object to a height (Δh) is equal to **gravitational potential energy** gained by the body (ΔU). Figure 4.6 provides an example.

gravitational potential energy: the energy stored in a body due to its position in a gravitational field

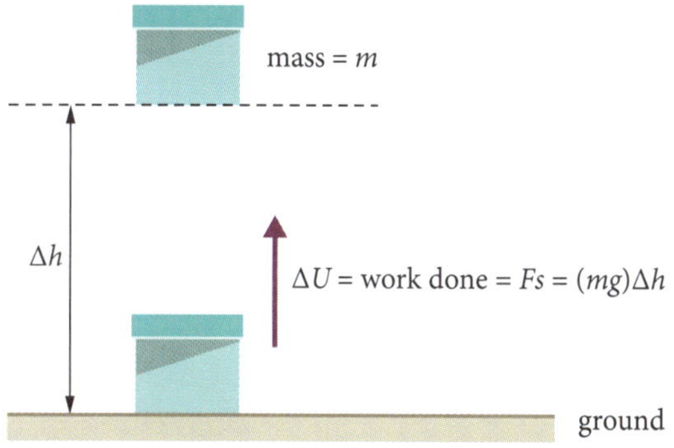

Figure 4.6 Work done to lift an object of mass m to a height h

The change in gravitational potential energy is therefore:

$$\Delta U = mg\Delta h$$

where m is the mass

g the acceleration due to gravity

Δh is the change in height

- The work done to accelerate an object from rest to velocity v is a measure of the energy contained by the moving object at velocity v. We call this the **kinetic energy** (K) of the object, demonstrated in the following example.

kinetic energy: the energy stored in a body due to its motion

Consider a truck of mass m accelerating from rest to velocity v, as shown in Figure 4.7.

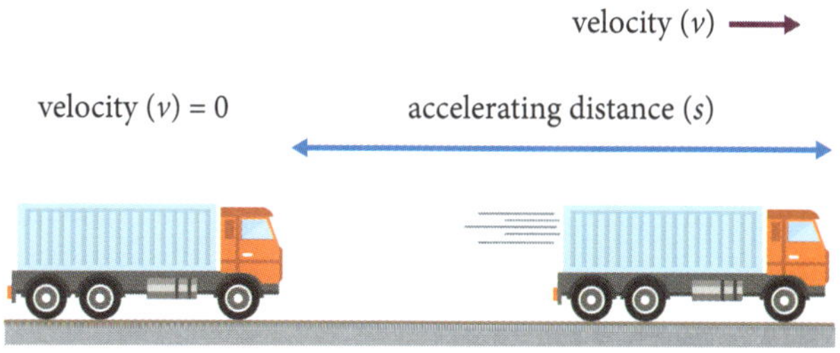

Figure 4.7 Deriving an expression for the kinetic energy of a moving object

Work done = $W = Fs$ and, as $F = ma$, we can write this as:

$W = mas$

From the equation of motion we have $v^2 = 2as + u^2$ and as $u = 0$ we can rearrange this equation as:

$as = v^2/2$

Substituting this into the first equation gives an expression for the kinetic energy:

$$K = \frac{1}{2}mv^2$$

where m is the mass
v is the velocity

- An object is said to have gravitational potential energy ($U = mgh$) when raised to a height h above the ground, while an object moving with velocity v is said to have kinetic energy ($K = \frac{1}{2}mv^2$).
- In the absence of air friction, as shown in Figure 4.8, the total energy of a body in freefall will be constant. Hence as it falls its gravitational potential energy (U) must be converted into kinetic energy (K) to ensure the total energy of the falling object remains constant. If a ball is thrown upwards it will slow down as its kinetic energy is transformed into gravitational potential energy.

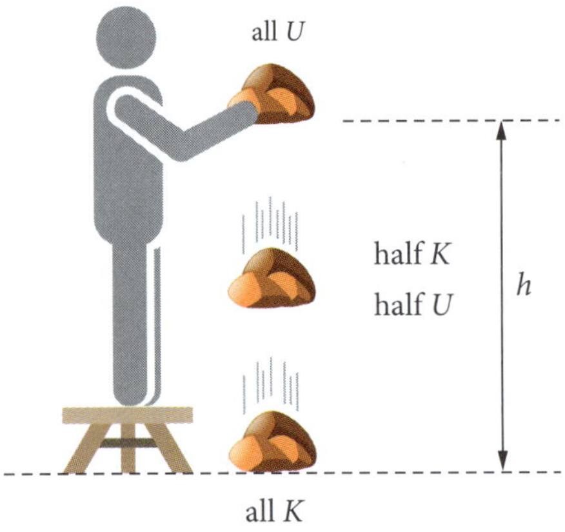

Figure 4.8 Energy of a falling object

At the top of the flight all the energy would be gravitational potential energy and, just before it landed, all the energy would be in the form of kinetic energy.

- **Power** (P) is the rate at which work is done or energy is transformed from one form to another. Hence:

$$P = \Delta E/t = \text{work}/t$$

where ΔE is the amount of energy transferred/ transformed
W is the work done
t is the time over which the work was done

Power is measured in watts, thus using unit analysis we can see that:

Watt = $Js^{-1} = Nms^{-1} = kgm^2s^{-3}$

power: the rate at which work is done or energy is transformed

- If a force is applied to an object and it moves at constant velocity, the force applied and the opposing frictional force must be equal. The work done against friction is therefore:

$W = F_{applied}\, s$

The power required to do this work is given by:

Power $P = W/t = F_{applied}\, s/t$

As the velocity is constant $v = s/t$ and this expression reduces to:

$$\text{Power} = P = F_{||}v = Fv\cos\theta$$

where F is the applied force
v the velocity

EXAMPLE 3

A fireman of mass 85 kg slides down a 10 m pole, as shown in Figure 4.9.

Figure 4.9 Forces on a fireman

If he was initially at rest and reached a velocity of 2 ms^{-1} just before he reached the ground, find:

- **a** **the initial gravitational potential energy of the fireman**
- **b** **the final kinetic energy of the fireman**
- **c** **the work done against friction as the fireman slid down the pole**
- **d** **the force of friction on the fireman as he slid down the pole (assume this friction force is constant).**

Answer:

a The initial gravitational potential energy of the fireman is:

Energy is conserved in a closed system

$\Delta U = mg\Delta h = 85 \times 9.8 \times 10 = 8330$ J

b The final kinetic energy of the fireman is:

$K = \frac{1}{2}mv^2 = \frac{1}{2} \times 85 \times 2^2 = 170$ J

c As energy is conserved in the system:

Initial energy = work done against friction + kinetic energy gained

Hence the work done against friction = initial energy − K gained = 8330 − 170 = 8160 J

d As work = Fs we can write

$F = \text{work}/s = \frac{8160}{10} = 816$ N upwards.

→ KEY QUESTIONS

3 **How are force, work, energy and power related to one another?**

4 **What do we mean when we say a body possesses kinetic energy or gravitational potential energy?**

5 **If air friction is neglected, how does the energy of an object change as it falls under the influence of gravity?**

Answers ➲ p. 52

3 Investigating force and motion

» Students investigate, describe and analyse the acceleration of a single object subjected to a constant net force and relate the motion of the object to Newton's second law of motion through the use of:

- qualitative descriptions
- graphs and vectors
- deriving relationships from graphical representations including $\vec{F}_{net} = m\vec{a}$ and relationships of uniformly accelerated motion.

FIRSTHAND INVESTIGATION

Newton's second law

We can verify Newton's second law by applying different forces to an object and measuring the resulting acceleration of the object. A simple experiment to do this is shown in Figure 4.10. In this experiment the mass of the object being accelerated is the trolley–mass system.

The force applied to the system is the weight of the mass carrier = mg. By moving the masses from the trolley to the mass carrier, the force (mg) acting on the system can be changed, while keeping the total mass of the system constant.

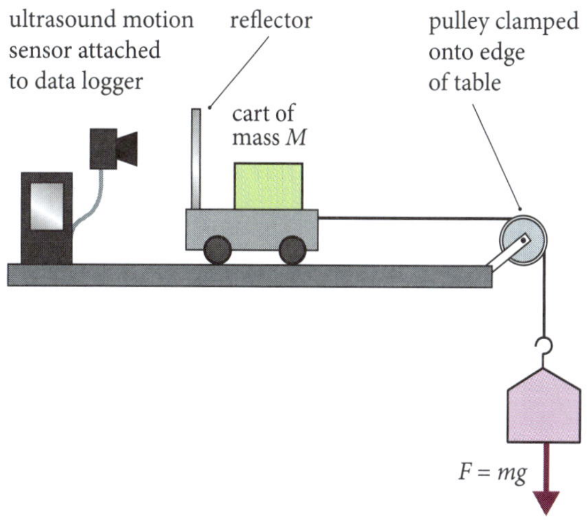

Figure 4.10 Newton's second law experiment

A ticker-tape timer or sonic ranger can be used to measure the acceleration produced for different applied forces. By plotting the acceleration against the applied force, as shown in Figure 4.11, we obtain a graph that represents the relationship $F = ma$.

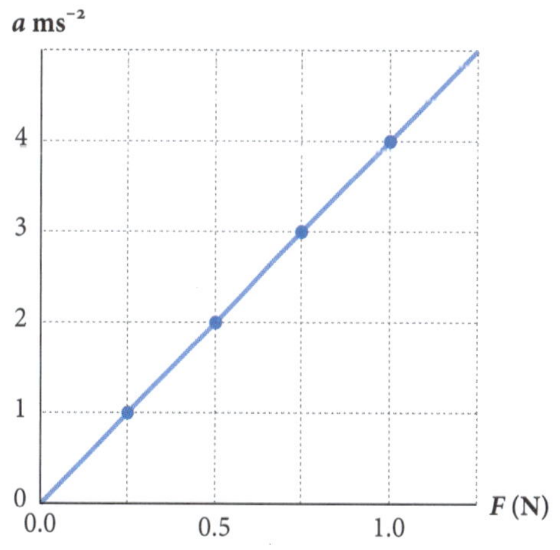

Figure 4.11 Typical results for Newton's second law experiment

The gradient gives the mass of the system, which in this case is $\frac{4}{1} = 4$ kg.

4 Investigating friction, work, energy and power

» Students conduct investigations over a range of mechanical processes to analyse qualitatively and quantitatively the concept of average power ($P = \frac{\Delta E}{\Delta t}$, $P = F_{\parallel}v = Fv\cos\theta$), including but limited to:
- uniformly accelerated rectilinear motion
- objects raised against the force of gravity
- work done against air friction, rolling resistance and friction.

SECONDARY-SOURCED INVESTIGATIONS

The following investigations enable students to gain a deeper understanding of friction, work and energy.

Using gravitational potential energy to produce and store electrical energy

Students can visit a hydroelectric power station or research how hydroelectric power stations use gravity to produce and store energy.

The gravitational potential energy of water changes into kinetic energy as it falls. Energy in falling water is used to turn electric generators to produce electrical energy in hydroelectric plants. During times when the grid has too much electricity, the excess can be used to pump water back up to the high-level reservoir where it can be used at a later time to generate electricity when demand is high. A schematic diagram showing how a hydroelectric system can produce and store energy is shown in Figure 4.12.

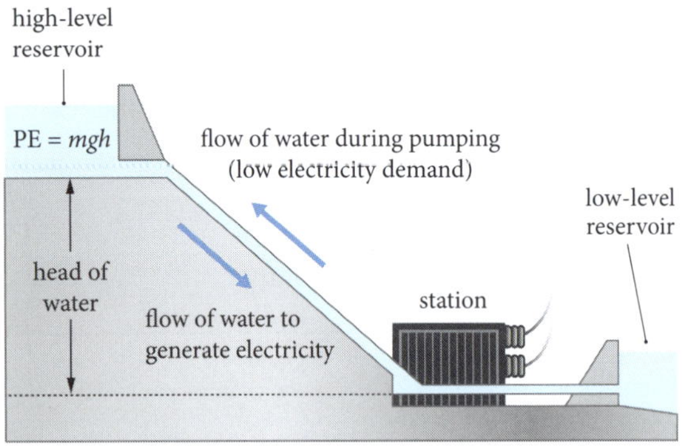

Figure 4.12 Hydroelectric energy production and storage system

The power of the electrical pump will determine how fast water is pushed up to the upper reservoir. Assuming 100% efficiency:

Power of pump = $P = \Delta E$ of the water/time = mgh/t

Hence the mass of water pumped to the upper reservoir in a time (t) would be:

$m = Pt/gh$

Measuring the coefficient of friction (μ)

This experiment investigates the resistance to moving an object from rest due to friction with the surface of contact (**static friction**) μ_s and the resistance to keeping an object moving (**kinetic friction**) μ_k.

static friction: the friction between the object and the surface it rests on that produces a force opposing an object moving from rest
kinetic friction: the friction between the object and the surface of contact when the object is moving

Consider the mass on an inclined plane in Figure 4.13.

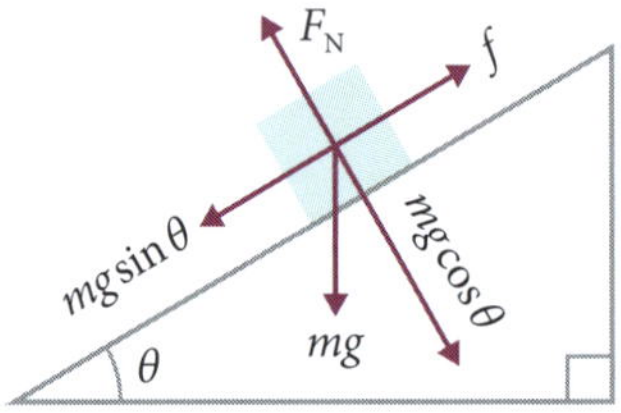

Figure 4.13 Forces acting on an object on an inclined plane

Static friction

We can measure static friction by slowly increasing the angle of the plane (θ) until the mass just starts to move. When motion starts, the force down the plane ($mg\sin\theta$) is equal to the static friction. We can use this value of the friction to calculate the coefficient of static friction for the surfaces:

$f_{\text{friction}} = mg\sin\theta = \mu_s F_N$

And as $F_N = mg\cos\theta$ then

$\mu_s = mg\sin\theta/F_N$
$= mg\sin\theta/mg\cos\theta$
$= \tan\theta$

Kinetic friction

To find the coefficient of kinetic friction we increase the angle of the plane (θ) until the mass continues to slide at constant velocity down the plane when it is given a slight push. The coefficient of kinetic friction is then given by:

$\mu_k = \tan\theta$

Constant acceleration down the plane

Once the coefficient of kinetic friction is known, the net force on the mass as it moves down the plane can be determined as follows:

Force down the plane = $mg\sin\theta$

$f_{friction} = \mu_k F_N = \mu_k(mg\cos\theta)$

Therefore $F_{net} = mg\sin\theta - \mu_k(mg\cos\theta)$ down the plane.

Students can conduct an experiment to verify this by measuring the acceleration of the body down the plane and finding the net force using $F = ma$.

Friction

Power required to do work against friction

If a force (F) has to be applied to an object for it move at constant velocity v, the power put into the object by the applied force is given by:

Power = $P = Fv$

If friction is opposing the motion this energy is converted into heat.

Static and kinetic friction

Generally it is harder to start an object moving than to keep it moving and, as consequence, there is a static coefficient of friction and kinetic coefficient of friction for surfaces rubbing together. Some typical values for the coefficient of friction are provided in Table 4.1.

Table 4.1 Coefficients of friction*

Surfaces	Coefficients of static friction, μ_s	Coefficients of kinetic friction, μ_k
Lubricated ball bearings	<0.01	<0.01
Metal on metal (lubricated)	0.15	0.07
Rubber on dry concrete	1.0	0.8
Rubber on wet concrete	0.7	0.5
Steel on steel (unlubricated)	0.7	0.6

* Approximate values only

Rolling resistance

Rolling resistance is the friction opposing the motion of rolling objects like wheels or balls. Rolling friction, like surface friction, depends on a constant coefficient of rolling friction μ_r for the object rolling and the normal reaction force on the object. Rolling friction is generally very much less than surface friction. It is represented by the equation:

Rolling friction = $\mu_r F_N$

For example, the rolling resistance of a pushbike is about 0.01.

rolling resistance: the frictional force that opposes the motion of a rolling object

Air friction

Unlike kinetic and rolling friction, **air friction** depends on the shape and size of objects and on the velocity at which the object travels. For small objects travelling at low speeds, air friction is proportional to the velocity:

Air friction = $f_{air} = bv$

where v is the velocity

b is a constant that depends on the dimensions of the moving object and air density

At high velocities, air friction is related to the square of the speed.

air friction: the drag force that opposes the motion of objects through the air

EXAMPLE 4

Sketch a velocity versus time graph for an object falling in the atmosphere.

The force exerted by air friction increases as an object moves faster

Answer:

Calling the downwards velocity positive, the velocity would change, as shown in Figure 4.14.

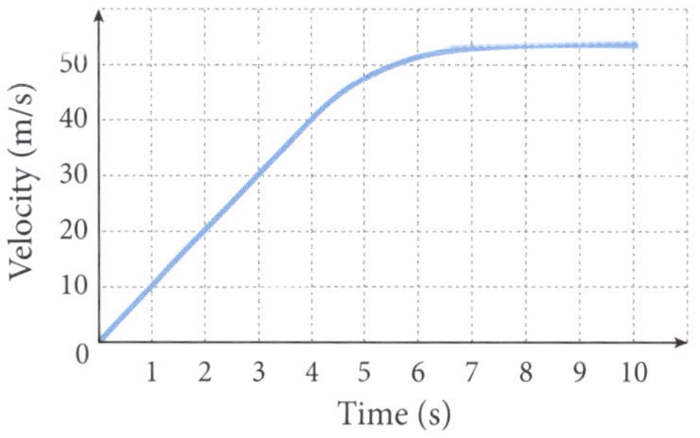

Figure 4.14 Velocity versus time graph for a falling object in air

Note that the gradient (acceleration) in initially 9.8 ms^{-2} but as the object falls its velocity increases and this in term increases air friction, which lowers the net force on the object. Eventually the velocity increases to a point when the air friction equals the weight and the net force on the object is zero. From this point it falls at constant velocity (terminal velocity) and the gradient is zero.

→ KEY QUESTIONS

6 How does kinetic friction depend on velocity and mass?

7 Is air friction constant? If not, what does it depend on?

8 Why do cars use more power to travel at greater velocities?

9 Would a spacecraft use power as it travelled at high speed in space?

Answers ➲ p. 52

CHAPTER SYLLABUS CHECKLIST

Are you able to answer these questions from the syllabus for this chapter? Tick each question as you go through the checklist if you are able to answer it. If you cannot answer a question, turn to the relevant page in the study guide to find the answer. For NESA key word meanings, go to www.educationstandards.nsw.edu.au and search 'key words'.

FOR A COMPLETE UNDERSTANDING OF THIS TOPIC:		PAGE NO.	✓
1	Can I use the coefficient of friction to determine the force of friction on an object of known mass?	41	
2	Can I use Newton's laws to explain motion that involves friction?	42	
3	Can I define and calculate the work that a force does when it moves an object?	43	
4	Can I use conservation of energy and the concept of work to derive expressions for the change in an object's gravitational potential energy and its kinetic energy?	43–44	
5	Can I use the special case of conservation of mechanical energy to predict transformations of gravitational and kinetic energy for a free-falling object?	44	
6	Can I use conservation of energy to solve problems such as simple pendulums, motion down inclined planes, and freefalling objects?	44	
7	Can I relate power to the rate at which energy is transformed?	44	
8	Can I use power to explain energy transformations involved when bodies accelerate and when bodies move with constant velocity against a friction?	44	
9	Can I conduct an investigation and produce graphical evidence to prove Newton's second law?	45	
10	Can I compare surface friction to rolling friction and air friction?	47	
11	Can I explain how the acceleration of a falling object changes due to air friction as it falls?	47	
12	Can I relate the terminal velocity of a falling object to the forces acting on the object?	47	

YEAR 11 EXAM-TYPE QUESTIONS

Objective-response questions (1 mark each)

1 A man pushes and pulls a rock at constant velocity, as shown in Figure 4.15. The rock has a mass M and its motion is resisted by surface friction between the rock and the ground. Which statement correctly compares the forces (F and F') applied to the rock by the man in each case?

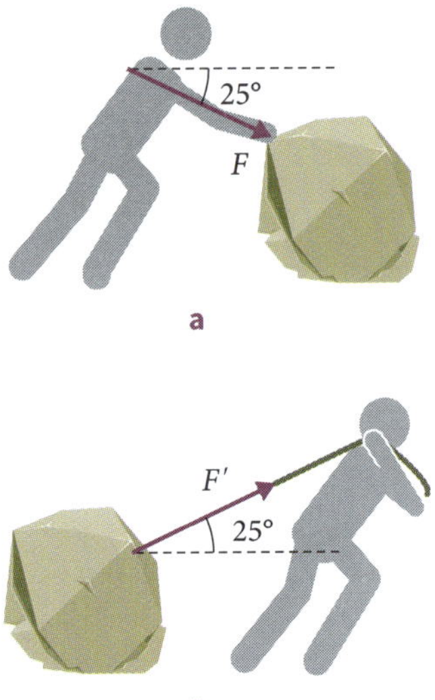

Figure 4.15 Using two different forces to move a rock

A $F > F'$

B $F < F'$

C $F = F'$

D It is impossible to compare the forces from the information provided.

2 A skier experiences a constant frictional force F when she slides down a hill. She skies straight down, along a path of length d. If the height of the hill is h, which of the following expressions shows how energy is conserved in this situation?

A $mgh = \frac{1}{2}mv^2$

B $mgh = Fd$

C $mgh = \frac{1}{2}mv^2 - Fd$

D $mgh = \frac{1}{2}mv^2 + Fd$

3 When dropped from a great height a cricket ball will attain a greater terminal velocity than a tennis ball. How do the forces on the balls compare when the balls reach terminal velocity?

A The net force on the cricket ball is greater.

B The force of air friction on the cricket ball is greater.

C The force of air friction on the tennis ball is greater.

D The net force on the tennis ball is greater.

4 Which of the following statements best describes why one team wins a tug of war, like the one shown in Figure 4.16?

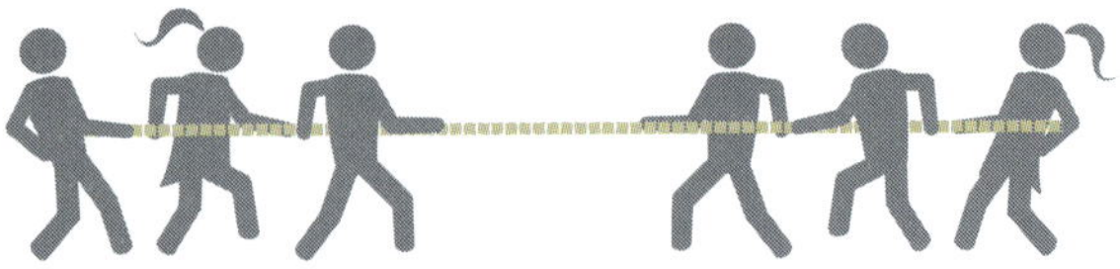

Figure 4.16 Students having a tug of war

A The winning team applies a greater force to the rope, which produces a net force on the system.

B The net force on the system is zero but the winning team applies a greater force to the rope.

C The net force on the system is zero but the winning team applies a greater force to the ground than the losing team.

D The winning team applies a greater force to the ground than the losing team, which produces a net force on the system.

5 It is easier to push a wheelbarrow up an inclined slope with a small gradient than it is to push a wheelbarrow up an inclined plane with a steep slope. Which of the following statements best explains why this occurs?

A Less work has to be done if the gradient is less steep.

B The same work is done but less force is required to push a wheelbarrow up a less-steep gradient.

C If the wheelbarrows take the same time to be pushed up the inclined planes, less power is required for the less-steep plane.

D Each barrow is just as difficult to push because the work done and force required are the same in each case.

Extended-response questions

6 A crate of mass 40 kg slides down a smooth inclined plane onto a rough horizontal surface, as shown in Figure 4.17.

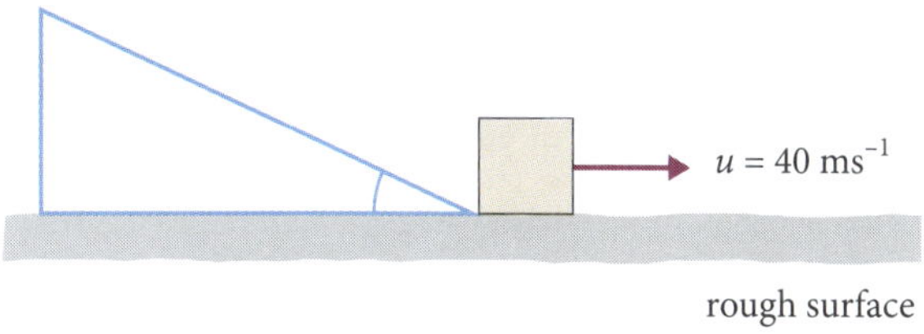

Figure 4.17 Crate moving on a rough horizontal surface

The crate leaves the ramp with a velocity of 4 ms^{-1} and the coefficient of friction between the crate and the horizontal surface is 0.02.

- a Find the force of friction experienced by the crate on the horizontal surface. (1 mark)
- b Find the rate at which the crate will decelerate when it is on the horizontal surface. (1 mark)
- c Find the distance the crate will travel on the horizontal surface before it comes to rest. (1 mark)
- d Find the work done by the crate against the frictional force on the horizontal surface. (1 mark)
- e Explain the energy transformations that occur as the crate slides down the inclined plane and comes to rest on the horizontal surface. (2 marks)

7 A student sets up the experiment shown in Figure 4.18, where a mass m_1 = 2 kg is pulled by a mass m_2. When m_2 is 1 kg the system moves with constant velocity.

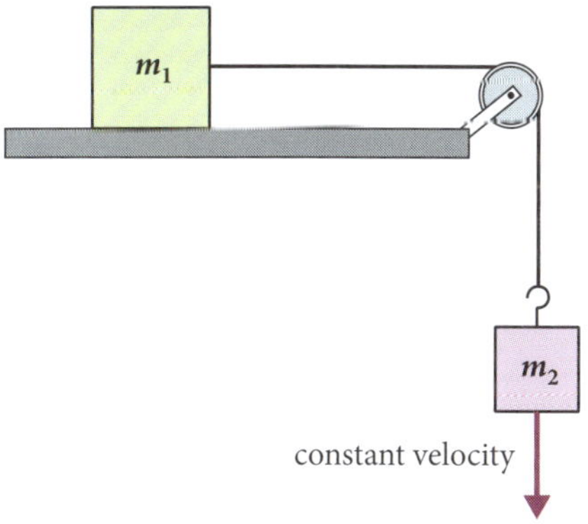

Figure 4.18 Mass experiment

- a Find the net force on the system. (1 mark)
- b Find the tension in the string. (1 mark)
- c Find the frictional force on m_1. (1 mark)
- d Find the coefficient of friction between m_1 and the surface. (1 mark)
- e If a larger mass m_2 was used, would the system continue to move at constant velocity? Justify your answer. (2 marks)

8 A car of mass 2000 kg coasts along a level road with a constant velocity of 80 kmh^{-1}. The coefficient of friction for the car is 0.08. Find:

- a the frictional force on the car (1 mark)
- b the power used by the engine to maintain this constant velocity (1 mark)
- c the energy used by the car to travel 10 km. (1 mark)

9 A man pushes a box of mass 35 kg up an inclined plane, as shown in Figure 4.19. The coefficient of friction is 0.5 and the man pushed the box at a constant velocity of 0.5 ms^{-1} up the slope.

Figure 4.19 Pushing a box up a rough inclined plane

Find:

- a the force friction on the box (1 mark)
- b the force applied to the box by the man (1 mark)
- c the power used by the man as he pushed the box (1 mark)
- d the total work done by the man to push the box up the slope (1 mark)
- e the potential energy gained by the box when it was moved from the bottom to the top of the slope. (1 mark)

10 A rubber ball of mass 0.2 kg is released 2 m above the ground. If the ball loses 20% of its kinetic energy every time it bounces, and air friction can be neglected, find:

- a the initial potential energy of the ball (1 mark)
- b the maximum velocity achieved by the ball (1 mark)
- c the impulse applied to the ball from the ground during the first bounce (2 marks)
- d the height reached by the ball after the second bounce. (1 mark)

ANSWERS

KEY QUESTIONS

Key questions ➲ p. 42

1 For an object to remain at rest, the sum of the forces acting on the body must be zero.

2 Surface friction is a force that opposes motion. Surface friction is the product of the normal reaction force (usually the weight) and the coefficient of friction for the surfaces involved.

Key questions ➲ p. 45

3 Work is energy being transformed or transferred to a different object and is equal to the product of force and the distance over which the force is applied. Power is the rate at which work is done.

4 Kinetic energy is energy of motion, while gravitational potential energy is the energy stored in a body due to its position in a gravitational field. A body with energy can do work.

5 The potential energy decreases and kinetic energy increases. This ensures the total energy remains constant.

Key questions ➲ p. 48

6 Kinetic friction, to a good approximation, is independent of the velocity, but it is proportional to the normal reaction force. (On a horizontal surface the normal reaction force is equal in magnitude to the weight of the object.)

7 No, air friction increases with velocity and depends on the shape of the moving object.

8 Air friction increases with speed and hence at higher velocities cars have to do more work against friction.

9 No. As there is no friction in space, travelling at constant velocity would require no force to be applied and hence no power. Accelerating to a high speed would require thrust, but Newton's first law tells us that once at the high speed the spacecraft would continue to move at constant velocity with no power input.

YEAR 11 EXAM-TYPE QUESTIONS

Objective-response questions

1 **A.** The horizontal component is the same for both forces and must equal friction as the rock moves at constant velocity. Because F' has an upwards component it reduces the friction as $F_{friction} = \mu F_N$. **B** is incorrect because pushing down would increase the friction. **C** is wrong because the friction will be different in each case. **D** is not true as the relative strength of the forces can be determined.

2 **D.** The U is converted into K and work against friction (heat) as she goes down the hill. **A** and **B** are incorrect because they do not take the work done against friction into account. **C** is incorrect because it implies that the force of friction gives the girl energy as she slides down the hill.

3 **B.** At terminal velocity the balls are not accelerating. Hence the net force is zero and as $mg = F_{air}$ the force of air friction must be greater on the heavier ball. **A** and **D** are incorrect because the ball will be travelling at constant velocity after it reaches terminal velocity and hence there would be no net force on the ball. **C** is incorrect because at the terminal velocity the weight force would equal the force due to air friction and, as the weight of the tennis ball is less than the cricket ball, the frictional force required to balance the weight would also be less.

4 **D.** The net force on the system is the sum of the reaction forces produced by the team pushing the ground. Hence the team that pushes the ground with the greatest force will win. **A** is incorrect because the tension in the rope requires both teams to experience equal and opposite forces from the rope. **B** and **C** are incorrect because if the net force on the system was zero the system would not accelerate in one direction.

5 **B.** The component of the weight down the plane is $mg\sin\theta$ and hence it requires less force to push the barrow if the gradient is less (θ smaller). Note that the same amount of work is done in either case. **A** is incorrect as the same work is done because the barrow is lifted to the same height (same change in gravitational potential energy). **C** is incorrect because if the barrow takes the same time to do the same amount of work it uses the same power. **D** is incorrect because the barrows will require a different force ($mg\sin\theta$) to push.

Extended-response questions

6 EM This question requires students to understand how to use the coefficient of friction to calculate friction, apply the equations of motion to a system slowing due to friction, and understand how conservation of energy is applied to an object slowing due to friction.

- a $f_{friction} = \mu F_N = \mu(mg) = 0.2(40 \times 9.8) = 78.4$ N ✓
- b $a = F/m = \dfrac{78.4}{40} = 1.96$ ms^{-2} ✓
- c $v^2 = u^2 + 2as$ and, as $v = 0$, we can write $s = -u^2/2a = \dfrac{4^2}{2\times1.96}$ $= 4.08$ m ✓
- d Work $= Fs = 78.4 \times 4.08 = 320$ J ✓
- e U is converted to K as it slides down the inclined plane ✓ and then the K is converted into heat due to friction as it comes to rest on the horizontal surface. ✓

7 EM This question requires students to apply Newton's laws to the whole system and part of a two-body system that is influenced by gravity and friction.

- a Net force on the system is zero as it moves with constant velocity. ✓
- b The net force on m_2 must also be zero as it moves at constant velocity and hence m_2g down = tension up or $1 \times 9.8 = T = 9.8$ N. ✓
- c Friction must balance the weight of m_2 for the system to move at constant velocity. Thus friction $= m_2g = 1 \times 98 = 9.8$ N to the left. ✓
- d $f_{friction} = \mu F_N = \mu m_1 g$ and hence $\mu = f_{friction}/m_1 g = \dfrac{9.8}{2\times9.8}$ $= 0.5$ ✓
- e No. ✓ The system would accelerate as the applied force m_2g would be greater than friction, resulting in a net force on the system. ✓

8 EM This question requires students to understand the relationship between friction, the applied force and the power input for an object moving with constant velocity. Students also have to utilise the relationship between power and energy.

- a $f_{friction} = \mu F_N = \mu mg = 0.08 \times 2000 \times 9.8 = 1568$ N ✓
- b Power $= Fv = \dfrac{1568 \times 80 \times 1000}{60 \times 60} = 34\,844 = 34.8$ kW ✓
- c Work $= Fs = 1568 \times (10 \times 1000) = 15.7$ MJ ✓

9 **EM** This question requires students to combine their understanding of Newton's first law, components of an object's weight on an inclined plane, the coefficient of friction and the conservation of energy.

a $f_{friction} = \mu F_N = \mu mg\cos\theta = 0.5 \times 35 \times 9.8 \times \cos 20°$ $= 161.16$ N down the ramp ✓

b As the box moves with the constant velocity the net force is zero and the man must be pushing with a force equal to the friction $= 161.16$ N up the ramp. ✓

c Power $= Fv = 161.16 \times 0.5 = 80.58$ watts ✓

d Length of inclined plane is given by $s = \frac{3}{\sin 20°} = 8.77$ m

Now work $= Fs = 161.16 \times 8.77 = 1413.37$ J ✓

e The change in $U = mgh = 35 \times 9.8 \times 3 = 1029$ J ✓

10 **EM** This question requires students to apply conservation of energy and the concept of impulse to a non-elastic collision between a ball and the ground.

a $U = mgh = 0.2 \times 9.8 \times 2 = 3.92$ J ✓

b Just before the ball reaches the ground before the first bounce, the K will equal the initial U and hence $\frac{1}{2}mv^2 = U$ or

$v = \sqrt{\frac{2U}{m}} = \sqrt{\frac{2 \times 3.92}{0.2}} = 6.26\ \text{ms}^{-1}$ downwards. ✓

c If the ball lost 20% of its K, the velocity after the first bounce must be 80% of the K that it hit the ground with and hence the new $K = 0.8 \times 3.92 = 3.14\ \text{J} = \frac{1}{2}mv^2$ or

$v = \sqrt{\frac{2K}{m}} = \sqrt{\frac{2 \times 3.14}{0.2}} = 5.6\ \text{ms}^{-1}$ upwards. ✓ Now if we call upwards positive and downwards negative, then impulse $= \Delta p = mv - mu = 0.2(5.6 - (-6.26)) = 2.37$ Ns upwards. ✓

d After the first bounce the energy of the ball is 3.14 J and after the second bounce it will be 80% of this $= 0.8 \times 3.14 = 2.512$ J. Now this energy will be converted into U at the top of the bounce and hence $mgh = 2.512$ or $h = 2.512/mg = \frac{2.512}{0.2 \times 9.8} = 1.28$ m. ✓

CHAPTER 5 MOMENTUM, ENERGY AND SIMPLE SYSTEMS

INQUIRY QUESTION:

How is the motion of objects in a simple system dependent on the interaction between the objects?

Newton's laws apply to the objects within a system and the system as a whole. Objects within an isolated system may exert forces on one another. Although these forces change the momentum of the objects, the total momentum of the system always remains constant. In the special case of elastic interactions between objects, the sum of the momentum and sum of the kinetic energy remain constant.

1 Conservation of momentum and kinetic energy

» Students analyse quantitatively and predict the results of interactions in elastic collisions using the law of conservation of momentum $\sum m\vec{v}_{\text{before}} = \sum m\vec{v}_{\text{after}}$ and, where appropriate, conservation of kinetic energy $\sum \frac{1}{2}mv^2_{\text{before}} = \sum \frac{1}{2}mv^2_{\text{after}}$, the results of interactions in elastic collisions.

➔ **Momentum** (p) is a vector quantity defined as the product of the mass and velocity of an object:

$$p = mv$$

The units of momentum are Nm = kgms^{-2}.

momentum: a vector quantity equal to the product of an object's mass and its velocity

➔ A moving object possesses kinetic energy and momentum. The magnitude of the momentum of a moving object is related to its kinetic energy by:

$K = p^2/2m$

➔ A **closed** or **isolated system** is one in which no external forces act on the objects within the system.

closed or isolated system: an isolated system in which no external forces act on an object

➔ The vector sum of the momentum of the objects in a closed system always remains constant; that is, momentum is conserved in closed systems. **Conservation of momentum** can by expressed by the relationship:

$$\sum mv_{\text{before}} = \sum mv_{\text{after}}$$

conservation of momentum: the sum of the momentum in a closed system remaining constant

➔ In some interactions the sum of the momentum and the sum of the kinetic energy are conserved. We call these special interactions **elastic collisions**, and for such collisions:

$\Sigma mv_{\text{before}} = \Sigma mv_{\text{after}}$

and

$$\sum \frac{1}{2}mv^2_{\text{before}} = \sum \frac{1}{2}mv^2_{\text{after}}$$

elastic collision: a collision in which kinetic energy is conserved

➔ **Conservation of kinetic energy** only occurs in elastic collisions. Because none of the initial kinetic energy is lost in elastic collisions, none of the initial kinetic energy can be transformed to friction, sound or heat during the collision. Such collisions only involve forces that act over a distance such as the gravitational force or the electric or magnetic force. If a ball had an elastic collision with the ground, it would bounce back to the same height it was dropped from as no energy would be lost in the ball's collision with the ground.

conservation of kinetic energy: the sum of the kinetic energy in a closed system remaining constant; this only occurs in the special case of elastic collisions

→ Figure 5.1 shows a perfectly elastic and a perfectly inelastic collision (where K after = 0). In everyday situations collisions are generally somewhere between these two extremes.

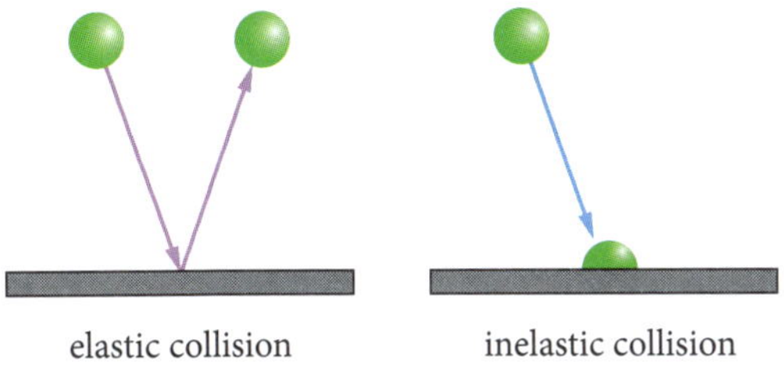

Figure 5.1 Perfectly elastic and perfectly inelastic collisions

EXAMPLE 1

A car and truck have a head-on collision, as shown in Figure 5.2. After the collision the vehicles become entangled and move as one.

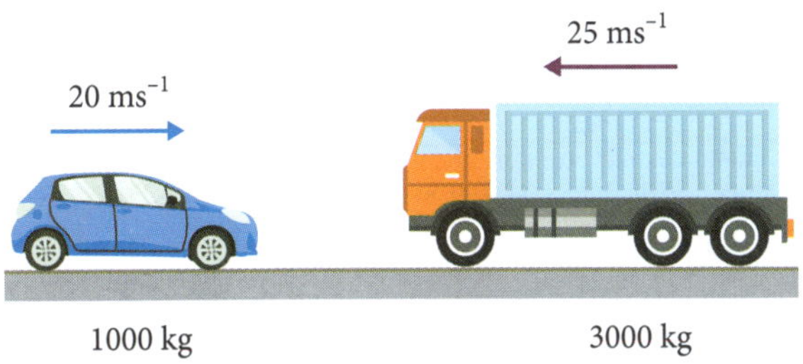

Figure 5.2 Showing a car and a truck just before they collide

a **Find the velocity of the entangled wreckage after the collision.**

b **Was the collision elastic? Justify your answer.**

Answer:

> Momentum is conserved in a closed system

a We can solve this problem using conservation of momentum. Calling motion to the left positive, the momentum of the vehicles before the collision is:

$p_{car} = mv = 1000(-20) = -20\,000$ Ns and

$p_{truck} = mv = 3000 \times 25 = 75\,000$ Ns

Now as $\Sigma mv_{before} = \Sigma mv_{after}$ and the mass of the wreck will be the sum of the masses:

$75\,000 - 20\,000 = (1000 + 3000)v_{after}$

$$v_{after} = \frac{55\,000}{4000}$$

$$= 13.75 \text{ ms}^{-1} \text{ to the left}$$

b To determine if the collision was elastic we must see if the sum of the kinetic energy remained constant. Note that as energy is a scalar we treat all energies as positive.

$$\Delta K = \Sigma \tfrac{1}{2} mv^2_{after} - \Sigma \tfrac{1}{2} mv^2_{before}$$

$$= \tfrac{1}{2}(4000)(13.75^2) - (\tfrac{1}{2}(1000)20^2 + \tfrac{1}{2}(3000)25^2)$$

$$= -7.6 \times 10^5 \text{ J}$$

This is the amount of the initial kinetic energy that was converted into other forms of energy during the collision.

EXAMPLE 2

Consider the collision between a white pool ball travelling with speed v_1 and a black ball at rest, as shown in Figure 5.3. Both balls have the same mass. Assuming the collision is elastic, find the velocity of each of the balls after the collision.

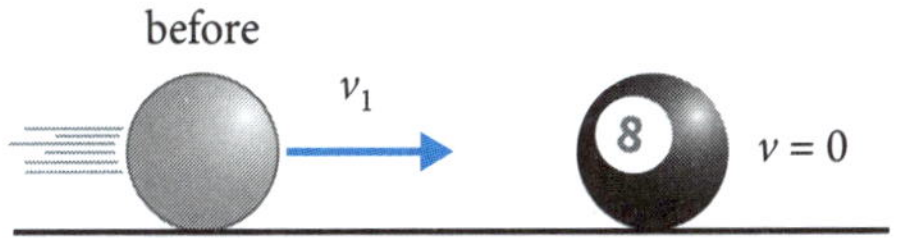

Figure 5.3 An elastic collision between pool balls

Answer:

> Momentum and kinetic energy are conserved in an elastic collision

As the collision is elastic, momentum and kinetic energy are conserved. If we call motion to the right positive and the velocity of the white ball after the collision v_1' and the velocity of the black ball after the collision v_2':

$\Sigma mv_{before} = \Sigma mv_{after}$

Or, for this example, $mv_1 = mv_1' + mv_2'$

$\Sigma \tfrac{1}{2} mv^2_{before} = \Sigma \tfrac{1}{2} mv^2_{after}$

Or $\tfrac{1}{2}mv_1^2 = \tfrac{1}{2}mv_1'^2 + \tfrac{1}{2}mv_2'^2$

Cancelling out the mass and the $\tfrac{1}{2}$ leaves us with two simultaneous equations:

$v_1 = v_1' + v_2'$ and

$v_1^2 = v_1'^2 + v_2'^2$

Substituting $v_1 - v_2'$ for v_1' into the second equation gives:

$v_1^2 - v_2'^2 = (v_1 - v_2')^2$

Hence $2v_1v_2' = 2v_2'^2$

$v_2' = v_1$

Substituting this into the first equation to remove v_2' gives $v_1' = 0$.

Thus the white ball stops after the collision, and the black ball moves to the right with the same velocity as the white ball had initially.

→ KEY QUESTIONS

1 What is a conservation law?

2 Under what circumstances is momentum conserved?

3 Under what circumstances is kinetic energy conserved?

Answers ➲ p. 64

2 Changing forces

» Students investigate the relationship and analyse information obtained from graphical representations of force as a function of time.

- When a net force is applied to an object, the object changes its momentum. We call this change in momentum the **impulse**.

impulse: the product of the applied force and the time over which it is applied; it is equal to the change of momentum ($\Delta p = p_{\text{final}} - p_{\text{initial}}$) that the force produces

- Newton's second law can be expressed in terms of impulse:

  $F = ma = m\Delta v/t = \Delta p/t$, and hence:

$$\text{Impulse} = \Delta p = F\Delta t$$

- The area under a graph of force acting on an object against time is equal to the **change in momentum** (Δp = impulse) of the object.

change in momentum: this generally refers to an increase or decrease in an object's velocity that results from a force being applied to the object over a time period; the change of momentum is equal to the impulse applied

EXAMPLE 3

A 10 kg hammer strikes a nail and applies a force that varies with time, as shown in Figure 5.4. The maximum force of 900 N is achieved after 0.04 s and begins to decrease 0.09 s after the nail was struck.

a Find the change of momentum of the hammer.

b Find the change of momentum of the nail.

c Assuming the hammer came to rest after 0.14 seconds, find the initial velocity of the hammer.

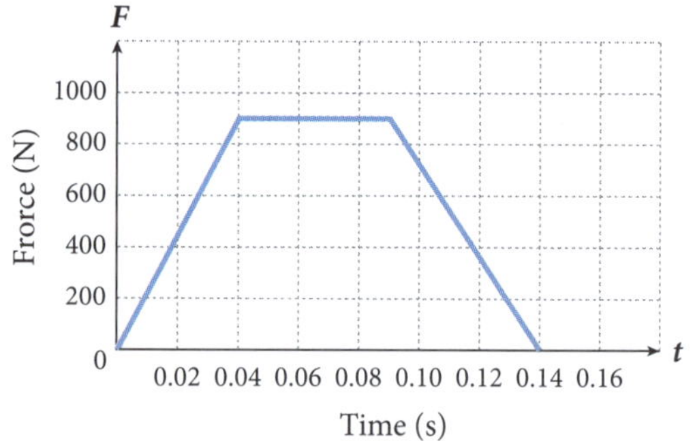

Figure 5.4 Force versus time graph for a nail hit by a hammer

Answer:

The area under a force against time graph is equal to the impulse

a The change in momentum is equal to the area under the graph:

Area under
$= \Delta p$
$= 0.5 \times 900 \times 0.04 + 900 \times 0.05 + 900 \times 0.05 \times 0.5$

Thus $\Delta p = 85.5$ Ns in the opposite direction to the initial velocity.

b Because the total change in momentum must be zero, the change of momentum of the nail will be equal to that of the hammer, but in the opposite direction. For the nail, $\Delta p = 85.5$ Ns in the direction of the initial velocity.

c $\Delta p = mv - mu$ and, as the final velocity for the hammer is zero, $v = 0$.

Thus we have $\Delta p = mu$ and hence:

$$u = \Delta p/m = \frac{85.5}{10} = 8.5 \text{ ms}^{-1}$$

3 Impulse in collisions

» Students evaluate the effects of forces involved in collisions and other interactions, and analyse quantitatively the interactions using the concept of impulse ($\Delta\vec{p} = \vec{F}_{\text{net}}\Delta t$).

- We can prove that momentum is conserved in collisions by applying Newton's third law and the concept of impulse.
- Consider the collision shown in Figure 5.5, in which ball A was moving to the right and ball B towards the left before they collided. When they collide ball A exerts a

force on ball B and, by Newton's third law, ball B exerts an equal and opposite force on ball A. Because the forces are applied for the same time, the impulse applied to ball A from ball B is equal and opposite to the impulse applied to ball B from ball A. As impulse is equal to the change in momentum of each ball we see that both change their momentum in equal and opposite amounts, and therefore the sum of the momentum for the system remains unchanged. This is why momentum is conserved in collisions.

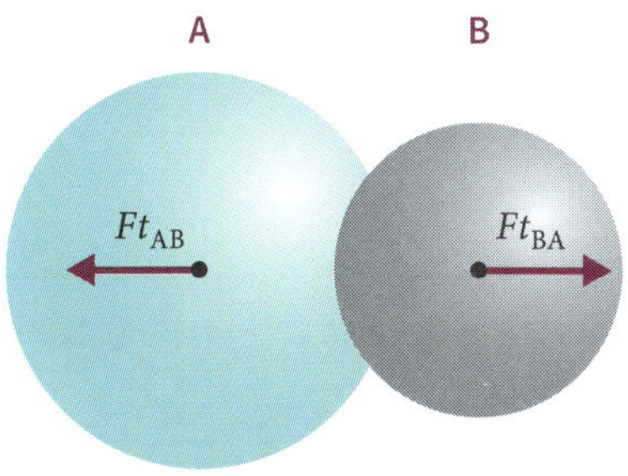

Figure 5.5 Newton's third law and conservation of momentum

➔ Another way to show this is to consider the equation $\Delta p = F\Delta t$ applied to the whole system. Provided it is a closed system, no external forces act on the system and as $F = 0$ the change of momentum of the system (Δp) must also be zero. Hence the sum of the momentum of all the objects in the system cannot change, which means that the momentum of the system is conserved.

EXAMPLE 4

A tennis racquet applies an impulse of 3.0 Ns to a ball of mass 56 g that was approaching the racquet at 19 ms^{-1}. The collision between the ball and the racquet is illustrated in Figure 5.6.

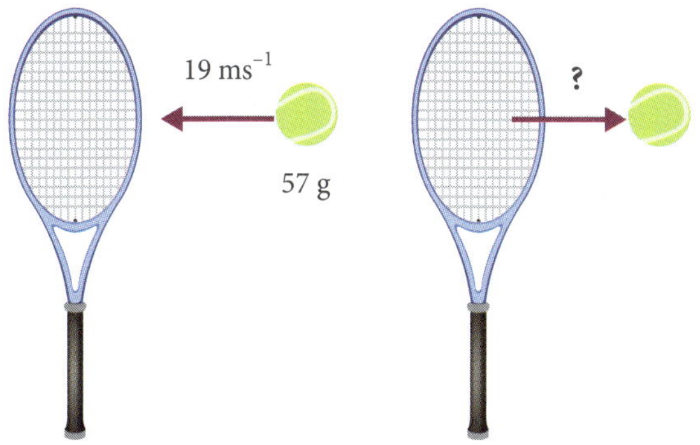

Figure 5.6 Collision between a ball and a tennis racquet

Given that the impulse was applied over a time of 0.015 seconds:

a find the average force applied to the ball by the racquet

b find the final velocity of the ball, assuming the racquet was perpendicular to the ball during the impact

c explain why the kinetic energy of the ball increases when it collides with the racquet.

Answer:

a As $\Delta p = F\Delta t$ the average force would be given by:

Impulse is equal to the change in momentum

$$F = \Delta p/\Delta t = \frac{3.0}{0.015} = 200 \text{ N towards the right}$$

b As $\Delta p = mv - mu$, and calling the direction of the initial velocity negative, then $\Delta p = mv - mu$. Substituting in the values gives:

$$3.0 = 0.056v - 0.056 \times (-19)$$

$$v = \frac{3.0 - 1.06}{0.056}$$

$$= 34.6 \text{ ms}^{-1} \text{ to the right}$$

c The kinetic energy increases because the movement of the racquet imparts some of the racquet's energy to the ball during the collision.

➔ KEY QUESTIONS

4 **How can the momentum of an object be changed?**

5 **How can conservation of momentum in a collision be explained using Newton's third law and impulse?**

6 **How can the change in an object's momentum be determined?**

Answers ➲ p. 64

4 Elastic and inelastic collisions

» Students analyse and compare the momentum and kinetic energy of elastic and inelastic collisions.

➔ While momentum is conserved in all collisions, kinetic energy is only conserved in elastic collisions. In **inelastic collisions** some of the initial kinetic energy is transformed into other types of energy such as heat and sound.

inelastic collision: a collision in which momentum is conserved but kinetic energy is not conserved

➔ We can determine if a collision is elastic or inelastic by comparing the sum of the kinetic energy before and after the collision. If the sum of the kinetic energy is unchanged, the collision is elastic.

➔ By calculating the change in kinetic energy during a non-elastic collision we can determine the amount of energy that goes into other forms of energy in the collision.

EXAMPLE 5

Consider the head-on collision between a large and small truck travelling at the same speed, shown in Figure 5.7.

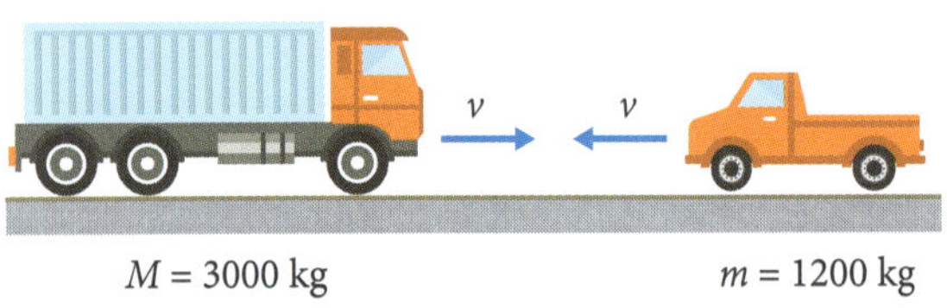

Figure 5.7 Collision between a large and small truck

Both vehicles were initially travelling at 72 kmh^{-1} and, immediately after the collision, an eye witness claimed the small truck moved to the right at 90 kmh^{-1}.

a **Find the momentum of each truck before the collision.**

b **Find the velocity of the large truck after the collision.**

c **Determine the total loss of kinetic energy during the collision.**

d **Was this an elastic collision? Justify your answer.**

Answer:

> Momentum is a vector quantity

a First changing to SI units:

$$72\ kmh^{-1} = 72 \times \frac{1000}{60 \times 60} = 20\ ms^{-1},$$

and 90 kmh^{-1} = 25 ms^{-1}

Calling motion to the right positive, then:

$p_M = Mv = 3000 \times 20 = 60\ 000$ Ns

$p_m = mv = 1200 \times (20) = -24\ 000$ Ns

b Using conservation of momentum,

$\Sigma mv_{before} = \Sigma mv_{after}$:

$60\ 000 - 24\ 000 = 3000v + 1200(25)$

Therefore $v = 2\ ms^{-1}$ to the right.

c The change in kinetic energy is given by:

$$\Delta K = \Sigma \frac{1}{2} mv^2_{after} - \Sigma \frac{1}{2} mv^2_{before}$$

$$\Delta K = (0.5 \times 3000 \times 2^2 + 0.5 \times 1200 \times 25^2) - (0.5 \times 3000 \times 20^2 + 0.5 \times 1200 \times 20^2)$$
$$= -459\ 000\ J$$

d This collision is not elastic as 459 000 J of the initial kinetic energy is converted into other forms of energy in the collision.

5 Investigating collisions

» Students conduct an investigation to describe and analyse one-dimensional (collinear) and two dimensional interactions of objects in simple closed systems.

FIRSTHAND INVESTIGATION 1

Conservation of momentum in one dimension

In this investigation students examine conservation of momentum in rectilinear motion.

Figure 5.8 shows how spring-loaded dynamics carts can be used to investigate conservation of momentum in one dimension.

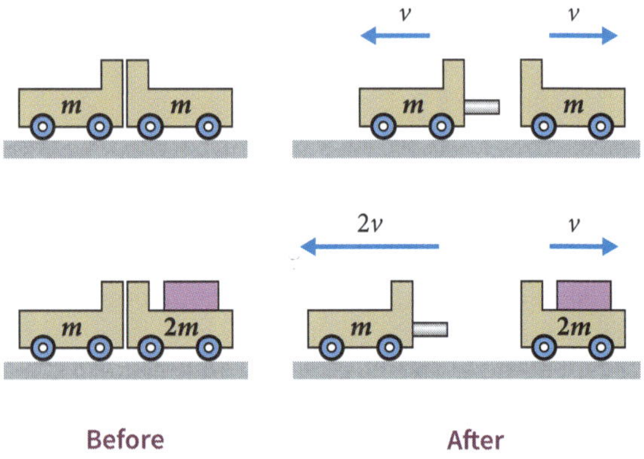

Figure 5.8 Conservation of momentum experiment

Before the trolleys are released, the momentum of the system is zero. To conserve momentum after the spring pushes the trolleys apart, the sum of the momentum must also be zero. The velocity of the dynamics carts after the release can be measured with sonic rangers or timing gates. Placing different masses on the trolleys changes their relative velocities after, but the sum of the momentum of the trolleys always remains zero.

Conservation of momentum in two dimensions

In this investigation students examine conservation of momentum in two dimensions.

Consider an experiment that involves a collision between two 50 g pool balls. The white ball is initially moving to the right with a velocity of 4ms^{-1} when it strikes the stationary black ball. After the collision the black ball has a velocity of 2 ms^{-1} NE, as shown in Figure 5.9.

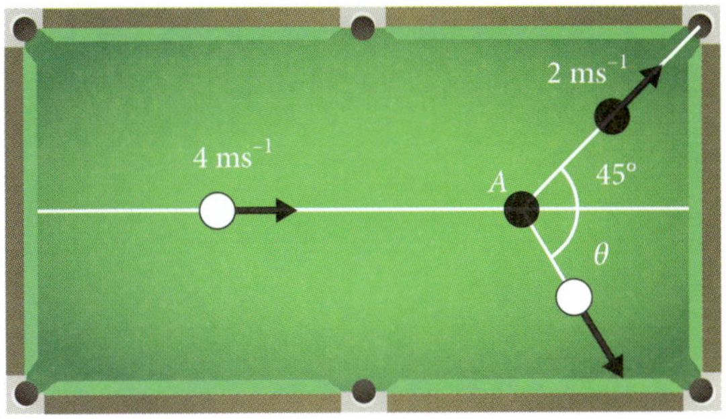

Figure 5.9 Two-dimensional collision

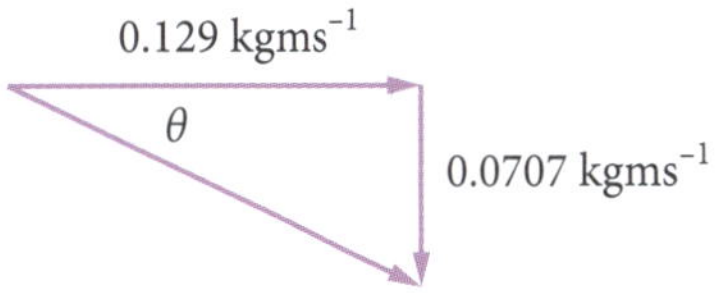

$$p_{\text{white}} = \sqrt{0.1292 + 0.07072} = 0.147 \text{ kgms}^{-1}$$

$$\theta = \tan^{-1}\left(\frac{0.0707}{0.129}\right) = 28.7°$$

Figure 5.10 Finding the momentum of the white ball after the collision

We can use conservation of momentum to find the velocity of the white ball after the collision. This problem can be solved in two ways, outlined below.

Method 1

The total momentum and the easterly and northerly components of the momentum are conserved.

The easterly component of momentum before the collision was simply the momentum of the white ball:

$$mv = 0.05 \times 4$$
$$= 0.2 \text{ kgms}^{-2} \text{ east}$$

After the collision the black ball has an easterly component of:

$$mv \cos 45° = 0.05 \times 2 \times \cos 45°$$
$$= 0.0707 \text{ kgms}^{-2} \text{ east}$$

As the sum of the momentum after must equal the momentum before in the easterly direction, the easterly component of the white ball's momentum can be calculated:

$$\Sigma mv_{\text{before}} = \Sigma mv_{\text{after}}$$
$$0.2 = p_{\text{white}} + 0.0707$$

Therefore $p_{\text{white}} = 0.2 - 0.0707$
$= 0.129 \text{ kgms}^{-1}$ east

As the northerly component of the momentum was zero before the collision, the sum of the northerly components of momentum after the collision must be zero:

$$\Sigma mv_{\text{before}} = \Sigma mv_{\text{after}}$$
$$0 = 0.05 \times 2 \sin 45° - p_{\text{white}}$$

Therefore $p_{\text{white}} = -0.05 \times 2 \sin 45°$
$= 0.0707 \text{ kms}^{-1}$ south

Combining the two components, as shown in Figure 5.10, gives the final momentum:

$$p_{\text{white}} = 0.147 \text{ kgms}^{-1} \text{ N118.7°}$$

$$v = p/m = \frac{0.147}{0.05}$$
$$= 2.95 \text{ ms}^{-1} \text{ N118.7°}$$

Method 2

We can also solve this problem using vector addition in two dimensions.

Draw a vector diagram, as shown in Figure 5.11, to demonstrate that the sum of the momentum before is equal to the sum of the momentum after:

$$p_{\text{black (after)}} = mv = 0.05 \times 2$$
$$= 0.1 \text{ kgms}^{-1}$$

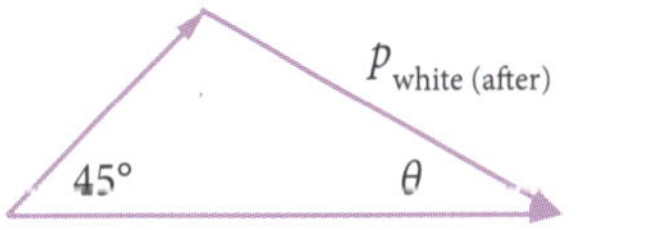

$$p_{\text{before}} = mv = 0.05 \times 4 = 0.2 \text{ kgms}^{-1}$$

Figure 5.11 Equating the sum of the momentum before and after the collision

Using the cosine rule ($a^2 = b^2 + c^2 - 2bc \cos A$) to solve for the momentum of the ball gives:

$$p^2 = 0.1^2 + 0.2^2 - 2 \times 0.1 \times 0.2 \times \cos 45°$$
$$p = 0.147 \text{ kgms}^{-1}$$

The sin rule gives $0.147/\sin 45° = 0.1/\sin\theta$ and

hence $\theta = \sin^{-1}\left(\frac{0.1 \sin 45°}{0.147}\right) = 28.7°$

Thus $v_{\text{white}} = p/m = \frac{0.147}{0.05}$
$= 2.95 \text{ ms}^{-1} \text{ N118.7°}$

6 Investigating changing forces

» Students investigate the relationship and analyse information obtained from graphical representations of force as a function of time.

FIRSTHAND INVESTIGATION 2

Impulse

In the following investigation students examine how the impulse produced by a changing force can be determined.

A player kicks a soccer ball of mass 0.43 kg that was initially at rest. Figure 5.12 shows how the force exerted on the ball changes with time.

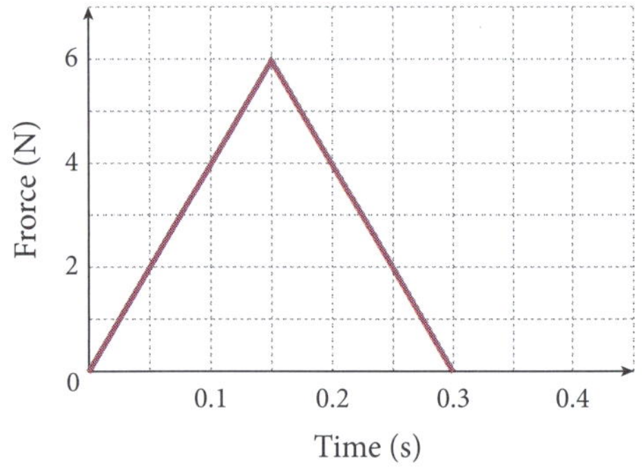

Figure 5.12 Force versus time graph

We can use Figure 5.12 to find the impulse applied to the ball by the force, and the velocity of the ball when it leaves the player's boot.

The area under the graph gives the change in momentum and hence the change of momentum is

$\Delta p = 6 \times \frac{0.3}{2} = 0.9$ Ns

Now as the change in momentum is

$\Delta p = p_{\text{final}} - p_{\text{initial}}$ and $p_{\text{initial}} = 0$

$\Delta p = 0.9 = p_{\text{final}} - 0$ and $p_{\text{final}} = 0.9$ kgms^{-1}

Thus the final velocity $v = p/m = \frac{0.9}{0.43} = 2.1$ ms^{-1}

➔ KEY QUESTIONS

7 If the force applied to an object is not constant, how can the change in momentum of the object be determined?

8 What energy transformations occur when an object is in free flight?

9 If the same change of momentum occurred over a shorter time, would the force on the object be different?

Answers ➲ p. 64

CHAPTER SYLLABUS CHECKLIST

Are you able to answer these questions from the syllabus for this chapter? Tick each question as you go through the checklist if you are able to answer it. If you cannot answer a question, turn to the relevant page in the study guide to find the answer. For NESA key word meanings, go to www.educationstandards.nsw.edu.au and search 'key words'.

FOR A COMPLETE UNDERSTANDING OF THIS TOPIC:		PAGE NO.	✓
1	Can I calculate the momentum of a moving object?	54	
2	Can I differentiate between an open and closed system?	54	
3	Can I differentiate between elastic and inelastic collisions?	54	
4	Can I use conservation of momentum to solve problems involving interactions and collisions in one and two dimensions?	54–55	
5	Can I calculate the change in kinetic energy involved in an inelastic collision?	55, 58	
6	Can I apply conservation of momentum and conservation of kinetic energy to solve problems involving one- and two-dimensional elastic collisions?	55	
7	Can I calculate the change in momentum of an object (impulse), given the initial and final momentum or the net force on the body and time over which it acts?	56–57	
8	Can I find the force applied from the change in momentum of an object and the time over which it occurs?	56–57	
9	Can I use a force against time graph to determine the impulse imparted to an object from a changing force?	56, 60	

YEAR 11 EXAM-TYPE QUESTIONS

Objective-response questions

(1 mark each)

1 **Two cars with the same mass travelling in opposite directions with the same velocity collide head on, as shown in Figure 5.13.**

Figure 5.13 Head-on collision between cars of equal mass

Which statement best describes the physics of this situation?

A The change in momentum of each car is equal.
B The kinetic energy is conserved in the collision.
C The vector sum of the change in momentum of each car is zero.
D The impulse applied to each car is equal.

2 **Why is the force imparted from a hammer to a nail so large?**

A The hammer transfers all its kinetic energy to the nail.
B The hammer has a very large change of momentum.
C The hammer has a large change of momentum that occurs over a very short time.
D The change of impulse of the nail is larger than the change of impulse of the hammer.

3 **If a ball travelling with velocity v collides with a ball with the same mass at rest, which of the following possible outcomes is least likely to occur?**

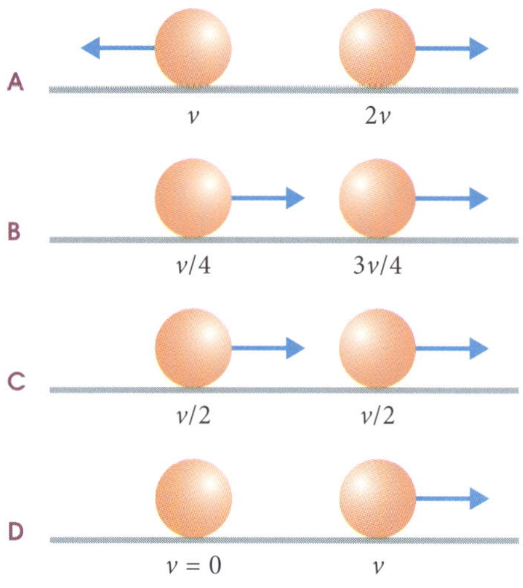

4 **Why is it better for a gymnast to bend her knees when she lands after jumping from a high bar?**

A Bending her knees would decrease the force on her because it increases her change in momentum.
B Bending her knees would decrease the force on her because it increases the time over which the change in momentum occurs.
C Bending her knees would decrease the force on her because it decreases the gymnast's momentum.
D Bending her knees would decrease the force on her because it decreases the impulse applied to the gymnast by the floor.

5 **Which of the following statements best describes what happens when a ball is thrown upwards into the air, if we neglect air friction?**

A The ball's kinetic energy is conserved throughout the flight.
B The ball's momentum is conserved throughout the flight.
C The force on the ball changes throughout the flight.
D The total energy of the ball remains constant throughout the flight.

Extended-response questions

6 **An ice hockey goalkeeper of mass 60 kg at rest catches a 0.15 kg puck travelling at 45 ms^{-1}, as shown in Figure 5.14.**

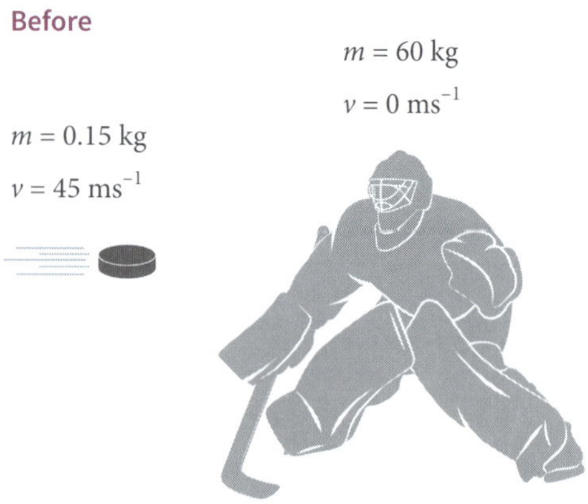

Figure 5.14 Ice hockey player about to catch a ball

a **Find the velocity of the goalkeeper after he has caught the puck.** (2 marks)
b **Is this an elastic collision? Justify your answer.** (2 marks)

7 Tarzan steps off from a branch 10 m above Jane, as shown in Figure 5.15. He swings down and picks up Jane as he passes.

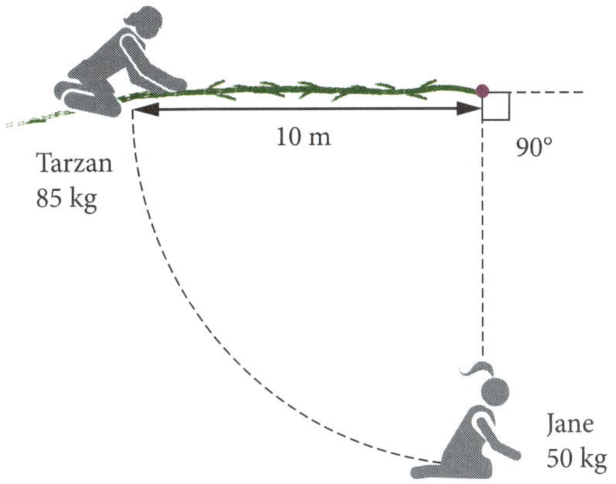

Figure 5.15 Tarzan swinging on a vine about to catch Jane

Given that Tarzan has a mass of 85 kg and Jane has a mass of 50 kg, find:

a Tarzan's initial gravitational potential energy (1 mark)

b Tarzan's kinetic energy just before he grabs Jane (1 mark)

c Tarzan's velocity just before he grabs Jane (1 mark)

d Tarzan's velocity immediately after he grabs Jane (1 mark)

e the maximum height Tarzan will swing to with Jane in his arms. (1 mark)

8 In a high-school chemistry demonstration a small piece of sodium metal with a mass of 5 grams was placed on the surface of water in a beaker. The piece of sodium reacted immediately and separated into three pieces that skimmed across the top of the water, as shown in Figure 5.16.

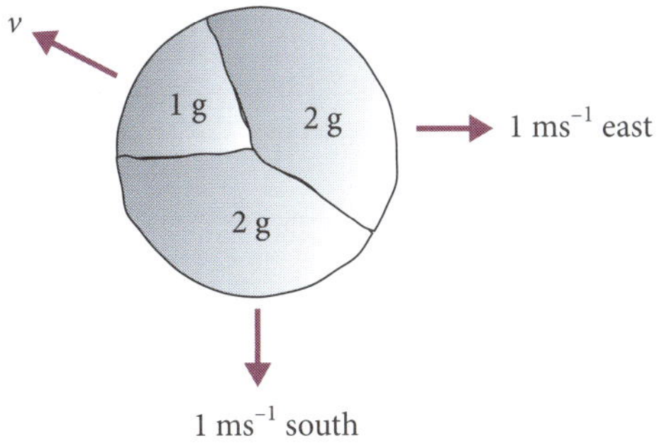

Figure 5.16 Small piece of sodium exploding into three pieces

a Find the sum of the momentum of the two 2 g masses immediately after the reaction. (2 marks)

b Find the momentum of the 1 g mass immediately after the reaction. (2 marks)

c Find the velocity of the 1 g mass immediately after the reaction. (1 mark)

9 The force applied to a tennis ball of mass 56 g travelling initially west towards the racquet at 20 ms^{-1} is graphed in Figure 5.17.

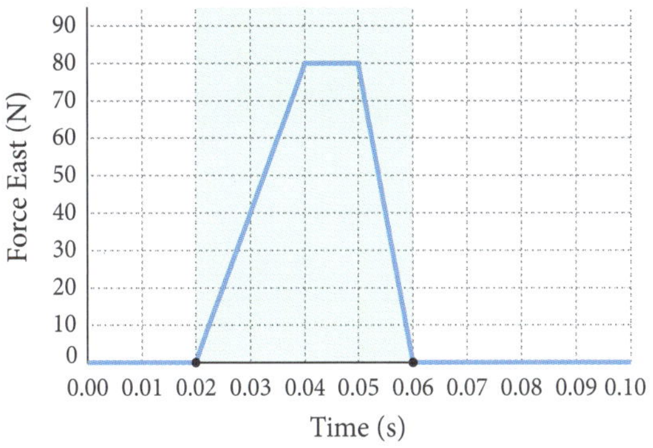

Figure 5.17 Force exerted on a tennis ball by a racquet

a Use Figure 5.17 to determine the impulse applied to the ball by the racquet. (1 mark)

b Calculate the change of velocity of the ball. (1 mark)

c Determine the final velocity of the tennis ball after it is hit. (2 marks)

d Does the ball gain or lose energy in the collision? Justify your answer. (2 marks)

10 Consider the collision shown in Figure 5.18. The ball has a mass of 200 g and approaches and leaves the wall 0.1 seconds later with a velocity of 10 ms^{-1}.

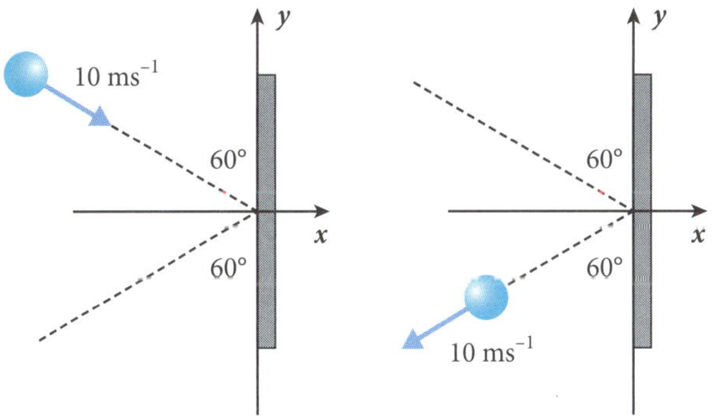

Figure 5.18 A ball colliding with a solid wall

a Is this an elastic collision? Justify your answer. (2 marks)

b Find the change in momentum of the ball during the collision. (3 marks)

c Find the average force the wall applied to the ball during the collision. (1 mark)

KEY QUESTIONS

Key questions ➲ p. 56

1 A conservation law reflects a specific symmetry in nature that results in the sum of a measurable quantity such as momentum or energy not changing in a closed system.

2 The vector sum of the momentum of the objects in a closed system always remains constant (i.e. is conserved).

3 Kinetic energy is only conserved in elastic collisions.

Key questions ➲ p. 57

4 Applying a net force over a time (i.e. an impulse) to an object will change its momentum.

5 We can use Newton's third law to show that the change in momentum of objects involved in a collision is always equal and opposite, which means the sum of the momentum of the system remains constant. Hence conservation of momentum is implied by Newton's laws of motion.

6 The change in momentum can be calculated by subtracting the final momentum of the object from the initial momentum of the object ($mv - mu$), or by multiplying the net force applied to the object by the time over which it was applied (i.e. $\Delta p = F\Delta t$).

Key questions ➲ p. 60

7 The area under the force against time graph is equal to the impulse. This method must be used to find the impulse when the force is not constant.

8 The energy transformations involved depend on whether air friction is taken into account or not:

a If air friction is neglected then energy is transformed between gravitational potential energy and kinetic energy but the total energy of the object (i.e. the sum of the K and U) remains constant.

b If air friction is taken into account, energy is still transformed between gravitational potential energy and kinetic energy but the total energy of the object decreases as it moves because the object does work on the air by opposing air friction.

9 A greater force is required to produce the same change of momentum that occurs over a shorter time interval (i.e. $F = \Delta p/\Delta t$).

YEAR 11 EXAM-TYPE QUESTIONS

Objective-response questions

1 **C.** Momentum of the system before the crash is zero and hence for the total momentum of the system to remain zero, the sum of the changes of momentum in the system must be zero. **A** and **D** are incorrect because the change in momentum (impulse) of each car is in an opposite direction. (Impulse and Δp are vectors.) **B** is incorrect because the collision is clearly not elastic as the collision creates permanent deformation to the vehicles and makes sound, heat, etc.

2 **C.** $F = \Delta p/\Delta t$. Hence to produce a large force the change of momentum must occur over a short time. **A** is incorrect because the collision would not be elastic as it makes noise, heat, etc. **B** is not correct because a large change in momentum over a long time would not produce a large force. **D** is incorrect because the change in impulse of the nail and hammer are equal and opposite.

3 **A.** Momentum is conserved in all of the collisions but in answer **A** the sum of the kinetic energy of the balls after the collision is greater than the kinetic energy before the collision. **B**, **C** and **D** are possible as they all conserve momentum and the sum of the kinetic energy is less after each collision than before.

4 **B.** As $F = \Delta p/\Delta t$, increasing the time over which the change in momentum occurs will reduce the force required to change the momentum. **A** is incorrect because the change in momentum is the same in each case. **C** is incorrect because bending her knees would not change her momentum ($p = mv$). **D** is incorrect because the impulse is the same in each case as it is equal to the change in momentum.

5 **D.** The ball's $K + U$ remains constant due to conservation of energy. **A** and **B** are incorrect because the ball's velocity and hence its kinetic energy and momentum change as it goes up and falls down. **C** is incorrect because the only force acting on the ball is gravity, which is constant throughout the flight.

Extended-response questions

6 **EM** This question requires students to apply conservation of momentum to a simple collision and to find the change in kinetic energy in a collision.

a Using conservation of momentum: $\Sigma mv_{\text{before}} = \Sigma mv_{\text{after}}$

Therefore $0.15 \times 45 = (60 + 0.15)v$ ✓ and hence $v = \dfrac{0.15 \times 45}{60 + 0.15}$

$= 0.112\ \text{ms}^{-1}$ in the direction the ball was initially travelling. ✓

b This is not an elastic collision ✓ as kinetic energy was not conserved. $K_{\text{before}} = \frac{1}{2}mv^2 = \frac{1}{2}(0.15 \times 45^2) = 152$ J, and after it was $K_{\text{after}} = \frac{1}{2}mv^2 = \frac{1}{2} \times 60.15 \times 0.112^2 = 0.377$ J ✓

7 **EM** This question requires students to apply conservation of energy to a falling body (i.e. total energy = kinetic energy + gravitational potential energy) and conservation of momentum to a collision of a falling body.

a $U = mgh = 85 \times 10 \times 9.8 = 8330$ J ✓

b As energy is conserved, $U_{\text{top}} = K_{\text{bottom}} = 8330$ J ✓

c $K = \frac{1}{2}mv^2$ and hence $v = \sqrt{\dfrac{2K}{m}} = \sqrt{\dfrac{2 \times 8330}{85}} = 14\ \text{ms}^{-1}$ ✓

d Using conservation of momentum $\Sigma mv_{\text{before}} = \Sigma mv_{\text{after}}$,

$85 \times 14 = (85 + 50)v$. Hence $v = \dfrac{85 \times 14}{85 + 50} = 8.81\ \text{ms}^{-1}$ ✓

e $U_{\text{top}} = K_{\text{bottom}}$ or $mgh = \frac{1}{2}mv^2$ and hence

$h = v^2/2g = \dfrac{8.81^2}{2 \times 9.8} = 3.96$ m ✓

8 **EM** This question requires students to demonstrate an understanding of conservation of momentum in two dimensions.

a Figure A5.1 shows how the momentum can be added.

Sum of vectors is:

$p = \sqrt{(0.002^2 + 0.002^2)}$

$= 0.00283$ Ns ✓ at N135° ✓

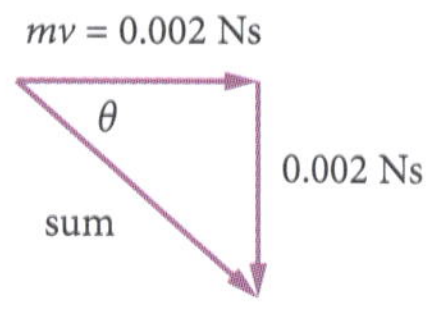

Figure A5.1
Adding momentum vectors

b As the initial momentum of the system was zero, it must remain zero and hence the momentum of the 1 g mass must be equal and opposite to the sum of the momentum of the other two masses. Therefore the momentum of the 1 g mass must be a vector of the same length but in the opposite direction to the sum vector shown in Figure A5.1. Hence:

$p = 0.00283$ Ns ✓ at N315° ✓

c $p = mv$ and hence $v = p/m = \dfrac{0.002\,83}{0.001} = 2.83\ \text{ms}^{-1}$ at N315° ✓

9 EM This question requires students to understand how to extract the change in momentum (impulse) from a force against time graph and apply this to a collision. Although the problem is in one dimension, students have to take the vector nature of momentum into account to correctly solve the problem.

Calling the initial direction of the ball's velocity positive:

a Impulse is the area under the graph during the time of impact $= (\frac{1}{2} \times 0.02 \times 80 + 0.01 \times 80) + (\frac{1}{2} \times 0.01 \times 80) = -2$ Ns and this force would be in the opposite direction of the ball's initial velocity. ✓

b Impulse $= \Delta p = mv - mu = m\Delta v$ and hence:

$\Delta v = \Delta p/m = \dfrac{-2}{0.056} = -35.71\ \text{ms}^{-1}$ ✓

c $\Delta v = v - u$ and hence $-35.71 = v - 20$, therefore $v = 20 - 35.71 = -15.71\ \text{ms}^{-1}$ or $15.71\ \text{ms}^{-1}$ ✓ in the opposite direction of the ball's initial velocity. ✓

d The ball has lost energy ✓ as its final kinetic energy is less than its initial kinetic energy ✓ (because the final velocity is less than the initial velocity).

10 EM This question requires students to be able use vector analysis to determine the change in an object's momentum and the impulse applied to change the momentum in a two-dimensional collision.

a Yes, ✓ it is elastic because the kinetic energy is the same before and after the collision. ✓

b $\Delta p = mv - mu$, but in this case we have a vector subtraction, as shown in Figure A5.2.

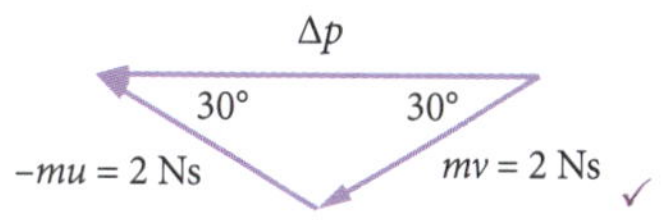

Figure A5.2 Finding the change in momentum

We can solve for Δp using the cosine rule or by using components. As the vertical components sum to zero, $\Delta p = 2 \times$ horizontal components $= 2 \times 2\cos 30° = 3.46$ Ns ✓ at N270°. ✓

c $F = \Delta p/\Delta t = \dfrac{3.46}{0.1} = 34.6$ N at N270° ✓

CHAPTER 6 WAVE PROPERTIES

INQUIRY QUESTION:

What are the properties of all waves and wave motion?

Waves are periodic disturbances of a medium or field that carry energy. All waves can be described using the following wave properties: velocity, frequency, wavelength, amplitude and intensity. Figure 6.1 illustrates some wave properties.

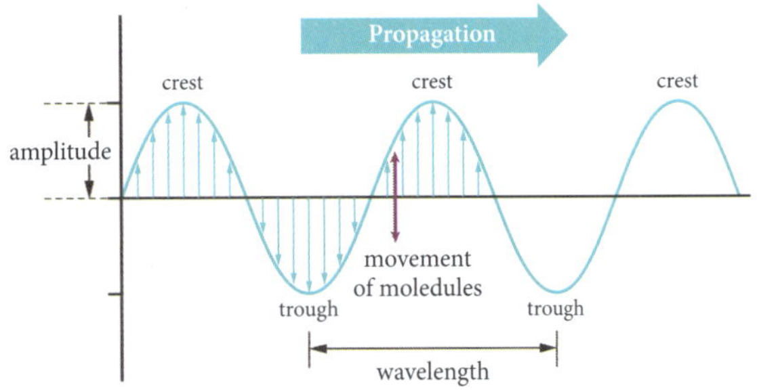

Figure 6.1 Wave diagram showing the transverse movement of the medium as the wave passes

1 Graphing wave motion

» Students construct and/or interpret graphs of displacement as a function of time and as a function of position of transverse and longitudinal waves, and relate the features of those graphs to the following wave characteristics:
- velocity
- frequency
- period
- wavelength
- wave number
- displacement and amplitude.

➔ A wave is a periodic disturbance of a medium or field that carries energy through the medium or field. If the wave moves through a *material* medium we call it a **mechanical wave**, and if the wave moves through a *field* we call it a **non-mechanical wave**. Water waves and sound waves are examples of mechanical waves, while electromagnetic waves and gravity waves are examples of non-mechanical waves.

mechanical wave: a wave that requires a material medium to propagate; for example, sound is a mechanical wave

non-mechanical wave: a wave that oscillates a field and hence can move through a vacuum; for example, electromagnetic waves are non-mechanical waves

➔ If the **oscillation** of the medium or field is *perpendicular* to the wave velocity we call the wave a **transverse wave**, but if the oscillation is *parallel* to the wave velocity we call it a **longitudinal wave**. Water waves and electromagnetic waves are transverse waves, while sound waves are longitudinal waves. The difference between a transverse wave and a longitudinal wave is shown in Figure 6.2.

oscillation: a repeated movement back and forth in a regular rhythm

transverse wave: a wave in which the medium oscillates in a direction perpendicular to the wave velocity; for example, water waves and electromagnetic waves

longitudinal wave: a wave in which the medium oscillates in a direction parallel to the wave velocity; for example, sound waves are longitudinal waves

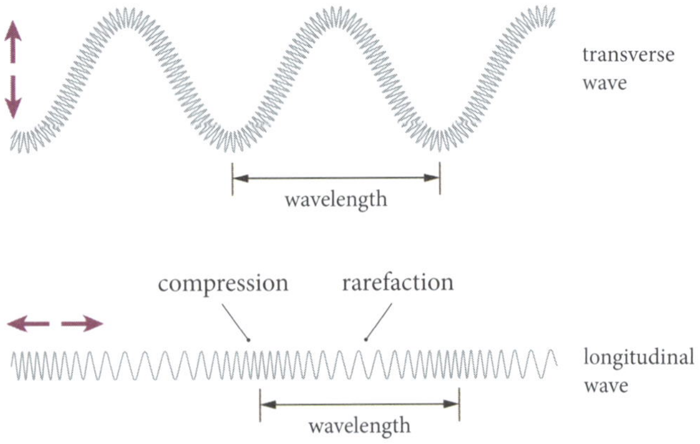

Figure 6.2 Comparing transverse and longitudinal waves in a spring

➔ Table 6.1 lists the measurable quantities (variables) that are used to describe waves.

Table 6.1 Wave quantities

Wave property	Symbol	SI unit	Definition	Relationship
Velocity	v	ms^{-1}	The speed at which the wave travels; we can think of this as the speed of a specific crest or compression	$v = f\lambda$
Frequency	f	s^{-1} or Hz	The number of crests that pass a point per second or the number of complete oscillations per second	$f = 1/T$
Period	T	s	The time for one complete wave cycle	$T = 1/f$
Wavelength	λ	m	The distance between adjacent crests (compressions) or troughs (rarefactions)	$\lambda = vT = v/f$
Displacement	y	m	The distance from equilibrium position	
Amplitude	a	m	The maximum displacement from equilibrium position	
Intensity	I	Wm^{-2}	The power delivered by the wave per unit area; intensity is proportional to the square of the wave amplitude	$I \alpha a^2$

➔ KEY QUESTIONS

1. **What is a wave?**
2. **What is the difference between a mechanical wave and a non-mechanical wave?**
3. **What is the difference between a transverse wave and a longitudinal wave?**

Answers ➲ p. 77

➔ We can graphically represent waves in two ways:

- We look at the wave at a specific instant in time, as if we are taking a photograph of the wave. This produces a graph of the wave's displacement (y) against *distance* in the direction of the wave velocity (x). The distance between adjacent crests on this graph is the **wavelength** (λ), and the maximum displacement from the equilibrium position is the **amplitude** (a) of the wave. See Figure 6.3(a).

wavelength: the distance between adjacent crests (compressions) or between adjacent troughs (rarefactions); the lowercase letter from the Greek alphabet *lambda* (λ) is the symbol used for wavelength

amplitude: the maximum displacement of the medium from the equilibrium position

- Alternatively, we use a graph to show how the displacement of the medium at a specific point (x) changes as the wave moves past. This will produce a graph of displacement (y) against *time* (t), and the distance between adjacent crests on this graph would be the time for one complete oscillation, called the **period** (T). See Figure 6.3(b).

➔ The **phase** of a wave describes how far through a cycle the wave is at the point or time of interest. Because waves can be described by sine or cosine functions, phase is often specified in degrees or radians. Note that 360° = 2π radians.

period: the time for one complete cycle; period is the inverse of the frequency (i.e. $T = 1/f$)

phase: a term used to represent how far a wave is through its cycle of oscillation; a phase is usually specified in radians as one complete cycle is equivalent to 2π radians

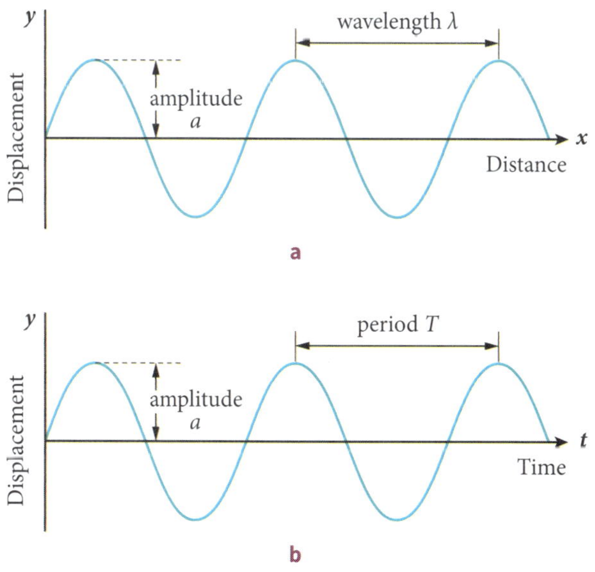

Figure 6.3 Graphs for **a** displacement against distance and **b** displacement against time for a wave

- We can graphically represent waves in two ways because the displacement changes with distance and with time. By holding one of these variables constant we can examine the change in displacement that the other variable produces. Thus by looking at the wave at an instant in time we are keeping time (t) constant and seeing how the displacement changes at different distances. By looking at a single point along the wave we are keeping distance (x) constant and seeing how the displacement at a specific point changes as time passes.
- Once the wavelength and period are found from the graphs, the frequency (f) can be determined from the relationship:

$$f = \frac{1}{T}$$

where T is the period (time for one oscillation)

- Mathematically we say the wave displacement (y) is a function of two variables: time (t) and distance (x). The displacement of the wave can be expressed by:

  $y = a\sin(2\pi x/\lambda - 2\pi t/T)$

  This is often simplified to:

  $y = a\sin(kx - \omega t)$

  where a = amplitude
  $k = 2\pi/\lambda$
  $\omega = 2\pi/T$

- This equation tells us that a wave is a sine function that varies with time (t) and distance (x). This is why we can represent a wave by two sine graphs: one that changes with distance and the other that changes with time. Note that while these trigonometric relationships that represent the wave displacement are important, students will not be tested on them in the Year 11 course.
- Sometimes other properties of the wave that reflect the displacement are graphed. For example, the displacement of air molecules can be plotted against time or distance for sound waves, or the air pressure can be plotted against time or distance for sound waves. Examples of these types of graphs for sound waves are shown in Figure 6.4.

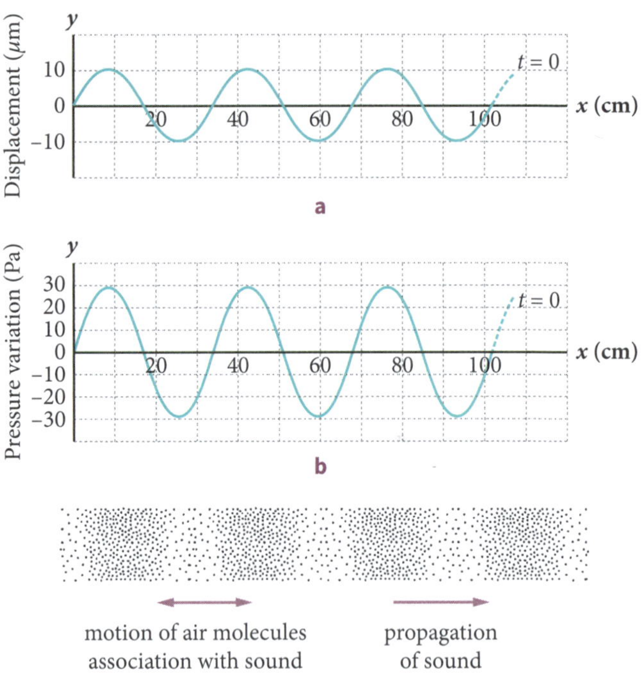

Figure 6.4 Sound wave represented as **a** displacement of molecules against distance and **b** as variation of air pressure against distance. The corresponding position of air molecules as the wave passes is represented in **c**

EXAMPLE 1

The displacement against distance graph for a wave in a stretched spring is shown in Figure 6.5.

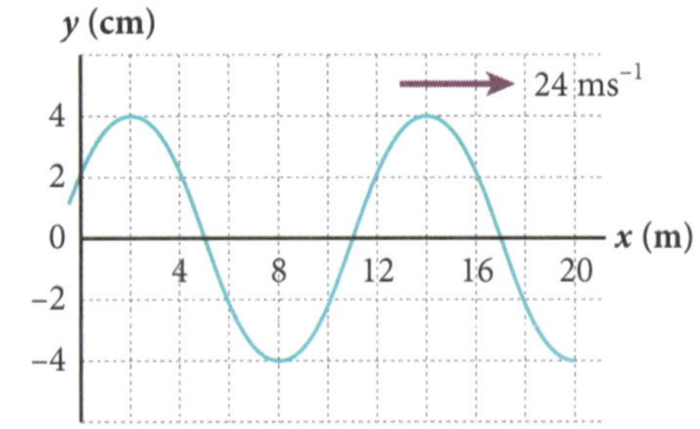

Figure 6.5 Graph of a wave travelling along a stretched spring

Given that the wave travels at 24 ms^{-1}, use the information in Figure 6.5 to find:

a **the wavelength of the wave**
b **the amplitude of the wave**
c **the frequency and period of the wave.**

Recall that the wave equation relates wavelength, frequency and velocity

Answer:

a The wavelength is the distance between adjacent crests, hence $\lambda = 12$ m.

b The amplitude is the maximum displacement from the equilibrium position, hence $a = 4$ cm.

c Using the velocity relationship stated in Table 9.1, $v = f\lambda$, we can write $f = v/\lambda = \frac{24}{12} = 2$ Hz.

The period of the wave is the inverse of the frequency, $T = 1/f = \frac{1}{2} = 0.5$ s.

EXAMPLE 2

Draw a displacement against time graph to illustrate $1\frac{1}{4}$ cycles of a wave with an amplitude of 15 cm and a frequency of 0.05 seconds.

Recall that the period is the inverse of the frequency

Answer:

The period $= T = 1/f = \frac{1}{0.05} = 20$ seconds.

Knowing the period is 20 seconds and the amplitude is 15 centimetres enables us to draw the graph shown in Figure 6.6.

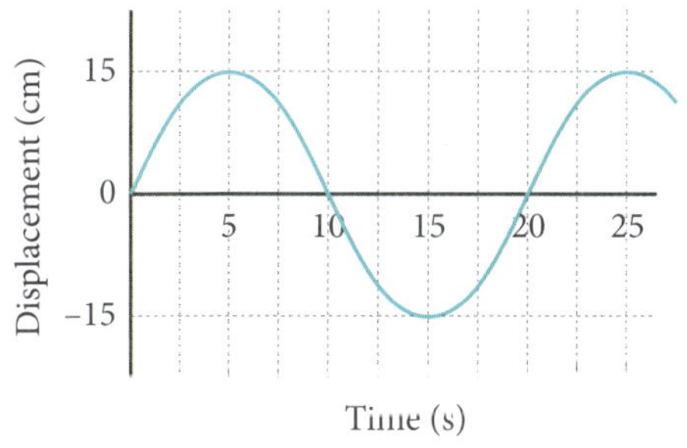

Figure 6.6 Displacement against time graph

EXAMPLE 3

Given that the speed of sound in air is approximately 340 ms^{-1}, use the graphs shown in Figure 6.4 to find:

a the wavelength of this sound wave

b the frequency of this sound wave

c the maximum sound pressure produced by the wave

d how the wavelength would change if the experiment was conducted on a hotter day. (Note that the speed of sound increases with air temperature.)

Recall that the wave equation relates frequency, velocity and wavelength

Answer:

a The wavelength is the distance between adjacent crests, hence $\lambda = 33$ cm $= 0.33$ m.

b Using the wave equation $v = f\lambda$ we can write:

$f = v/\lambda = \frac{340}{0.33} = 1030$ Hz

c The amplitude is the maximum displacement for the equilibrium position = 30 Pa.

d As $v = f\lambda$, then $\lambda = v/f$ and hence, if the velocity increased, the wavelength of sound wave with this frequency would also increase.

➔ KEY QUESTIONS

4 How can wave displacement be represented graphically?

5 How can we determine the wavelength of a transverse wave and a longitudinal wave?

Answers ➲ p. 77

2 The wave equation

» Students solve problems and/or make predictions by modelling and applying the following relationships to a variety of situations:

- $v = f\lambda$
- $f = \frac{1}{T}$

➔ The period (T) of a wave is the time for the wave to complete one full cycle, and the **frequency** (f) of the wave is the number of cycles that occur per second. Now if there were f cycles per second, the time for each cycle T would be given by $T = 1/f$ and hence:

$$f = \frac{1}{T}$$

frequency: the number of complete oscillations per second

➔ You will recall from Chapter 1 that velocity = rate of change of displacement (s/t). Applying this relationship to one cycle of a wave yields:

$$v = s/t = \lambda/T = f\lambda$$

We call this relationship the *wave equation*, written as:

$$v = f\lambda$$

where f is the frequency

λ is the wavelength of the wave

- The velocity of the wave depends on the type of wave and the medium through which it travels. Note that in a vacuum all electromagnetic waves travel at:

  $c = 3 \times 10^8$ ms^{-1}

  where c is the speed of light in a vacuum

  In dry air at sea level at 20 °C the speed of sound is:

  $v_{sound} = 343$ ms^{-1}

- The **wave number** ($k = 2\pi/\lambda$) is a useful concept when describing waves mathematically because the wave number (k) can be used to replace $2\pi/\lambda$ in the equation for the wave displacement. The wave number can be thought of as the number of radians per metre but is simply another measure related to the wavelength of the wave it describes.

wave number: the number of radians per unit length

EXAMPLE 4

Given that all electromagnetic waves travel at 3×10^8 ms^{-1} in a vacuum, complete the Table 6.2.

Table 6.2 Electromagnetic table

Type of electromagnetic wave	Frequency	Wavelength
Radio wave	105.7 MHz	
Microwave	2.45 GHz	
Red light		620 nm
Blue light		450 nm
X-ray	2×10^{18} Hz	
Gamma ray		2.5×10^{-12} m

Answer:

Converting to SI units and applying the wave equation $v = f\lambda$ in each case:

The wave equation can be rearranged to make the frequency, the wavelength or the velocity the subject of the equation

- Radio wave, $\lambda = v/f = \dfrac{3\times10^8}{105.7\times10^6} = 2.84$ m
- Microwave, $\lambda = v/f = \dfrac{3\times10^8}{2.45\times10^9} = 0.122$ m
- Red light, $f = v/\lambda = \dfrac{3\times10^8}{620\times10^{-9}} = 4.84 \times 10^{14}$ Hz
- Blue light, $f = v/\lambda = \dfrac{3\times10^8}{450\times10^{-9}} = 6.67 \times 10^{14}$ Hz
- X-ray, $\lambda = v/f = \dfrac{3\times10^8}{2\times10^{18}} = 1.5 \times 10^{-10}$ m
- Gamma ray, $f = v/\lambda = \dfrac{3\times10^8}{2.5\times10^{-12}} = 1.2 \times 10^{20}$ Hz

EXAMPLE 5

Given that sound waves travel at 340 ms^{-1} in air and at 6100 ms^{-1} in steel, compare the wavelength of a sound wave with a period of 4 ms in air and steel.

Answer:

We first change the period to frequency by substituting into $f = 1/T = \dfrac{1}{4 \times 10^{-3}} = 250$ Hz.

Remove SI prefixes first

Then using the wave equation $v = f\lambda$:

- the wavelength in air is given by

  $\lambda = v/f = \dfrac{340}{250} = 1.36$ m

- the wavelength in steel is given by

  $\lambda = v/f = \dfrac{6100}{250} = 24.4$ m.

Note that the wavelength is much longer in the medium where the wave travels faster.

3 Investigating mechanical waves

» Students conduct a practical investigation involving the creation of mechanical waves in a variety of situations in order to explain:
- the role of the medium in the **propagation** of mechanical waves
- the transfer of energy involved in the propagation of mechanical waves.

propagation: refers to the process by which a wave is transmitted through the medium or field

- In the two investigations that follow students produce mechanical waves and examine their properties.

FIRSTHAND INVESTIGATION 1

Waves in springs

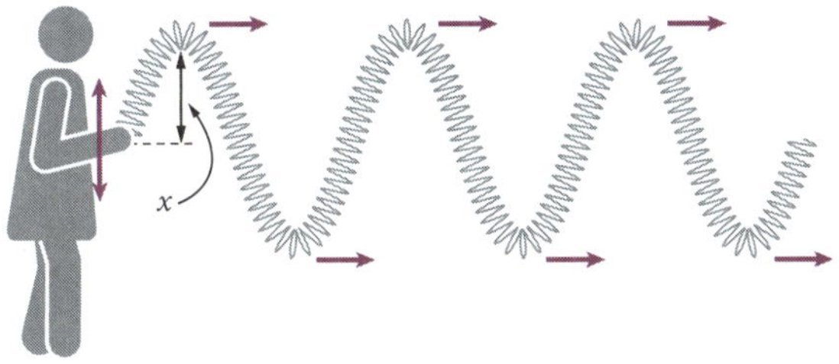

Figure 6.7 Student producing transverse waves on a long spring

As shown in Figure 6.7, students can produce transverse waves in long springs to investigate:

- the relationship between the frequency of oscillation and the wavelength of the wave produced, by measuring the wavelength as a function of the frequency at which the spring is oscillated
- the relationship between the velocity of the wave and the tension in the spring, by measuring the velocity as a function of tension (a spring balance can be used to measure the tension)
- the effect of the velocity of a wave on the wavelength for a wave oscillating at constant frequency, by measuring the wavelength produced for a specific frequency in springs at different tensions
- how the wave carries energy by allowing the wave in the spring to do work (e.g. allowing the wave to oscillate an object).

Typical results

Students will find the following results for both longitudinal and transverse waves:

- Increasing the frequency of oscillation decreases the wavelength.
- Increasing the tension in the string will increase the velocity of the wave. Careful measurement will show that velocity (v) is proportional to the square root of the tension:

  $$\left(t \propto \sqrt{\text{tension}}\right)$$

- Increasing the tension will increase the wavelength for a fixed frequency of oscillation.
- If a student holds the end of the spring, a wave propagating down the spring will cause their hand to move when it reaches them. This means that work is done on the person holding the spring, proving that waves carry energy. The greater the amplitude of the wave the more work it can do and hence the energy carried by the wave increases with the wave's amplitude.

FIRSTHAND INVESTIGATION 2

Water waves

Figure 6.8 A simple ripple tank to investigate water waves

A commercial ripple tank can be used (or students can make a simple ripple tank using a glass baking dish) and light, as shown in Figure 6.8. The transverse water waves produced can be used to investigate:

- what the effect of rate of oscillation on wavelength can be
- whether water waves carry energy
- how wave amplitude affects the speed of a water wave.

Typical results

Students will find the following results:

- The wavelength of the water waves depends on the frequency of the oscillation that creates the waves.
- Placing a cork in the water as the waves pass can be used to show the waves do work on the cork by oscillating the cork, proving that water waves carry energy. The bigger the amplitude of the wave the more it moves the cork, showing that the energy carried by the wave increases with amplitude.
- Students might also find that, unlike other forms of waves, the speed of a water wave is related to the amplitude of the wave. Larger water waves travel faster than smaller water waves, which is why surfers must be towed into huge waves because they cannot paddle fast enough to catch such fast-moving waves.

➔ KEY QUESTIONS

6 What determines the velocity of a wave?

7 How is the frequency of a wave and its wavelength related for waves in a specific medium?

Answers ➲ p. 77

4 Investigating different types of waves

» Students conduct practical investigations to explain and analyse the differences between:
- transverse and longitudinal waves
- mechanical and electromagnetic waves.

➔ In the two investigations that follow students examine the similarities and differences between transverse and longitudinal waves and between mechanical and non-mechanical waves.

FIRSTHAND INVESTIGATION 3

Comparing transverse and longitudinal waves in springs

The difference between transverse waves and longitudinal waves can be demonstrated by creating waves in stretched springs, as shown in Figure 6.9. The relationship between frequency and wavelength, and between the speed of the wave (tension of the spring) and wavelength, can be investigated for transverse and longitudinal waves.

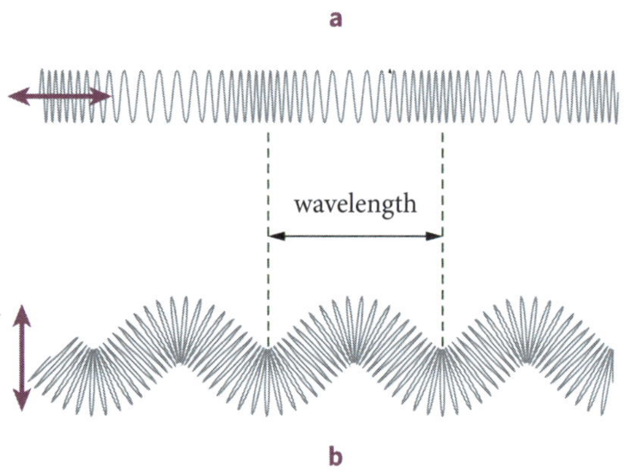

Figure 6.9 Producing transverse and longitudinal waves on a long spring

Typical results

Students will find that longitudinal waves behave in the same way as transverse waves. That is:
- Wavelength decreases as frequency increases.
- The speed of the wave increases with the tension of the spring.
- The wavelength is longer for a fixed-input frequency when the spring is more stretched (i.e. when the velocity of the wave is greater).

FIRSTHAND INVESTIGATION 4

Comparing mechanical waves and non-mechanical electromagnetic waves

Electromagnetic waves

Students could investigate **electromagnetic waves** by conducting experiments with the electromagnetic radiation that reaches us from the Sun. The fact that the electromagnetic radiation reaches the Earth through the vacuum of space proves that electromagnetic radiation can travel through a vacuum. The energy carried by the radiation reaching the Earth can be investigated by using a photovoltaic panel or by showing that the radiation will heat water in a solar oven (e.g. a black can).

electromagnetic wave (or electromagnetic radiation): a transverse wave that carries energy through the electromagnetic fields that permeate the universe; electromagnetic waves travel at the speed of light through a vacuum and consist of mutually perpendicular, oscillating, electrical and magnetic fields

Mechanical waves

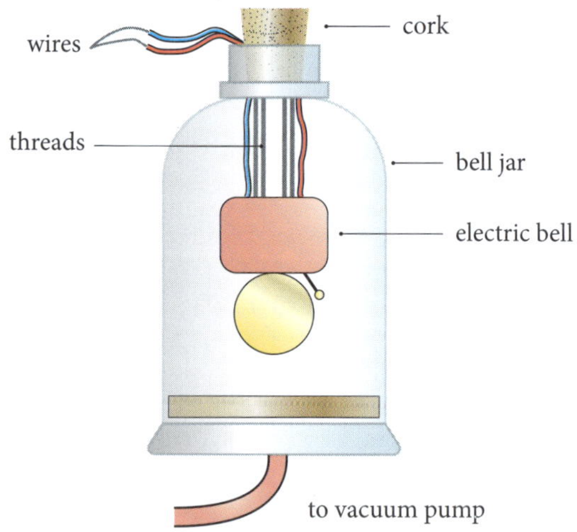

Figure 6.10 Experiment to show sound waves (mechanical waves) cannot travel in a vacuum

Students can show that mechanical waves cannot travel through a vacuum by placing a source of sound (such as an electric bell) in a gas jar and pumping the air out of the jar (see Figure 6.10). When the jar is evacuated the sound of the bell will not be heard outside of the jar.

The key differences between mechanical waves and electromagnetic waves is set out in Table 6.3.

Table 6.3 Difference between mechanical and electromagnetic waves

Mechanical waves	Electromagnetic waves
Require a medium to propagate	Can travel in a vacuum
Travel at a velocity less than the speed of light	Travels at the speed of light in a vacuum
Are produced by some form of oscillation that transfers energy to the medium	Produced by an oscillating charge transferring energy to the electric field around the charge
Detected by transferring energy to an oscillator (e.g. an eardrum or a boat on the ocean)	Detected by transferring energy to charges

→ KEY QUESTIONS

8 How are waves produced?

9 How are waves propagated through a mechanical medium?

Answers ➲ p. 77

CHAPTER SYLLABUS CHECKLIST

Are you able to answer these questions from the syllabus for this chapter? Tick each question as you go through the checklist if you are able to answer it. If you cannot answer a question, turn to the relevant page in the study guide to find the answer. For NESA key word meanings, go to www.educationstandards.nsw.edu.au and search 'key words'.

	FOR A COMPLETE UNDERSTANDING OF THIS TOPIC:	PAGE NO.	✓
1	Can I explain what a wave is and what it does?	66	
2	Can I explain the difference between mechanical and non-mechanical waves, and give an example of each?	66	
3	Can I explain the difference between longitudinal and transverse waves, and give an example of each?	66	
4	Can I relate the compressions and rarefactions of a longitudinal wave to the crests and troughs of a transverse wave?	66, 68	
5	Can I explain what the terms wavelength, period, frequency, amplitude, displacement and wave number mean, and how they can be applied to transverse and longitudinal waves?	67	
6	Can I recall the symbols and SI units used to represent wavelength, period, frequency, amplitude, displacement and wave number?	67	
7	Can I draw graphs of displacement (y) against distance (x), and displacement (y) against time (t), for waves?	67	
8	Can I use graphs of displacement (y) against distance (x) and displacement (y) against time (t) to determine the wavelength, wave number, period, frequency and amplitude of a wave?	69	
9	Can I solve problems using the relationships $T = 1/f$ and $v = f\lambda$?	69	
10	Can I conduct practical investigations to show how energy is transferred by mechanical waves through a medium?	71	
11	Can I conduct practical investigations to produce transverse and longitudinal waves?	72	
12	Can I use observations and simple experiments to illustrate the difference between mechanical waves and electromagnetic waves?	72	

YEAR 11 EXAM-TYPE QUESTIONS

Objective-response questions

(1 mark each)

1 How would a sound wave change if the frequency of the sound wave was doubled?

A The velocity would double.
B The velocity would halve.
C The wavelength would double.
D The wavelength would halve.

2 What is the amplitude and frequency of the wave shown in Figure 6.11?

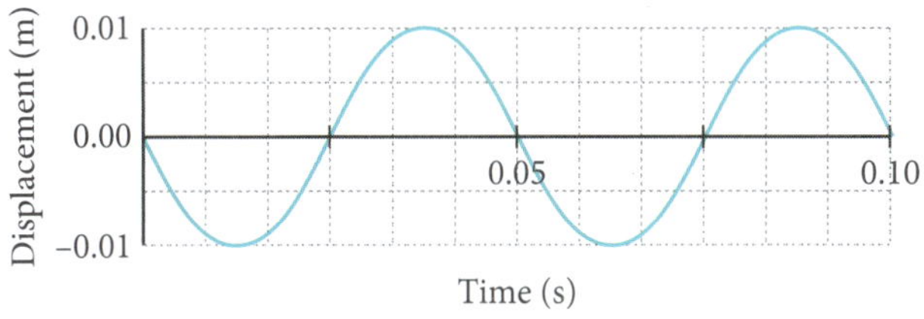

Figure 6.11 Wave graph

A amplitude: 0.02 m; frequency: 20 Hz
B amplitude: 0.02 m; frequency: 0.05 s
C amplitude: 0.01 m; frequency: 20 Hz
D amplitude: 0.01 m; frequency: 0.05 s

3 Which of the following are all mechanical waves?

A Water waves, seismic waves, sound waves in water, microwaves
B Waves in a string, sound waves in air, longitudinal waves in a spring, tidal waves
C Sound waves in steel, radio waves, water waves, seismic waves
D Transverse waves on a spring, water waves, x-rays, seismic waves

4 Given that sound travels at 340 ms^{-1}, what is the period of the sound wave shown in Figure 6.12?

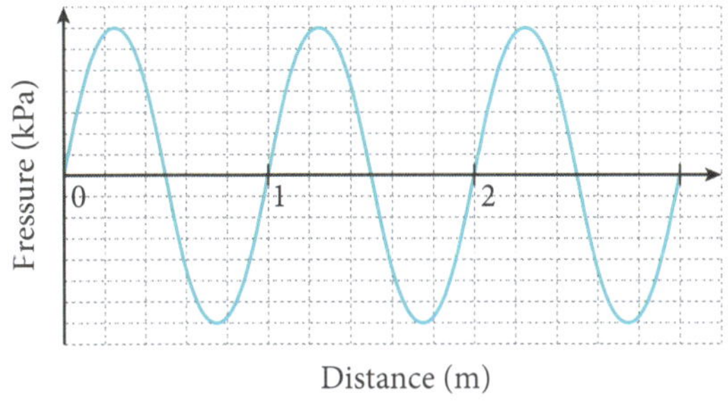

Figure 6.12 Graph of air pressure against distance for a sound wave

A 2.9 s
B 340 Hz
C 2.9 ms
D 0.0029 ms^{-1}

5 How could the wave in the Figure 6.13 be best described?

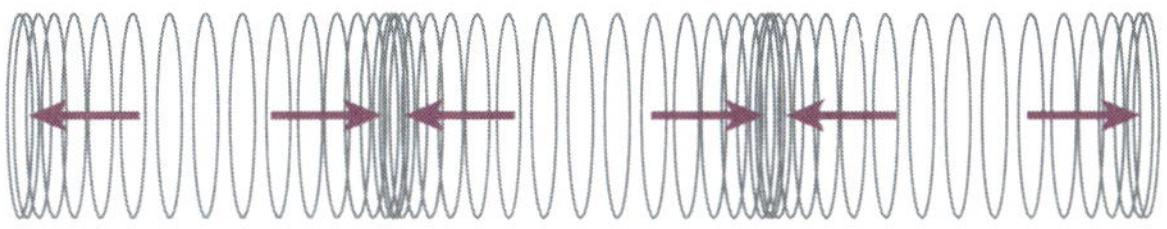

Figure 6.13 Waves on a spring

The wave is:

A a longitudinal wave on a spring showing 3 wavelengths.
B a transverse wave in a spring showing $1\frac{1}{2}$ wavelengths.
C a transverse wave on a spring showing 3 wavelengths.
D a longitudinal wave on a spring showing 4 wavelengths.

Extended-response questions

6 State three differences between a sound wave and a light wave. (3 marks)

7 Sound travels at 340 ms^{-1} on air and 1480 ms^{-1} in water. For a sound wave with a frequency 360 Hz:

a find the period of the sound wave (1 mark)
b find the wavelength of the sound wave in air (1 mark)
c find the wavelength of the sound wave in water (1 mark)
d suggest a reason why sound might travel faster in the water than in the air. (2 marks)

8 A boy sitting at the end of a wharf, as shown in Figure 6.14, observes 8 water waves passing per minute and measures the distance between the adjacent wave crests to be 25 m.

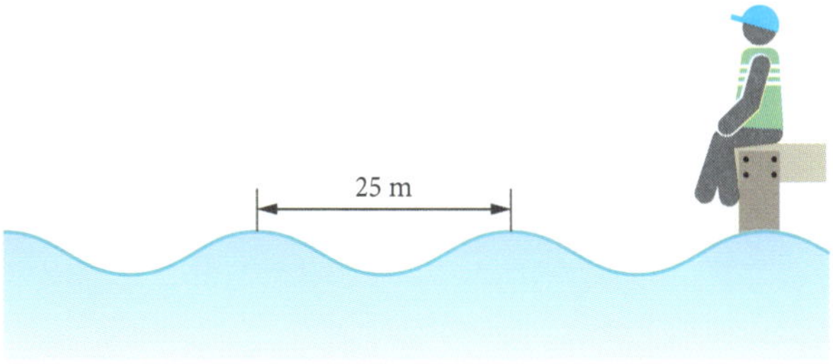

Figure 6.14 Water waves

a Find the period of the waves. (1 mark)
b Find the frequency of the waves. (1 mark)
c Find the velocity of the waves. (1 mark)
d What simple experiment could the boy do to prove that the waves were carrying energy? (2 marks)

9 Light travels at 3×10^8 ms^{-1} in air, while sound travels at 340 ms^{-1}. If a lightning flash takes 2 μs to reach an observer:

a find the distance from the observer to the lightning (1 mark)

b find the time after seeing the flash that the observer will hear the thunder clap from the lightning strike. (1 mark)

10 The motion of a point (position $x = 0$) on a stretched string is shown in Figure 6.15 as a function of time as transverse waves pass down the string.

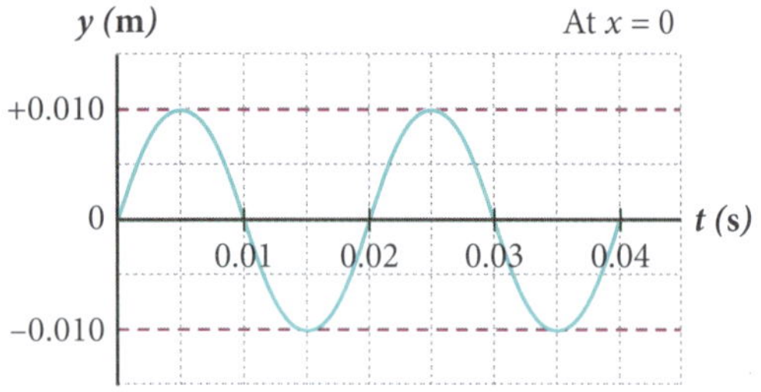

Figure 6.15 Waves on a stretched string

a What is the frequency of the wave? (1 mark)

b What is the amplitude of the wave? (1 mark)

c Given that the wave is travelling at 2 ms^{-1}, find the wavelength of the wave. (1 mark)

d Draw a graph of the displacement (y) against distance (x) graph for this wave, starting at $x = 0$. (3 marks)

ANSWERS

KEY QUESTIONS

Key questions ➲ p. 67

1 Waves are periodic disturbances of a medium or field that carry energy.

2 Mechanical waves require a material medium to propagate, while non-mechanical waves can pass through a vacuum. We can think of mechanical waves as an oscillation that moves through a material medium, and non-mechanical waves as an oscillation that moves through a field, such as an electric field or a gravitational field.

3 If the oscillation of the medium or field is perpendicular to the wave velocity we call the wave a transverse wave, but if the oscillation is parallel to the wave velocity we call it a longitudinal wave. Water waves and electromagnetic waves are transverse waves, while sound waves are longitudinal waves.

Key questions ➲ p. 69

4 We can represent displacement (y) graphically against either distance (x) or against time (t). The graph of displacement against distance represents the wave at a specific instant in time, while the time graph shows the displacement (y) at a specific point on the wave (e.g. at $x = 0$) as time passes.

5 The wavelength of a transverse wave is the distance between adjacent crests or between adjacent troughs. The wavelength of a longitudinal wave is the distance between adjacent compressions or between adjacent rarefactions.

Key questions ➲ p. 71

6 The medium through which the wave passes will determine the velocity of the wave. The velocity is usually independent of the amplitude and frequency of the wave, although in the special case of water waves, the velocity is related to the wave amplitude.

7 Wave velocity is constant in a specific medium and, as the product of the frequency and wavelength is equal to the velocity (i.e. $v = f\lambda$), the wavelength is inversely proportional to the frequency (i.e. $f = v/\lambda$ or $f \propto 1/\lambda$). This means that in a specific medium, increasing the frequency will decrease the wavelength.

Key questions ➲ p. 73

8 When an oscillation is coupled into a medium, energy from the oscillation is transferred to the medium. This energy moves away from the oscillator through the medium in the form of a wave.

9 Waves will only propagate through the medium if each part of the medium is coupled/connected to the adjacent part of the medium. When one part of the medium moves it therefore pulls or pushes the adjacent part of the medium. This coupling transmits the motion and energy through the medium.

YEAR 11 EXAM-TYPE QUESTIONS

Objective-response questions

1 **D.** $\lambda = v/f$. Note the velocity does not change as it is determined by the medium and independent of the frequency and wavelength. **A** and **B** are incorrect because the velocity depends on the medium and is independent of the frequency. **C** is incorrect because the period and frequency are inversely related, not proportional to each other.

2 **C.** The amplitude is the maximum displacement for the equilibrium position and the frequency is the inverse of the period. The period from the graph = 0.05 s and hence $f = 1/T = \frac{1}{0.05} = 20$ Hz.
A and **B** are incorrect because they use the peak-to-peak distance to determine amplitude rather than the maximum displacement from the equilibrium position. **D** is incorrect because it gives the period for the frequency.

3 **B.** All of these waves require a material medium to propagate. Note that radio waves, microwaves and x-rays are all electromagnetic (non-mechanical) waves. **A** is incorrect because microwaves are electromagnetic waves and hence non-mechanical. **C** is incorrect because radio waves are electromagnetic waves and hence non-mechanical. **D** is incorrect because x-rays are electromagnetic waves and hence non-mechanical.

4 **C.** Reading the wavelength for the graph gives $\lambda = 1$ m. Now as $v = f\lambda$ and $f = 1/T$ we can write $v = \lambda/T$ or $T = \lambda/v = \frac{1}{340} = 0.0029$ s or 2.9 ms. **A** is incorrect because it is 1000 times too large. **B** and **D** are incorrect because they have the wrong units.

5 **A.** As it is a longitudinal wave, the medium moves parallel to the velocity and, as the distance between compressions is one complete wavelength, there are three complete cycles of the wave. **B** and **C** are incorrect as it is not a transverse wave. **D** is incorrect as there are four compressions and hence only three wavelengths are shown.

Extended-response questions

6 EM This question requires students to recognise differences between sound and light waves.

- a Sound wave are longitudinal waves, while light waves are transverse waves. ✓
- b Sound waves are mechanical waves, while light waves are non-mechanic or electromagnetic waves. (Or sound waves require a mechanical medium to propagate but light does not.) ✓
- c Sound waves travel much more slowly than light waves. ✓

7 EM This question tests students' ability to apply the wave equation and suggest a plausible explanation for how density changes the velocity of a sound wave.

- a $T = 1/f = \frac{1}{360} = 2.78$ ms ✓
- b $v = f\lambda$ and hence $\lambda = v/f = \frac{340}{360} = 0.94$ m ✓
- c $\lambda = v/f = \frac{1480}{360} = 4.11$ m ✓
- d The molecules in water are much closer together than in air ✓ and hence the motion of one molecule is much more quickly transmitted to the next in water. ✓

8 EM This question tests students' understanding of period and frequency, SI units and applying the wave equation to solve a real-world problem. In addition, the question requires students to design a firsthand investigation to test a hypothesis about the energy carried by a wave.

- a As 8 waves pass per minute, the time for 1 wave would be $\frac{60}{8} = 7.5$ s. ✓

b $f = 1/T = 0.133$ Hz ✓

c $v = f\lambda = 0.133 \times 25 = 3.3\ ms^{-1}$ ✓

d The boy could drop something that floats into the water ✓ and show that the waves do work on it by oscillating the object. If the wave can do work it must possess energy because it would be transferring energy to the object. ✓

9 EM This question tests the ability of students to apply kinematics to electromagnetic waves and sound waves. An understanding of SI prefixes is also required.

a Distance $= vt = 3 \times 10^8 \times 2 \times 10^{-6} = 600$ m ✓

b Time for thunder (which travels at the speed of sound) would be given by $t = s/v = \frac{600}{340} = 1.76$ s. ✓ Note that you do not have to subtract the time for the lightning to reach the observer because this time is negligibly small compared to the time for the sound to reach the observer.

10 EM This question requires students to demonstrate an understanding of the two ways in which waves may be represented graphically and the relationship between the two types of graphs. Students are required to demonstrate proficiency at applying the wave equation.

a Frequency $= f = 1/T = \frac{1}{0.02} = 50$ Hz ✓

b Reading directly from the graph the maximum displacement from the equilibrium position is 0.01 m. ✓

c As $v = f\lambda$ the wavelength will be given by $\lambda = v/f = \frac{2}{50} = 0.04$ m. ✓

d We have the amplitude $a = 0.01$ m ✓ and the wavelength $\lambda = 0.04$ m. ✓ We also know the displacement will increase after time $t = 0$ at position $x = 0$. To do this the displacement must be negative in front of the position $x = 0$ at time $t = 0$ s, as shown in Figure A6.1. As the wave moves to the right, the displacement at position $x = 0$ will begin to increase, as we predicted it would.

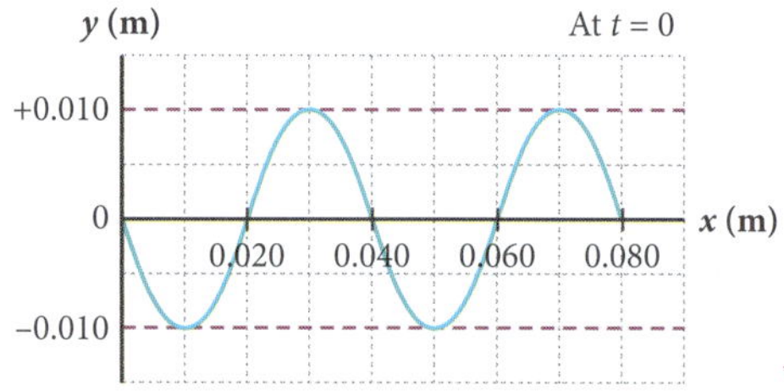

✓

Figure A6.1 Wave displacement against time

CHAPTER 7 WAVE BEHAVIOUR

MODULE 3
WAVES AND THERMODYNAMICS

INQUIRY QUESTION:

How do waves behave?

All types of waves display characteristic behaviour that can be explained by a small number of laws, theories and principles. For example, waves in the same medium add together (superimpose), waves reflect from boundaries, waves change direction (refract) when they change velocity and waves spread out (diffract) when they encounter small gaps, slits or obstacles.

1 Phenomena of wave behaviour

» Students explain the behaviour of waves in a variety of situations by investigating the phenomena of:
- reflection
- refraction
- diffraction
- wave superposition.

Reflection

→ A light beam from a ray box can be used to investigate the laws of reflection.

→ We can picture waves by drawing ray diagrams, as shown in Figure 7.1. A **ray** is a directed line drawn in the direction of the wave's velocity (or the direction that energy moves through the medium).

ray: a directed line drawn in the direction of the wave velocity (or the direction of energy transfer)

→ Sometimes wave fronts are drawn to illustrate the wave crests. Wavefronts are always drawn perpendicular to the ray and the separation between wavefronts is proportional to the wavelength of the wave.

→ **The normal** is a line drawn perpendicular to the surface or boundary between two different mediums.

the normal: a line drawn perpendicular to the surface or boundary; the angles of incidence, reflection and refraction are measured between the ray and the normal

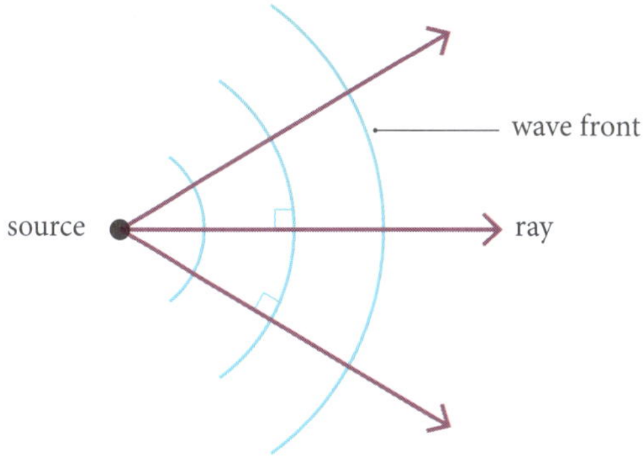

Figure 7.1 Rays and wavefronts emanating from a point source of waves

→ Reflected waves always obey the following laws of reflection (see Figure 7.2):

1 The angle between the *incident* ray and the normal (i) is equal to the angle between the *reflected* ray and the normal (r).

2 The incident ray, the reflected ray and the normal drawn from the surface at the point of incidence are all in the same plane.

3 The incident ray and reflected ray are on opposite sides of the normal.

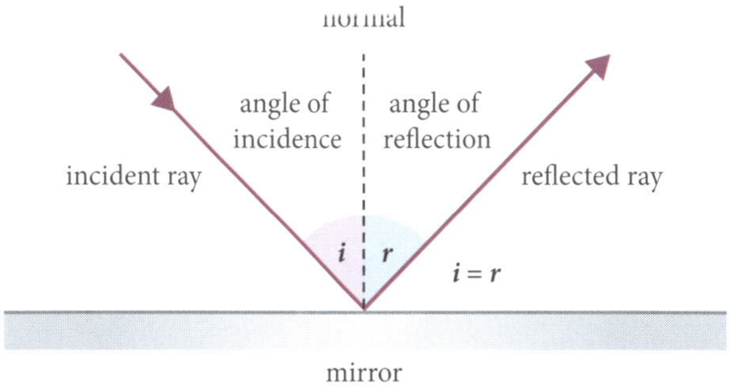

Figure 7.2 The laws of reflection. Note the incident ray, reflected ray and the normal are in the one plane (i.e. the plane of the page)

→ Reflection from a smooth surface, like a mirror, is called **specular reflection**. It is specular reflection that allows us to see our images in a flat mirror.

specular reflection: reflection from a smooth surface

- Reflection from a rough surface will produce reflected rays that move in many different directions. This type of reflection is called **diffuse reflection**. Diffuse reflection allows us to see everyday objects like this book. Both types of reflection are illustrated in Figure 7.3.

diffuse reflection: reflection from a rough surface

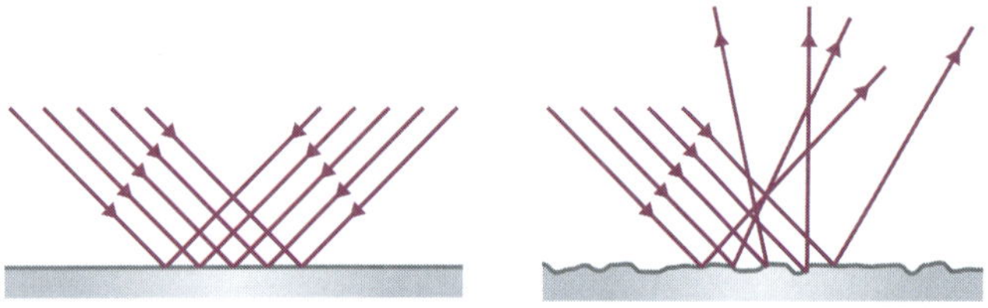

Figure 7.3 Types of reflection

- Curved mirrors, as shown in Figure 7.4, can be used to focus or disperse rays. Concave mirrors are used in torches, satellite dishes and car headlights, while the wide-angle view provided by convex mirrors is used in assisting drivers to see around blind corners.

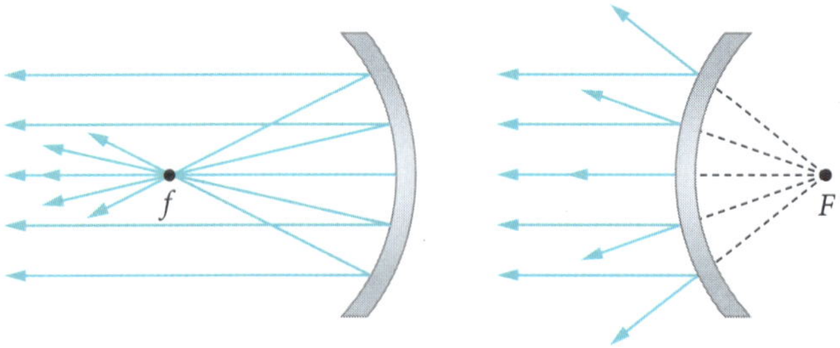

Figure 7.4 Ray diagrams for curved mirrors. Rays reflected from a concave mirror converge to a point, while rays from a convex mirror spread out or diverge

EXAMPLE 1

Consider a ray incident on a right-angled corner mirror, as shown in Figure 7.5. Given the angle of incidence is 65°, find the angles A, B, C and D.

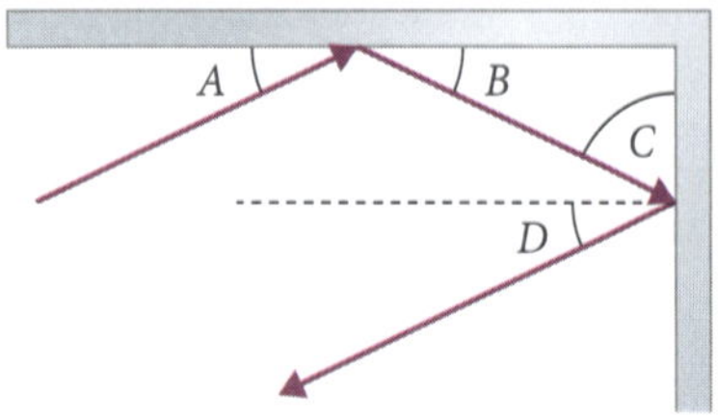

Figure 7.5 Reflection from a corner reflector

Answer:

If the angle of incidence is 65°, the angle $A = 90° - 65° = 25°$.

> Remember the angle of incidence and reflection refer to the angles between the rays and the normal

By the first law of reflection, the angle of reflection is also 65° and hence $B = 90° - 65° = 25°$.

As B and C are angles in a right-angled triangle, $C = 90° - 25° = 65°$

Angle D will be the same as the angle of incidence for this refection: $D = 90° - 65° = 25°$

→ KEY QUESTIONS

1. **Does the frequency or wavelength change when a wave is reflected?**
2. **What is the relationship between the direction that energy is transmitted by a wave through the medium and the wavefronts of the wave?**

Answers ⊃ p. 90

Refraction

- **Refraction** refers to the change in direction of the wave velocity that occurs when the wave moves from one medium to another. A refracted light ray is illustrated in Figure 7.6.

refraction: the change in direction of propagation of a wave when it moves from one medium to another

- If the wave moves slower in the second medium the ray is refracted *towards* the normal and, if the wave moves faster in the second medium, the ray is refracted *away* from the normal.

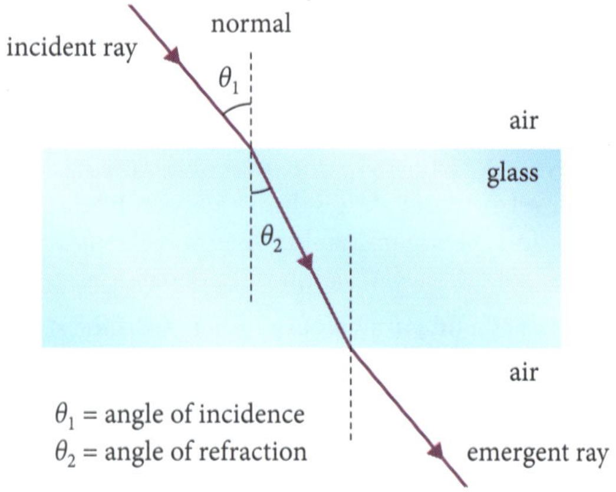

Figure 7.6 Refraction of a ray of light entering and emerging from a block of glass

➔ Note that because the wave must be continuous at the boundary, the frequency of the waves must be the same in each medium. This means that if the velocity changes the wavelength must also change. We can see this from the wave equation $v = f\lambda$. Hence, if the velocity halved in the new medium, the wavelength would also halve.

EXAMPLE 2

When a sound wave moves from air to water, some of the wave energy is reflected and some is transmitted into the water. On a diagram similar to Figure 7.7, sketch the reflected and refracted rays and three wavefronts of the reflected and refracted waves.

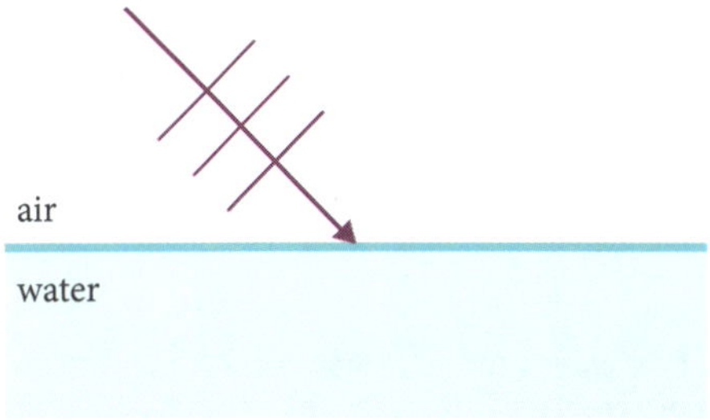

Figure 7.7 A sound wave approaching an air–water boundary

Answer:

Remember that sound travels faster in a denser medium

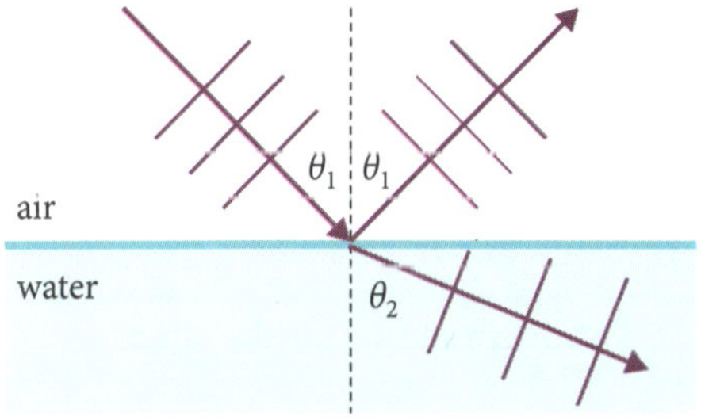

Figure 7.8 A sound wave reflected and refracted at an air–glass boundary

As shown in Figure 7.8, the angle of reflection would be the same as the angle of incidence and the reflected ray would have the same wavelength as the incident ray (i.e. the spacing between the wavefronts does not change).

The angle of refraction is greater than the angle of incidence because the sound wave would travel faster in the water than in the air. The wavelength is greater in the water than it is the air because the frequency is constant and hence the ratio of the wavelengths in each medium is equal to the ratio of the velocities in each medium, $v_1/v_2 = \lambda_1/\lambda_2$.

➔ KEY QUESTIONS

3 **What wave properties can change when a wave moves into a different medium?**

4 **Would a light ray and a sound ray be refracted in the same way if they passed into a denser medium?**

Answers ➲ p. 90

Diffraction

➔ When waves encounter a small object or gap, as shown in Figure 7.9, they spread out behind the object or gap. This spreading out of waves is called **diffraction**.

diffraction: the way waves spread out when they pass through a gap or encounter a small obstacle

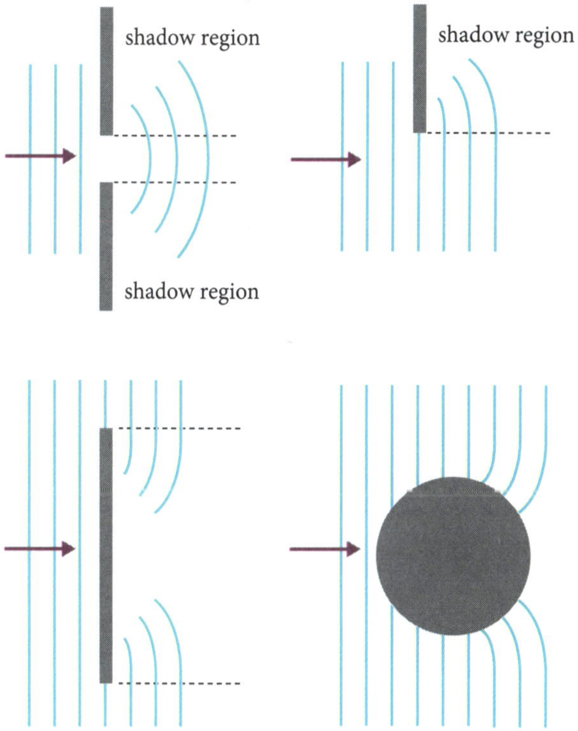

Figure 7.9 Examples of wave diffraction

➔ As shown in Figure 7.10, the smaller the gap or object is, the more the wave is diffracted. If the gap is much larger than the wavelength, the spreading out will be barely noticeable. But if the gap has similar dimensions to the wavelength, the wave will spread out significantly.

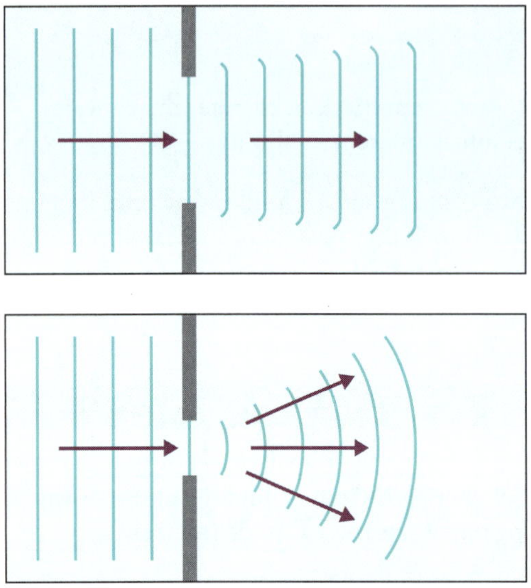

Figure 7.10 Effect of gap (or object) size on diffraction. Note that diffraction increases as the gap size is reduced

- The effect of gap size and wavelength (frequency) on diffraction can be readily investigated and confirmed for:
  - water waves using a ripple tank
  - sound waves using sound waves of different frequencies diffracted through openings such as doors
  - light waves using a laser pointer and narrow single slits.

Superposition

- When to two or more waves occur in the same place in a medium, the displacement produced by each of the waves is added together in a process called **superposition**.

superposition: the way the displacement of waves travelling in the same medium at the same time adds together

- Waves can pass through one another, superimpose during the interaction and then continue through the medium unchanged by the interaction. A good way to illustrate this is by passing pulses down a stretched spring, as shown in Figure 7.11.
- The superposition of two or more waves in the same medium at the same time is called **interference**. The amplitude of the resulting wave will depend on the relative phase of the interfering waves. As in Figure 7.12, if the waves have the same phase, their crests and troughs align and the resulting wave will have twice the amplitude of the interfering waves. But if the waves are a half-cycle out of phase, the waves will cancel out one another. When waves reinforce one another we say the interference is *constructive*, but when superposition causes the waves to cancel out each other, we call the interference *destructive*.

interference: occurs when two or more waves are superimposed

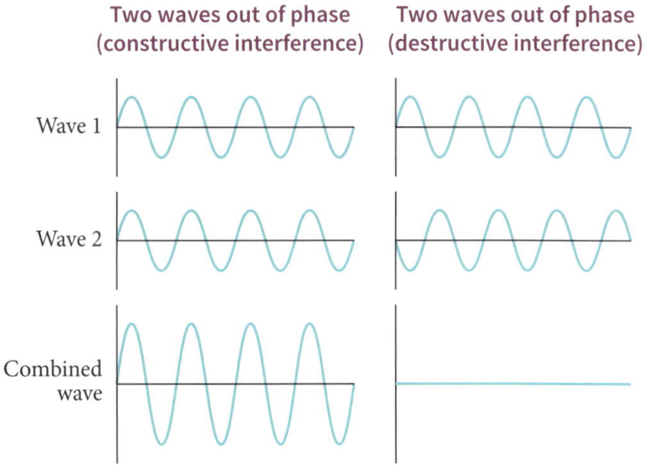

Figure 7.12 Constructive and destructive interference

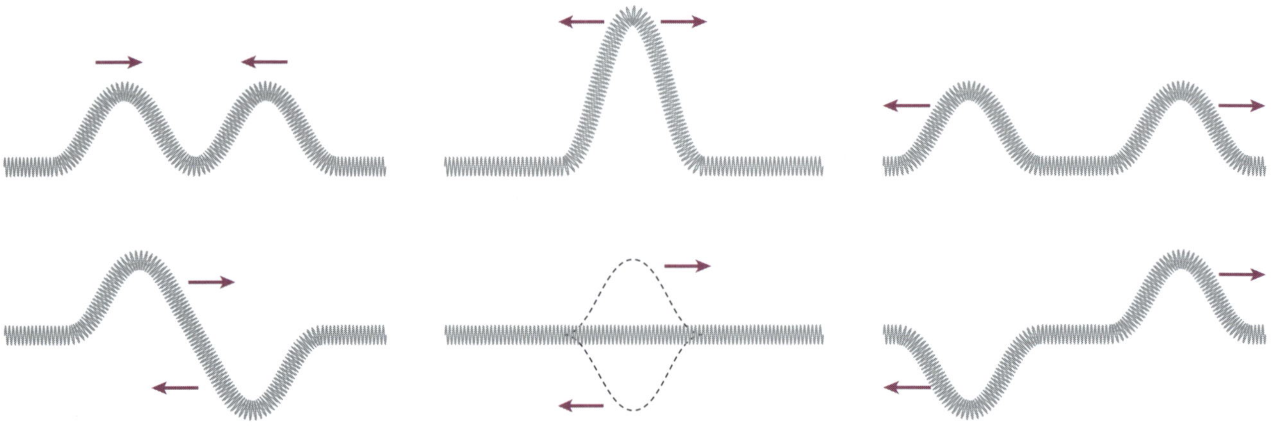

Figure 7.11 Superposition of pulses in a spring, showing the effects of positive and negative displacements on the superimposed waveform

→ KEY QUESTIONS

5 How does the size of a gap influence the amount of diffraction that results when a wave passes through a slit?

6 What happens when two waves pass through each other?

Answers ➲ p. 90

EXAMPLE 3

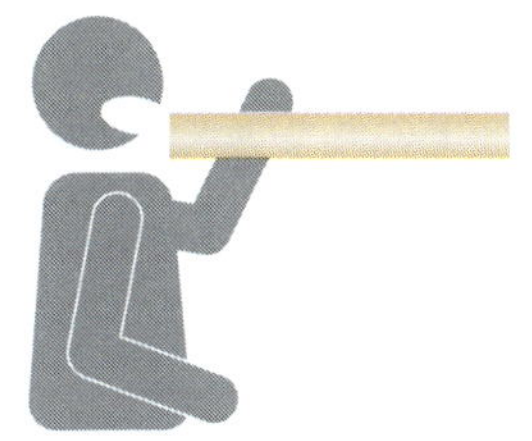

Figure 7.13 Shouting through a narrow tube

Refer to Figure 7.13. Will shouting through a narrow tube produce a thin column of sound? Justify your answer.

This is similar to a wave passing through a gap

Answer:

No, a column of sound will not be produced. This is because the sound waves will spread out (diffract) as they leave the tube because the wavelength of the sound waves is similar to or larger than the diameter of the tube shown. The diffracted waves are illustrated in Figure 7.14.

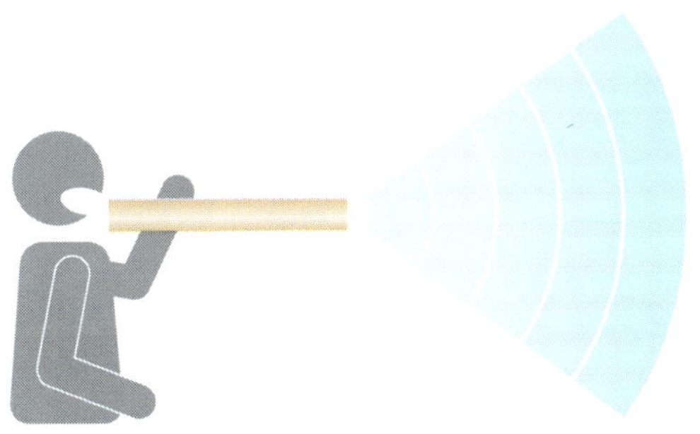

Figure 7.14 Diffraction of a sound wave leaving a narrow tube

EXAMPLE 4

Sketch the waveform that would result from the superposition of the two waves shown in Figure 7.15.

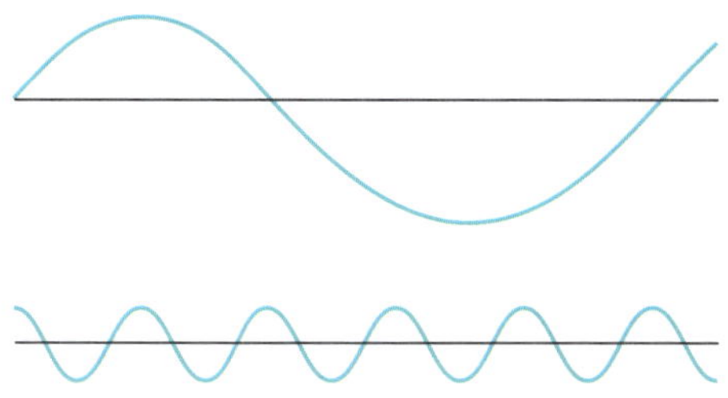

Figure 7.15 Two waves with different wavelengths and amplitudes

Answer:

Adding the two waves would produce the resultant shown in Figure 7.16.

For the displacement of superimposed waves, simply add them together

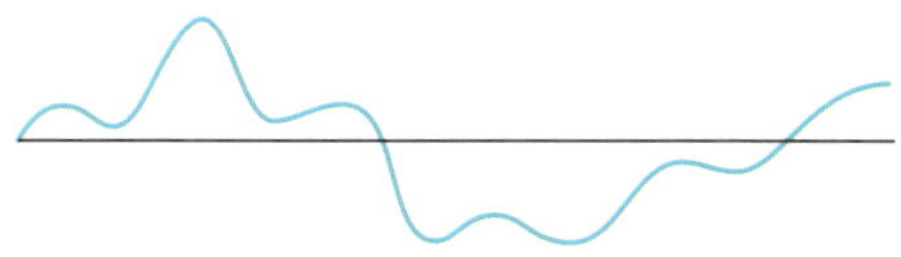

Figure 7.16 Superimposed waveform

2 Progressive waves and standing waves

» Students conduct an investigation to distinguish between progressive and standing waves.

→ A **progressive wave** is a travelling wave which has crests or compressions that travel through the medium or field at the wave velocity.

progressive wave: a travelling wave that carries energy

→ In contrast, a **standing wave** contains energy but does not move energy through the medium. Standing waves are produced by two travelling waves with the same wavelength moving in opposite directions. When a guitar string is plucked, standing waves are set up on the string by the waves being reflected at each fixed end of the string.

- Unlike progressive waves, standing waves will have some points that do not change their displacement (**nodes**), and some points that a have a maximum change in displacement (**antinodes**). As shown in Figure 7.17, the distance between adjacent nodes is $\lambda/2$.

standing wave: a wave made up by the superposition of two progressive waves moving in opposite directions that appears to be stationary
node: the position on a standing wave where the displacement of the wave does not change. The distance between adjacent nodes is a half of the wavelength
antinode: the position on a standing wave where the change in displacement is maximum

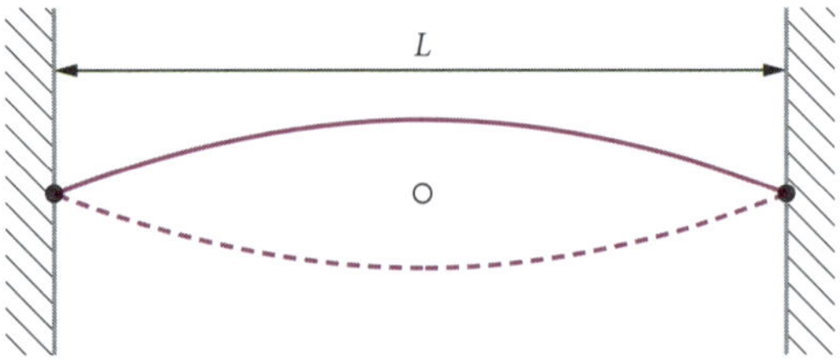

Figure 7.17 A standing wave on a string

- Standing waves can be demonstrated for:
  - water waves using a ripple tank, oscillator and barrier
  - waves on strings using any stringed instrument
  - waves on springs using long, stretched springs
  - sound waves using a sound source and open and closed pipes or a sonometer.
- Two-dimensional standing waves can be demonstrated by sprinkling fine sand on a drum skin and tapping the drum.
- By changing the frequency at which a standing wave system is driven, students can set up standing waves with different frequencies and wavelengths, as shown in Figure 7.18 (a) and (b).

EXAMPLE 5

A student finds that a standing wave with 5 nodes is set up when she oscillates a spring of length 6 m with a frequency of 4 Hz. Find the speed of the wave on the spring.

Answer:

Standing waves are produced by progressive waves moving in opposite directions. As the distance between adjacent antinodes is $\lambda/2$ there must be two complete wavelengths on the spring, as shown in Figure 7.19.

The distance between adjacent antinodes is $\lambda/2$

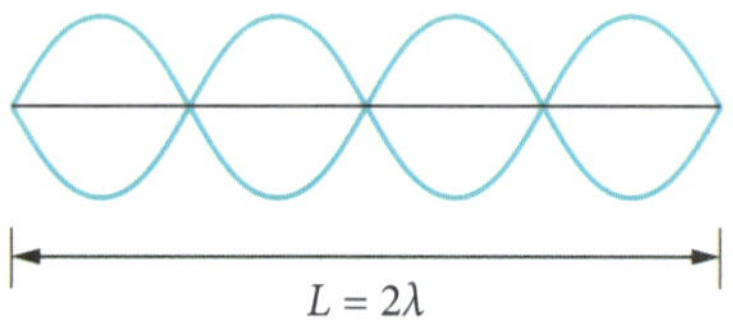

Figure 7.19 Standing waves on a string

As $L = 6$ m, $\lambda = L/2 = \frac{6}{2} = 3$ m

Applying the wave equation:

$v = f\lambda = 4 \times 3 = 12\ \text{ms}^{-1}$

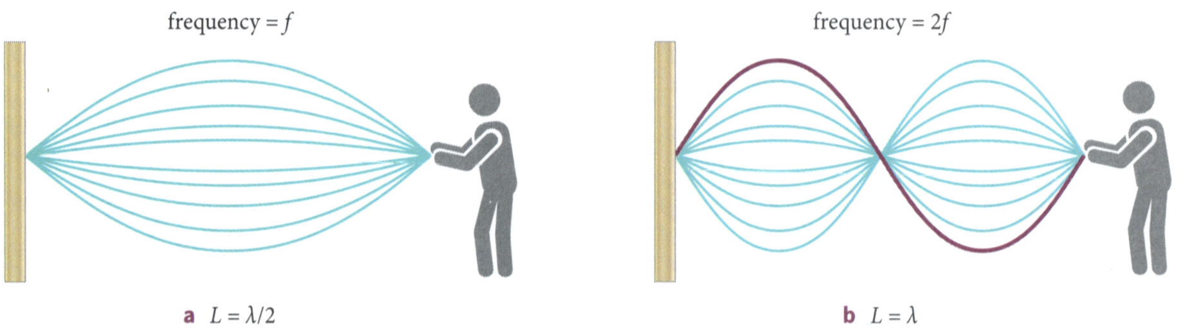

Figure 7.18 Setting up standing waves with different wavelengths and frequencies

→ KEY QUESTIONS

7 **Why do standing waves on a string only occur at specific frequencies?**

8 **Why does increasing the tension in a spring change the frequency of a specific standing wave on the spring?**

Answers ➲ p. 90

3 Resonance in mechanical systems

» Students conduct an investigation to explore resonance in mechanical systems and the relationships between:
- driving frequency
- natural frequency of the oscillating system
- amplitude of motion
- transfer/transformation of energy within the system.

→ When an oscillating system is displaced from its equilibrium position it will oscillate at one or more specific frequencies. For example, a child's swing will oscillate at one frequency but a guitar string can oscillate at several specific frequencies determined by the number of antinodes in the standing wave-pattern set up. We call these specific frequencies the **natural frequencies** of the oscillating system.

→ When an oscillating system is pushed or pulled we say it is being driven and, if the **driving frequency** is the same as one of the natural frequencies of the system, energy will be coupled into the oscillating system very effectively. When this occurs we say the system is being driven at its resonant frequency.

→ When a system is driven at its resonant frequency, energy is coupled into the system efficiently and the amplitude of the oscillations will increase rapidly. **Resonance** is the phenomena of driving an oscillating system at its natural (resonant) frequency.

→ We can investigate resonance in many ways. The examples that follow show how resonance can be investigated in two simple firsthand investigations.

natural frequency: the frequency (or frequencies) that an oscillating system vibrates at when it is displaced from its equilibrium position

driving frequency: the frequency with which an oscillating system is pushed or pulled by an external force

resonance: refers to a system being driven at its natural frequency

FIRSTHAND INVESTIGATION 1

Resonance of coupled pendulums

Students can demonstrate resonance using coupled pendulums, illustrated in Figure 7.20.

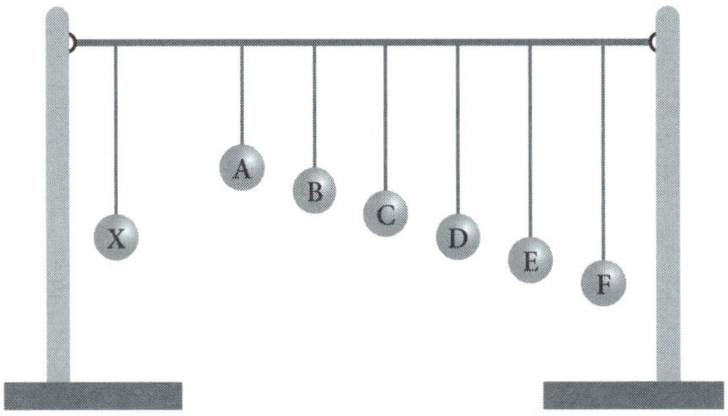

Figure 7.20 Resonance and coupled simple pendulums

The top string will transfer the motion of one pendulum to the other pendulums. That is, the top string will drive the other pendulums. Now as the frequency of a pendulum depends only on the length of the pendulum, setting pendulum *X* oscillating will drive the other pendulums at the frequency of *X* and will cause resonance in the pendulum with the same length as *X*. The pendulum with the same length of *X* will acquire energy from *X* and begin to oscillate with large amplitude, while the other pendulums will not because they are not being driven at their natural frequency.

FIRSTHAND INVESTIGATION 2

Resonance in a stretched string

Students can set up a ticker-tape timer or oscillator to drive a stretched string, as shown in Figure 7.21.

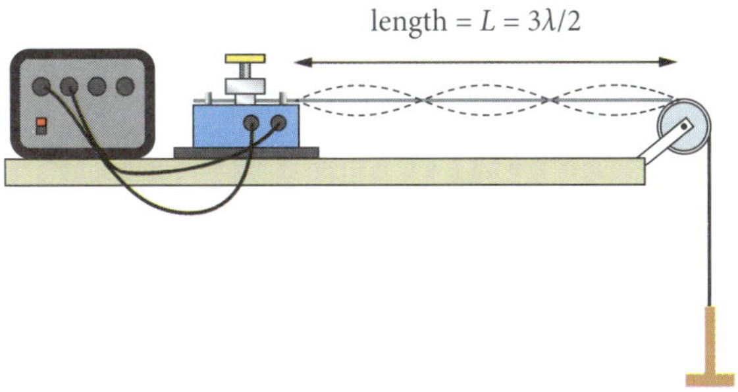

Figure 7.21 Experimental set up to investigate standing waves on a stretched string driven by a 50 Hz ticker-tape timer

The ticker-tape timer or oscillator can be slid along, to change the length (L) of string, until a standing wave appears on the string. The length of the string can then be readjusted to increase or decrease the number of wavelengths. For example, $L = \lambda/2$ and $L = \lambda$ or $L = 3\lambda/2$ (shown in Figure 7.21).

Note that the amplitude of the waves on the string only increases when the system is being driven at its resonance frequency, because only then is energy coupled into the system effectively.

Adding mass to the mass carrier will change the velocity of waves in the string, which will change the natural frequencies of the system and the wavelength of the standing waves that will occur at that tension. The length (L) of the string will have to be changed (by moving the oscillator) to set up standing waves on the string each time the tension is changed.

The velocity of the waves in the string as a function of the tension ($W = mg$) can be calculated from the wavelength of standing waves at that tension by using the wave equation, $v = f\lambda$. This velocity can be compared to the velocity predicted by theory:

$$v = \sqrt{\frac{T}{\mu}}$$

where T is the tension (mg) in the string

μ is the mass per unit length of the string

→ KEY QUESTIONS

9 **How are natural frequency, amplitude and driving frequency related for an oscillating system?**

10 **What is the most efficient way to couple energy into an oscillating system?**

Answers ➲ p.90

CHAPTER SYLLABUS CHECKLIST

Are you able to answer these questions from the syllabus for this chapter? Tick each question as you go through the checklist if you are able to answer it. If you cannot answer a question, turn to the relevant page in the study guide to find the answer. For NESA key word meanings, go to www.educationstandards.nsw.edu.au and search 'key words'.

FOR A COMPLETE UNDERSTANDING OF THIS TOPIC:		PAGE NO.	✓
1	Can I draw a normal to a surface as well as wave diagrams using rays and wavefronts?	79	
2	Can I recall and apply the laws of reflection to different types of waves?	79	
3	Can I predict the direction of refraction by the velocity of the wave in each medium?	80–81	
4	Can I predict the change in wavelength that will occur when a wave moves into a different medium and explain why this change occurs?	80	
5	Can I differentiate between diffraction and refraction?	80–81	
6	Can I explain how diffraction is related to wavelength or gap/slit size?	81–82	
7	Can I plot the resultant wave produced by the superposition of two component waves?	82	
8	Can I explain the difference between a standing wave and a progressive wave?	83	
9	Can I explain and give everyday examples of resonance in mechanical systems?	84	
10	Can I define natural frequency and driving frequency?	85	
11	Can I relate energy transfer and wave amplitude to driving frequency and natural frequency?	85	

YEAR 11 EXAM-TYPE QUESTIONS

Objective-response questions (1 mark each)

1 **Which of the following is a law of reflection?**

A The angle between the incident ray and the reflecting surface is equal to the angle between the reflected ray and the surface.

B The wavelength of the reflected wave is less than the wavelength of the incident wave.

C The angle between the normal and the incident ray is equal to the angle between the reflected ray and the normal.

D The angle between the normal and the incident ray is equal to the angle between the refracted ray and the normal.

2 **If the wavelength of a refracted wave is greater than the wavelength of the incident wave, which of the following statements would be true?**

A The velocity of the wave decreased when it entered the new medium.

B The frequency of the wave decreased when it entered the new medium.

C The velocity of the wave increased when it entered the new medium.

D The frequency of the wave increased when it entered the new medium.

3 **The girl shown in Figure 7.22 is standing 12 m from the tree and is oscillating the spring with a frequency f to set up standing waves. What would the wavelength of the standing waves on the spring be if she increased the driving frequency to $f/2$?**

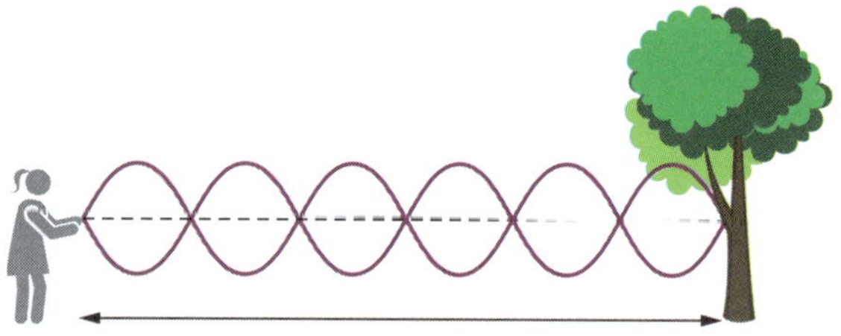

Figure 7.22 Setting up standing waves on a stretched spring

A 8 m
B 12 m
C 4 m
D 6 m

4 **If two identical waves are superimposed, what will the amplitude of the resultant wave be?**

A Twice the amplitude of the waves that make it up.

B Zero.

C Between zero and twice the amplitude of the waves that make it up.

D Between twice and four times the amplitude of the waves that make it up.

5 **Which of the following statements best describes the energy transfer that occurs when an oscillating system is driven at its natural frequency?**

A Energy will very effectively flow from the driving mechanism to the oscillating system.

B Energy will be transferred very effectively from the system to the driving mechanism.

C Energy will transfer backwards and forwards between the driving system and oscillator throughout each cycle.

D Because the system is not in resonance, no energy will flow between the driver and the oscillating system.

Extended-response questions

6 **Consider a ray of light moving from air to glass with three different angles of incidence, as shown in Figure 7.23.**

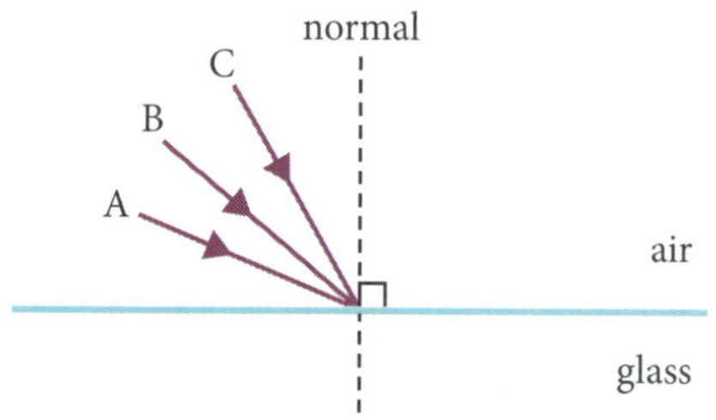

Figure 7.23 Rays of light approaching an air–glass boundary

a **Sketch and label the refracted ray for rays A, B and C in the glass.** (2 marks)

b **Would the velocity, frequency and wavelength of the wave change when it entered the glass slab and, if so, in what way would they change?** (3 marks)

7 **The displacement against distance graph for two waves in the one medium is shown in Figure 7.24.**

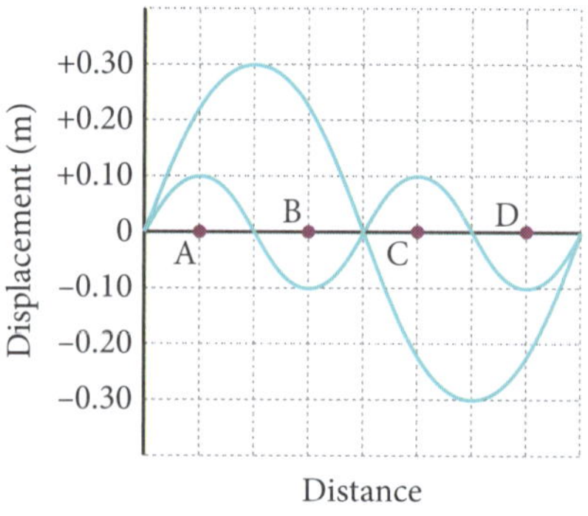

Figure 7.24 Two waves in the one medium

a **Determine the displacement of the resultant wave in the medium at each of the positions A, B, C and D marked on the graph.** (4 marks)

b Sketch the resultant wave onto Figure 7.24. (1 mark)

c Given that the distance between A and B was 2 m, find the wavelength of the resultant wave. (1 mark)

8 Consider two simple pendulums suspended from a single string tied to two retort stands, as shown in Figure 7.25. Note that the period of oscillation of a simple pendulum is related to length of the pendulum and the acceleration is due only to gravity (g).

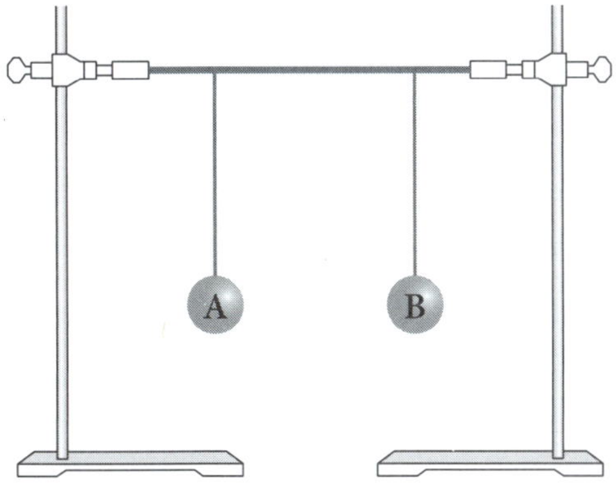

Figure 7.25 Coupled pendulums

If pendulum A is set swinging back and forth, the amplitude of its oscillation gradually decreases and B starts swinging and gradually increases the amplitude of its oscillation. This continues until pendulum A has stopped swinging. The process then repeats as B starts to swing less and A starts to oscillate.

Explain why the pendulums start and stop swinging in this way? (3 marks)

9 An 80 cm guitar string is plucked and the longest wavelength possible on the string is set oscillating with a frequency at 300 Hz, as shown in Figure 7.26.

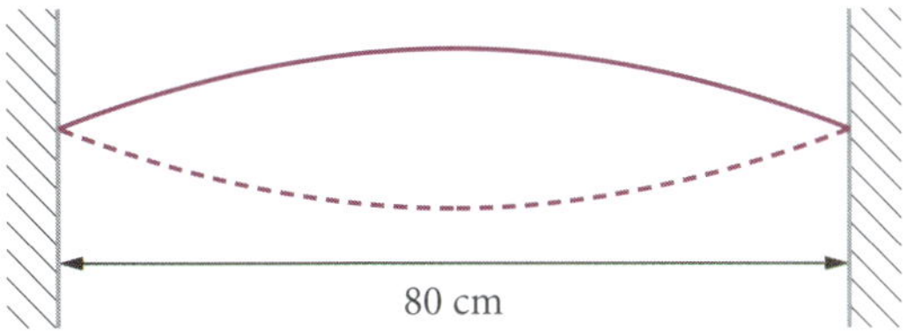

Figure 7.26 Standing wave on a guitar string

a Find the velocity of the waves in the string. (1 mark)

b Explain how the standing wave would change if the tension in the string was increased. (1 mark)

c Sketch the standing wave that would result if this string was oscillating at 900 Hz and find its wavelength. (2 marks)

10 a Given that light travels slower in glass than in air, redraw Figure 7.27 and sketch the path the ray would take through the glass block. (2 marks)

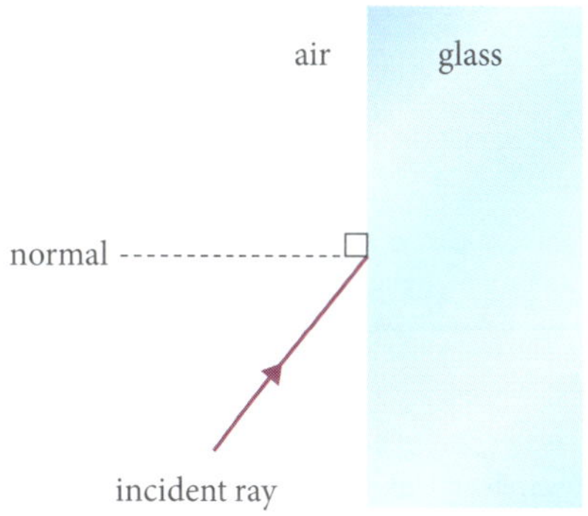

Figure 7.27 Light ray incident on an air–glass boundary

b Consider two mirrors placed at 90° to each another, as shown in Figure 7.28. The incident ray is 40° from the surface of mirror M_1. Sketch the path of the ray through the double-mirror system and include all relevant angles of incidence and reflection. (3 marks)

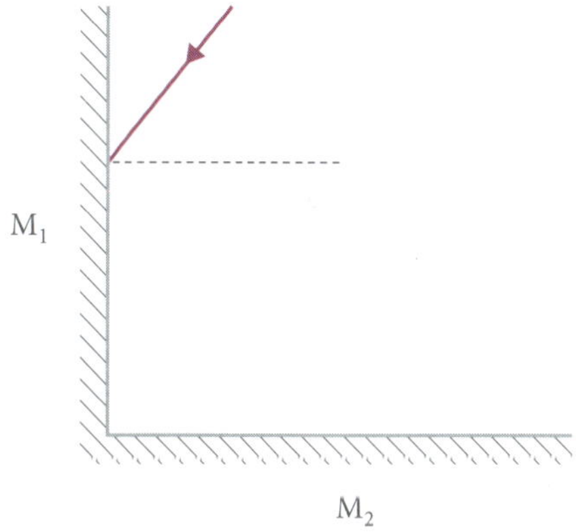

Figure 7.28 Reflection of light by a corner mirror

KEY QUESTIONS

Key questions ➲ p. 80

1 The frequency and the wavelength do not change when a wave is reflected.

2 Wave fronts are always drawn perpendicular the ray and hence wave fronts are perpendicular to the direction that energy is moved through the medium by the wave.

Key questions ➲ p. 81

3 When a wave moves into a different medium (is refracted) its velocity, wavelength and direction of propagation may change. Note that the wave frequency remains constant when a wave is refracted.

4 No, the light ray would be refracted towards the normal, and a sound ray would be refracted away from the normal. This occurs because light velocity decreases when it moves into a denser medium but the velocity of the sound wave increases when it moves into the denser medium.

Key questions ➲ p. 83

5 The smaller the gap encountered by a wave, the more it will be diffracted.

6 The waves will superimpose (add together) and, after passing through one another, will continue to propagate unaffected by the encounter.

Key questions ➲ p. 85

7 Because the ends of the string are fixed, the standing waves that can be set up on the string must have a node at each end of the string and an integer number of nodes in between. This limits the possible wavelengths (and hence the frequencies) of the standing waves that can be set up on the string.

8 Changing the tension changes the velocity of the waves on the string and, as the wavelength of the standing wave is constant, changing the velocity of the wave will change the frequency because $f = v/\lambda$.

Key questions ➲ p. 86

9 If a system is driven at its natural frequency, energy will be coupled into the system efficiently and the amplitude of the oscillations will increase significantly.

10 The most efficient way to couple energy into an oscillating system is to drive the system at its natural frequency of oscillation. The natural frequency is the frequency the system oscillates at when it is displaced from its equilibrium position and released.

YEAR 11 EXAM-TYPE QUESTIONS

Objective-response questions

1 **C**. The law of reflection states that the angle between the normal and the incident ray is equal to the angle between the refracted ray and the normal. **A** is incorrect because it refers to the angle between the ray and the surface. **B** is incorrect because the angles should be equal. **D** is incorrect because it refers to refraction not reflection.

2 **C**. Because the frequency remains constant when a wave is refracted, an increase in velocity results in an increase in wavelength as $\lambda = v/f$. **A** is incorrect because a decrease in velocity results in a decrease in wavelength. **B** and **D** are incorrect because frequency remains unchanged in refraction.

3 **A**. The initial wavelength is twice the distance between adjacent nodes and hence $\lambda = 4$ m. But if the driving frequency is halved the wavelength would double to 8 m, as $\lambda = v/f$. All the other answers are incorrect because they are not twice the initial wavelength.

4 **C**. If the crests of the two waves were aligned, the amplitude would double. But if the crests of one wave coincided with the troughs of the other wave, they could sum to zero. **A** is incorrect because it assumes the waves have their crests aligned (they are in phase). **B** is incorrect because it assumes the crest of one wave is aligned with the trough of the second wave (they are half-cycle out of phase). **D** is incorrect because adding two waves with the same amplitude cannot make a wave four-times bigger.

5 **A**. Driving an oscillator at its natural frequency will result in energy being transferred effectively to the oscillator. **B** is incorrect because the energy transfer is in the wrong direction. **C** is incorrect because the energy flows from the driver to the oscillator. It will not flow back and forth in one cycle of the oscillation. **D** is incorrect because if the system is driven at its natural frequency it is in resonance.

Extended-response questions

6 EM This question requires students to understand how a ray is refracted qualitatively and the changes a wave undergoes when it enters a different medium.

a Each ray will be refracted towards the normal, as shown in Figure A7.1.

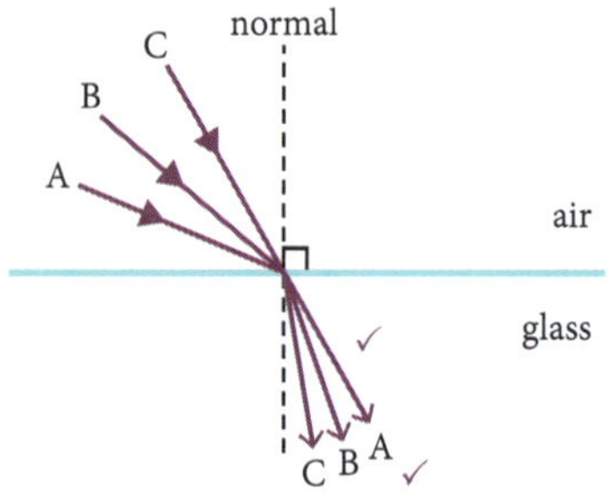

Figure A7.1 Refraction towards normal

b As the light rays are entering a denser medium their velocity ✓ and wavelength ✓ will decrease but their frequency will remain unchanged. ✓

7 EM This question requires students to have a sound understanding of superposition and the concept of wavelength.

a Displacement at A = 10 + 10 = 30 m ✓

Displacement at B = 20 − 10 = 10 m ✓

Displacement at C = 10 − 20 = −10 m ✓

Displacement at D = −20 − 10 = −30 m ✓

b Figure A7.2 shows the superposition of the two waves.

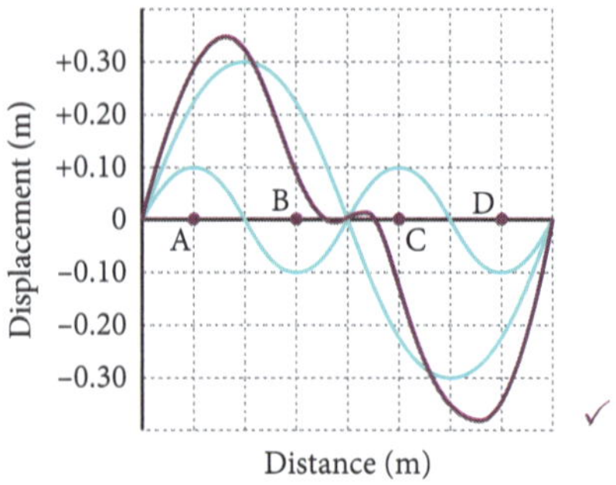

Figure A7.2 Superposition of two waves with different wavelengths and amplitudes c The distance is 8 m. ✓ As the distance between points A and B is 2 m, the distance between the vertical lines on the graph is 1 m and the wavelength of the superimposed wave would therefore be 8 m as it would take another 8 m to repeat the waveform.

8 EM This question tests students on their understanding of natural frequency and resonance.

Because the pendulums are the same length they will have the same natural frequency. ✓ When pendulum A is set oscillating, it will drive pendulum B at resonance and hence transfer its energy to pendulum B. ✓ When pendulum A has transferred its energy to pendulum B, B will act as the driver and transfer energy back to pendulum A. ✓

9 EM This question requires students to apply the wave equation to standing waves on a string.

a The distance between nodes is $\lambda/2$ and hence $\lambda = 1.6$ m.
Now $v = f\lambda = 300 \times 1.6 = 480\ \text{ms}^{-1}$. ✓

b If the tension was increased the velocity would increase. As the wavelength is constant, increasing the velocity would increase the frequency of the oscillation ✓ (as $f = v/\lambda$).

c Figure A7.3 shows the standing wave that would be produced on the string if it was driven at 900 Hz (i.e. third harmonic).

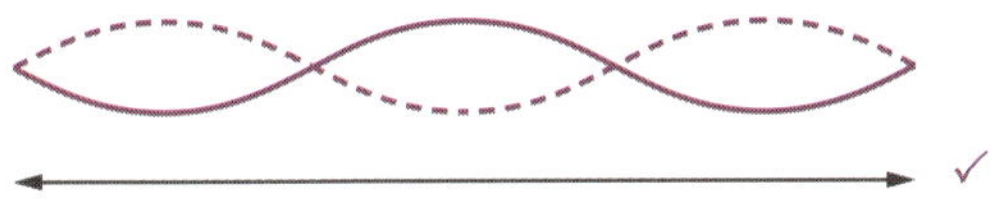

Figure A7.3 Standing wave that would result from a driving force of 900 Hz

We see from figure A7.3 that there are $\frac{3}{2}$ wavelengths in 80 cm, $\lambda = \frac{2 \times 0.8}{3} = 0.53$ m. ✓

10 EM This question requires students to demonstrate an understanding of the refraction of a ray moving to a medium in which it travels slower and a medium in which it travels faster. It also requires students to demonstrate an understanding of the laws of reflection.

a Figure A7.4 shows how a wave of light slows down as it passes through glass, which is denser than air, and thus it is refracted towards the normal.

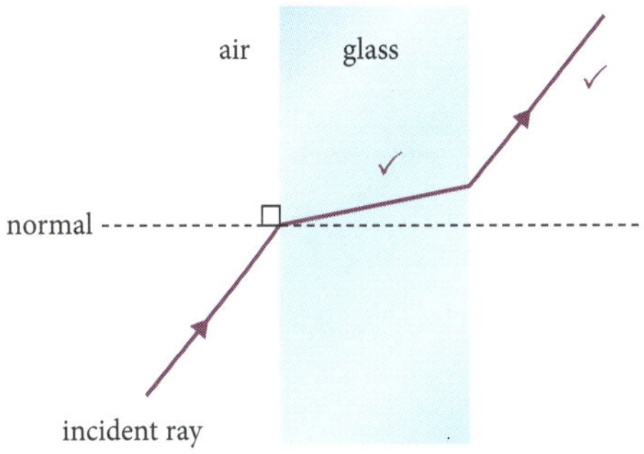

Figure A7.4 Refraction of a light ray as it travels through glass

b The correct angles of reflection are shown in Figure A7.5.

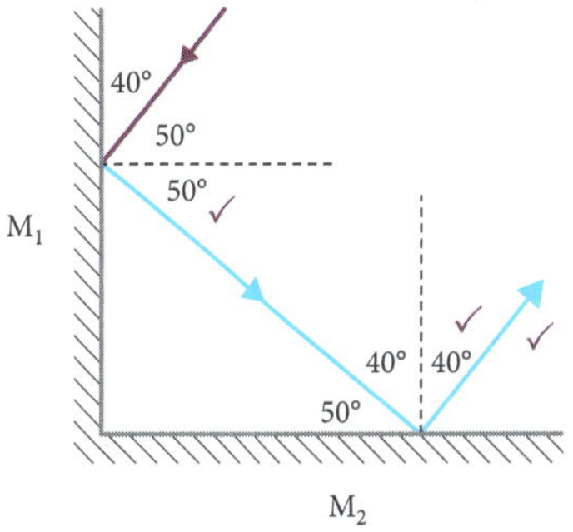

Figure A7.5 Path of the ray and angles of reflection for a 90° corner reflector.

CHAPTER 8 SOUND WAVES

MODULE 3
WAVES AND THERMODYNAMICS

INQUIRY QUESTION:

What evidence suggests that sound is a mechanical wave?

A sound wave is a longitudinal pressure wave. Experiments show that sound waves cannot propagate in a vacuum and hence sound waves are mechanical waves.

1 Pitch and loudness

» Students conduct a practical investigation to relate the pitch and loudness of a sound to its wave characteristics.

➔ The relationships between the frequency and pitch and between the amplitude and loudness of a sound wave are examined in the following investigations.

FIRSTHAND INVESTIGATION 1

Pitch and loudness

Students can investigate sound waves by using a microphone connected to a cathode ray oscilloscope (CRO) or to a computer or phone with software for displaying sound waves.

The microphone converts the sound wave (the air pressure changes) into a voltage that is displayed as a function of time on the CRO. A simple experiment of this type is shown in Figure 8.1.

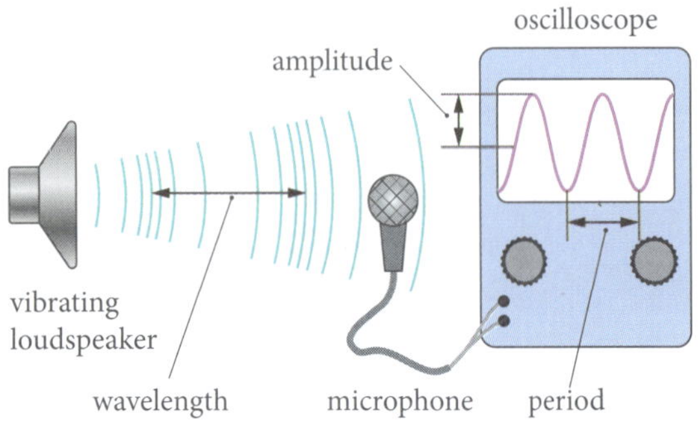

Figure 8.1 Investigating sound waves using a cathode ray oscilloscope (CRO)/computer or phone application

By using a wave generator connected to a loudspeaker, or by humming/whistling into the microphone, students will produce graphs like those shown in Figure 8.2. These show that:

- louder sounds produce a wave of greater amplitude (maximum change in air pressure)
- higher pitched sounds produce a wave with a smaller period (increased frequency)
- pure notes produce sine waves but noise or speech produce complex wave forms.

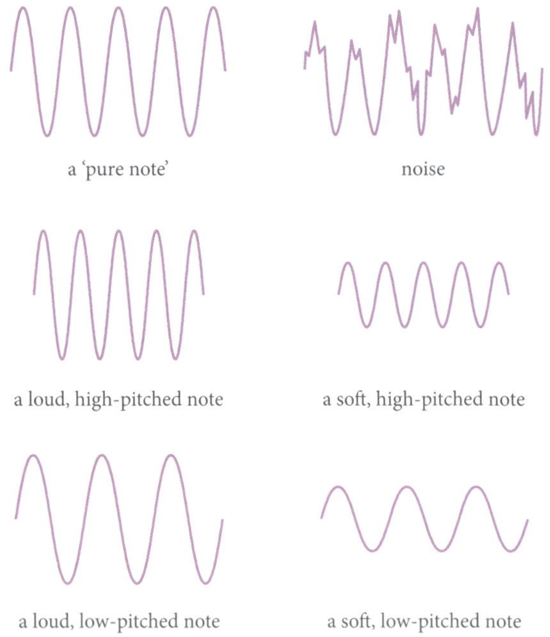

Figure 8.2 Different sound waves shown as pressure against time graphs

EXAMPLE 1

If the loudness of a sound was increased and its frequency decreased, describe how the sound wave would change.

Recall how loudness and frequency are related to the shape of a sound wave

Answer:

If the frequency is decreased, the wavelength would increase because $\lambda = v/f$. This means the compressions in the wave will move further apart. Increasing the loudness will cause the amplitude of the oscillations of the molecules to increase. For a sound wave this means that the variation in pressure produced by the wave will be greater.

2 Sound as a longitudinal pressure wave

» Students model the behaviour of sound in air as a longitudinal wave.
» Students relate the displacement of air molecules to variations in pressure.

- Investigations with tuning forks or wave generators and loudspeakers can be used to investigate the longitudinal nature of waves. As shown in Figure 8.3, both the tuning fork and loudspeaker involve an oscillation parallel to the wave velocity that will continually compress and rarefy the air in front of the tuning fork or loudspeaker. These compressions and rarefactions move away from the source of sound in the form of a longitudinal pressure wave.
- We hear sound because the change in pressure above the eardrum in our ears causes the eardrum to move back and forth parallel to the wave velocity.
- To produce changes in air pressure, air molecules must move back and forth parallel to the wave velocity. We can model the movement of molecules by using a longitudinal wave on a slinky spring, as shown in Figure 8.4. Note that pushing the spring or air in the direction of the wave produces (a) a **compression**, while moving the spring or air in the opposite direction of the wave velocity will produce (b) a **rarefaction**.

compression: a region of high pressure
rarefaction: a region of reduced pressure

- As the wave passes a point the air molecules move back and forth parallel to the wave velocity. This can be shown by tying a piece of ribbon onto a point on a spring and watching the ribbon move back and forth as a longitudinal wave passes along the spring.
- Pressure is measured in newtons per square metre (Nm^{-2}) or **pascals** (Pa).

pascal: the SI unit of pressure, the fundamental units of which are Nm^{-2}

- Atmospheric pressure is 101 kPa. A sound wave is a small pressure change on top of atmospheric pressure.

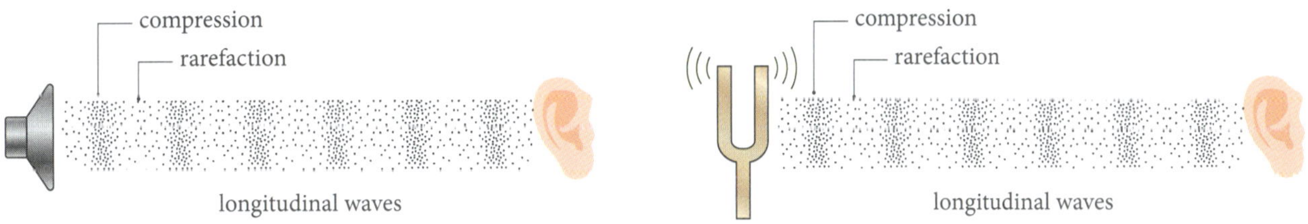

Figure 8.3 Sound waves produced by a loudspeaker and by a tuning fork

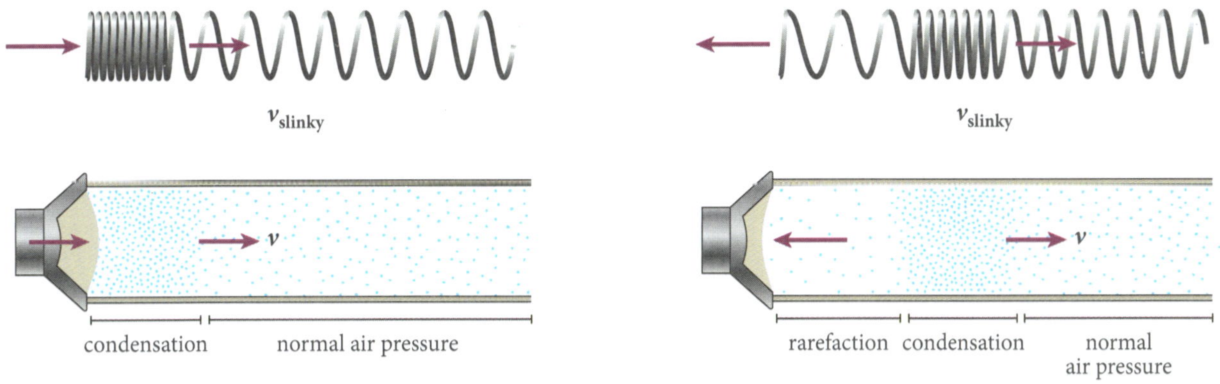

Figure 8.4 Using a slinky spring to model the propagation of sound

The change in pressure produced by a sound wave 1 m from someone speaking normally is about 0.02 Pa, while the pressure change 5 m from a busy road would be about 0.2 Pa.

EXAMPLE 2

When a sound wave *X* passes a point in air, the molecules oscillate backwards and forwards parallel to the velocity of the wave with an amplitude *A* and frequency *f*. Outline how *A* and *f* would change if the loudness of the sound wave was reduced and the pitch was increased.

Answer:

Recall that the pressure change produced by a sound wave is related to the amplitude of the oscillation of the molecules

Reducing the loudness would decrease the amplitude (A) of the molecule's oscillation, and increasing the pitch would increase the frequency (f) at which the particles oscillated.

→ KEY QUESTIONS

1 **What wave characteristics determine pitch and loudness?**

2 **How does the oscillation of air molecules produce a region of high pressure in a sound wave?**

3 **Explain why sound waves travel faster in denser materials.**

Answers ➲ p. 105

3 Intensity and distance

» Students investigate quantitatively the relationship between distance and intensity of sound.

→ Theory suggests that the **intensity** (I) of a spherical wave will decrease in proportion to the inverse square of the distance (r) from the source:

$$I \propto 1/r^2$$

where I is the light intensity

α (Greek letter *alpha*) means 'proportional to'

r is the distance from the source

intensity: the energy carried by a wave per second through a unit area perpendicular to the wave's velocity (i.e. Wm^{-2})

→ This because the power emitted from a points source is spread over the surface of a sphere that surrounds the source. As shown in Figure 8.5, the surface area increases in proportion to the radius squared and therefore the intensity (power ÷ area) must decrease as $1/r^2$ as we move further from the source.

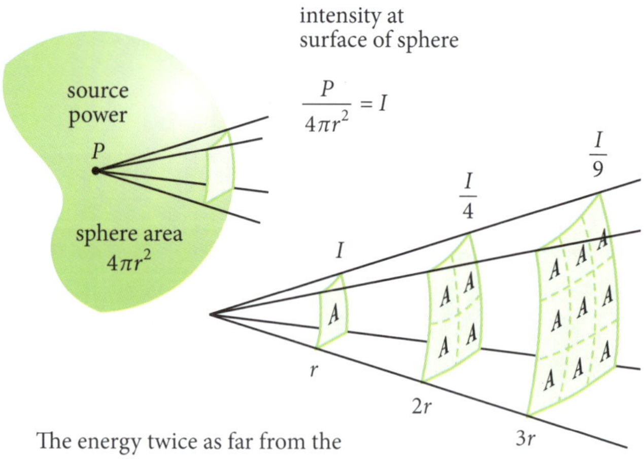

Figure 8.5 Inverse square law of intensity

→ The intensity of a sound at two different distances (r_1 and r_2) from a source can be related by taking the ratio of the two intensities, $I_1 \propto 1/r_1^2$ and $I_2 \propto 1/r_2^2$:

$$I_1/I_2 = r_2^2/r_1^2$$

where I_1 refers to the intensity at a distance r_1

I_2 refers to the intensity at a distance r_2

FIRSTHAND INVESTIGATION 2

Distance and intensity

To verify the inverse square law for sound intensity we could set up a loudspeaker and microphone and connect both to an oscilloscope, as shown in Figure 8.6. Note we may first have to calibrate the amplitude of the CRO signal to the intensity of the signal by comparing the amplitude at different sound intensities.

Using the amplitude of the wave from the microphone on the CRO as a calibrated measure of the intensity, we find the intensity as a function of distance (r).

If the intensity is proportional to the inverse of the distance from the source, a graph of intensity against $1/r^2$ should produce a straight-line graph.

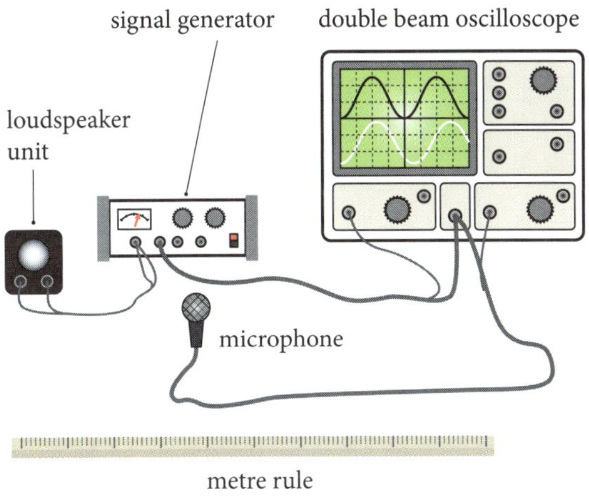

Figure 8.6 Experimental set up to demonstrate $I \alpha 1/r^2$

EXAMPLE 3

The intensity of a sound 5 m from a busy road is about 0.2 Wm^{-2}. Determine the intensity of sound 20 m from the road.

Answer:

> Recall that intensity is measured in Wm^{-2} and that the intensity of a wave decreased with the inverse square of the distance from the source

The new distance is four times further from the road and, as $I \alpha 1/r^2$, the new intensity will be the $\frac{1}{4^2} = \frac{1}{16}$ of the initial intensity. Hence:

$$I = \frac{0.2}{16} = 1.25 \times 10^{-2} \text{ Wm}^{-2}$$

4 Investigating the behaviour of sound waves

» Students conduct investigations to analyse the reflection, diffraction, resonance and superposition of sound waves.

- Sound waves exhibit many of the general wave behaviours we investigated in the previous chapter including reflection, refraction, diffraction and superposition.
- The following investigations examine a variety of wave behaviours for sound.

FIRSTHAND INVESTIGATION 3

Reflection of sound waves

The reflection of a sound wave is called an echo. We can confirm that sound waves obey the laws of reflection by setting up a simple experiment like the one shown in Figure 8.7. Note that a microphone or other sound-detecting device could replace the ear and a small loudspeaker could be used instead of the mechanical stopwatch shown.

> **echo:** a reflection of sound

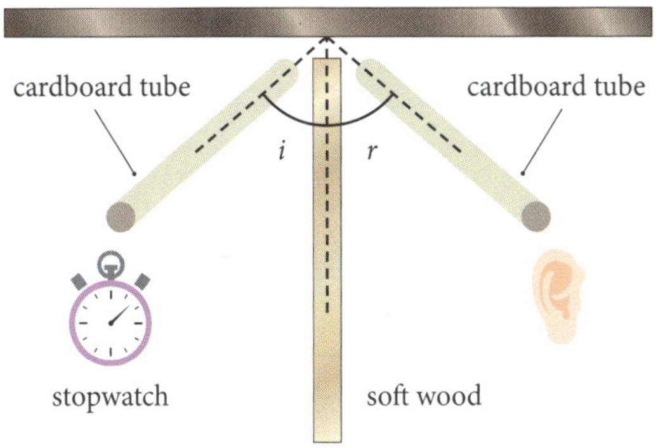

Figure 8.7 Experiment to investigate the reflection of sound waves

FIRSTHAND INVESTIGATION 4

Diffraction of sound waves

We can investigate the diffraction of waves in many ways. For example Figure 8.8 shows a simple experiment to test if sound waves diffract around corners. If a sound-measuring device was used rather than the boy, the relationship between the frequency and diffraction could also investigated. Typical results would show that low frequency sounds are diffracted more readily than high frequency sounds.

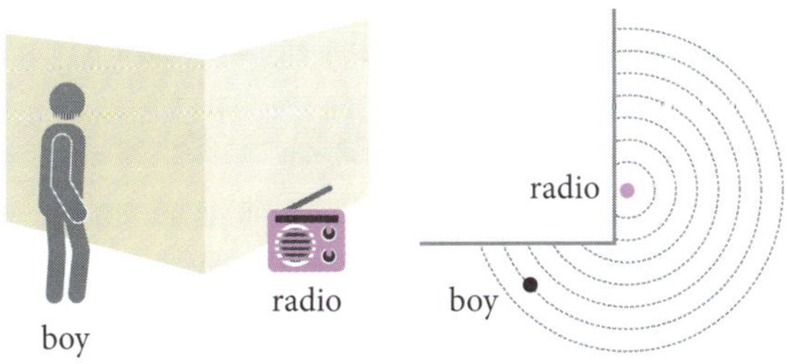

Figure 8.8 Investigating diffraction of sound waves

EXAMPLE 4

Sound barriers are often used to reduce the traffic noise from highways. With reference to Figure 8.9, compare the sound heard at point A to the sound that would be heard by the observer at point B.

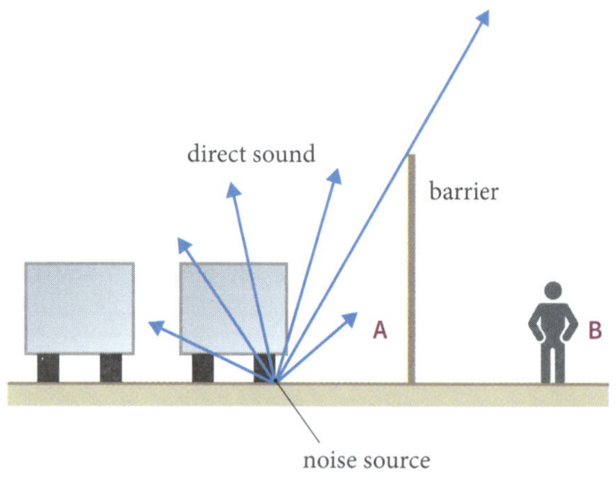

Figure 8.9 Effects of a highway sound barrier

Answer:

> Recall that longer wavelength waves are diffracted more readily than shorter wavelengths

The long wavelength (low frequency) sound waves would reach the observer at B, but the short wavelength sound waves may not be diffracted enough to reach the observer at B. Hence the sound heard at point A would be much louder and contain more high-pitched noise than the sound heard by observer B.

➔ KEY QUESTIONS

4. **How is the intensity of a sound wave related to the distance from the source of the sound?**
5. **Do sound waves obey the law of reflection?**
6. **How does diffraction of sound depend on the frequency of the sound?**

Answers ➲ p. 105

FIRSTHAND INVESTIGATION 5

Sound superposition

Superposition of sound waves to produce beats

Superposition of sound waves can be shown in many ways. Figure 8.10 shows how two tuning forks (one with a little plasticine attached to one of its forks to produce a slightly different frequency) can be used to demonstrate superposition using a microphone attached to a CRO. The tuning forks can be replaced with two loudspeakers attached to a signal generator to produce two slightly different frequencies.

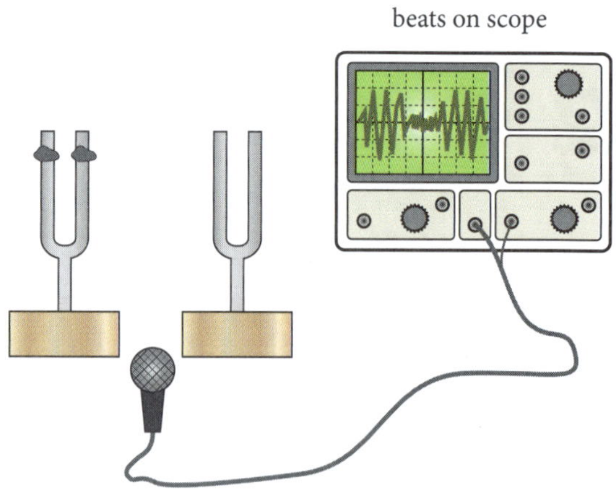

Figure 8.10 Simple experiment to demonstrate superposition of sound waves and the formation of beats

Because the two waves would initially have their crests aligned, but later the crest of one wave would be aligned with the trough of the other wave, the waves will sometimes add up to make a bigger wave (interfere constructively) and sometimes cancel each other out (interfere destructively), as shown in Figure 8.11. We call the periodic increases in amplitude produced by the superposition of two waves of slightly different frequencies **beats**. The period of these beats (T_{beat}) represents one complete cycle of the change of intensity of the superimposed wave form.

> **beat:** a periodic change in the amplitude of a wave

Superposition of sound waves to demonstrate interference

Another example of superposition of sound waves is shown in Figure 8.12. Here two loudspeakers are emitting waves of the same frequency. As the waves spread out from the speakers they superimpose and, because in some places the compressions of both waves are aligned (in phase), a louder sound results. At other places the compressions of one wave overlap with the rarefactions of the second wave, resulting in regions with no sound. An observer walking in front of the speakers will hear the sound get louder, softer, louder, softer, etc., as they walk along. The distance between two positions where the observer would hear a loud sound is marked as X in red on the diagram.

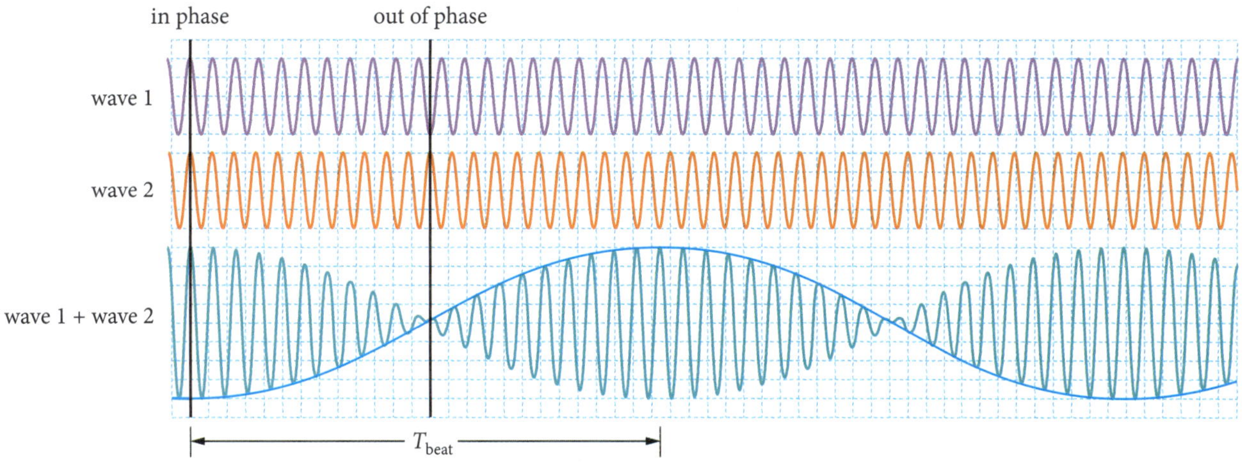

Figure 8.11 Two waves with slightly different frequencies superimposed to produce beats. The time for one complete cycle of the amplitude change of the superimposed wave form is called the beat period (T_{beat})

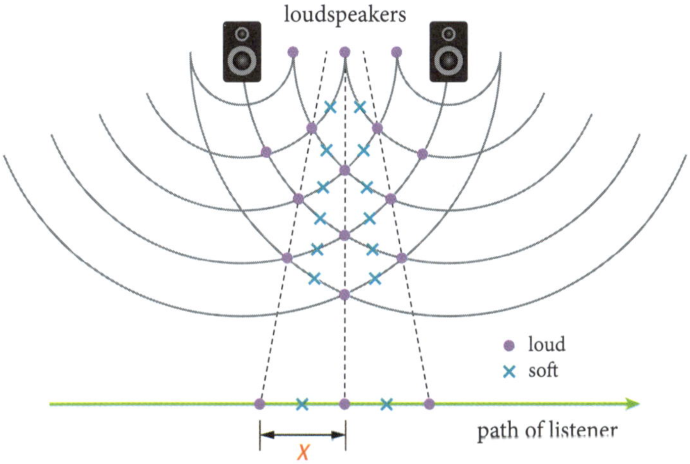

Figure 8.12 Superposition if sound waves from two loudspeakers

FIRSTHAND INVESTIGATION 6

Resonance of sound waves

A simple investigation to demonstrate resonance using sound waves is shown in Figure 8.13. Bottles filled with different levels of water will resonate at different frequencies because the standing waves set up in the bottles will have different wavelengths. Blowing across the top of the bottles creates noise of many frequencies, but only the particular frequency that coincides with the natural (resonance) frequency of the bottle will effectively drive the standing wave in the tube, resulting in a loud sound at the resonant frequency. A microphone and CRO or a mobile-phone application can be used to measure the frequency that is amplified by the bottle (the resonance frequency), and the relationship between the length of the air column and the resonance frequency could be investigated. Later we will see how the resonance frequency of a bottle depends on the length of the air column in the bottle.

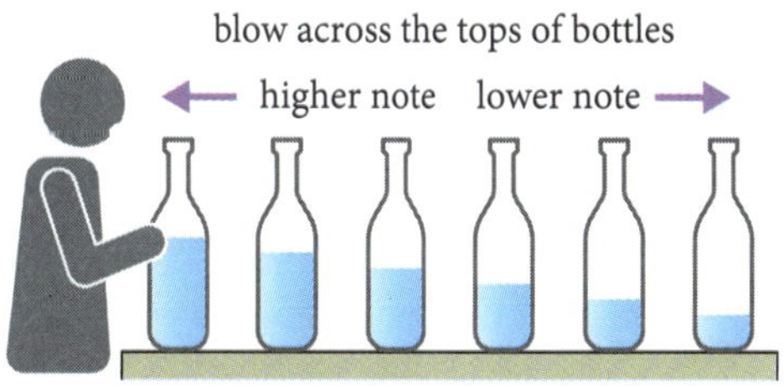

Figure 8.13 A simple experiment to demonstrate resonance of sound waves

5 Standing waves

» Students investigate and model the behaviour of standing waves on strings and/or in pipes to relate quantitatively the fundamental and harmonic frequencies of the waves that are produced to the physical characteristics (e.g. length, mass, tension, wave velocity) of the medium.

➔ Recall that we investigated standing waves in stretched strings in the previous chapter, and that **nodes** and **antinodes** refer to the positions on the wave where the displacement is zero and maximum, respectively.

node: those points on a standing wave where the displacement is zero
antinode: those points on a standing wave where the displacement is maximum

➔ Organ pipes, flutes and other wind instruments use the resonance of standing waves of sound in tubes to make sound. The type of standing waves set up depends on whether the end of the pipe is open or closed. Because the air molecules cannot move backwards and forwards parallel to the velocity of the wave, if the end of the tube is closed, there must be a node at the closed end. In contrast, an antinode must occur at an open end as the air molecules are free to move at an open end. Figure 8.13 shows the shape of the longest-wavelength standing wave (fundamental) that can exist on a pipe open at both ends.

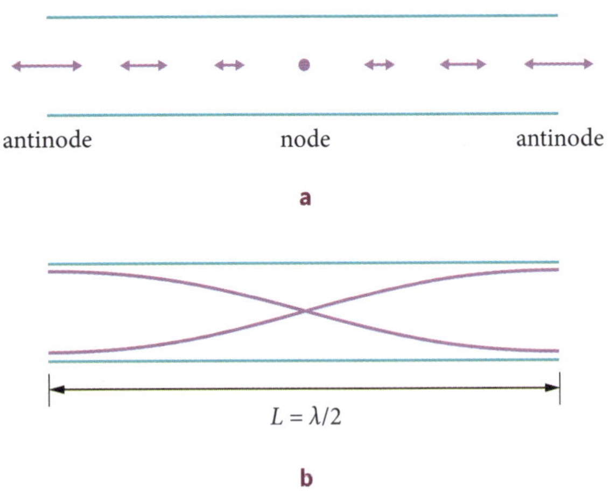

Figure 8.14 A fundamental standing wave in a pipe open at both ends. The molecular movement is shown in **a** and the change in air pressure involved is shown in **b**

➔ Because standing waves in a pipe open at each end will have an antinode at each end, and pipes closed at one end will have a node at the closed end, many standing waves with different wavelengths and frequencies can be set up in such pipes. Figure 8.15 shows some of the standing waves that can be set up in open and closed pipes. Using the observation that the distance between adjacent nodes is one half-wavelength ($\lambda/2$), we can write equations that relate the length of the pipe (L) to the wavelength of the standing waves that can be set up in each pipe. The relationship between the length of the pipe and the wavelength for each standing wave is shown in Figure 8.15.

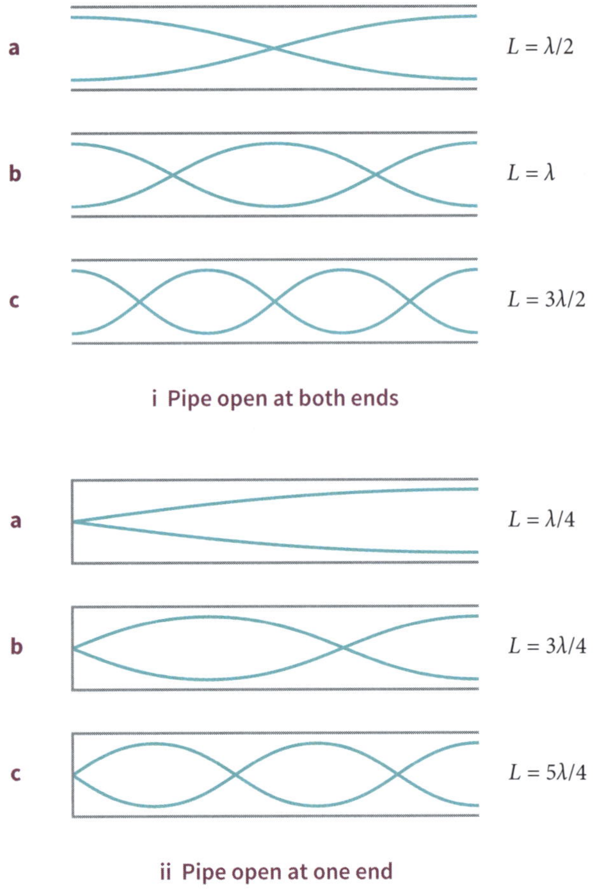

Figure 8.15 Fundamental **a**, first overtone **b** and second overtone **c** for standing waves in a pipe **i** open at both ends and **ii** one closed at one end

➔ Using the wave equation, we can write general equations for the standing waves that occur in pipes:

a For a pipe open at both ends:

$$f = \frac{nv}{2L}$$

b For a pipe closed at one end:

$$f = \frac{(2n - 1)v}{4L}$$

where $n = 1, 2, 3, 4,$ etc.

➔ The frequency of the longest wavelength standing wave that can be set up in a pipe is called the **fundamental frequency** or 1st harmonic. The next longest wavelength standing wave that can be set up is called the **1st overtone**. The 2nd **harmonic** refers to a standing wave with twice the frequency of the 1st harmonic. Stretched strings and pipes open at both ends can have standing waves with any multiple of the 1st harmonic, but a pipe closed at one end can only have odd harmonics.

fundamental frequency: the lowest frequency standing wave that can be produced in a system
1st overtone: the first frequency above the fundamental frequency that will produce a standing wave in the system
harmonic frequencies: multiples of the fundamental frequency that will produce standing waves in a system

EXAMPLE 5

A resonance box for a tuning fork consists of a wooden box open at one end and closed at the other, as shown in Figure 8.16. Placing a tuning fork on a resonance box makes the tuning fork sound much louder.

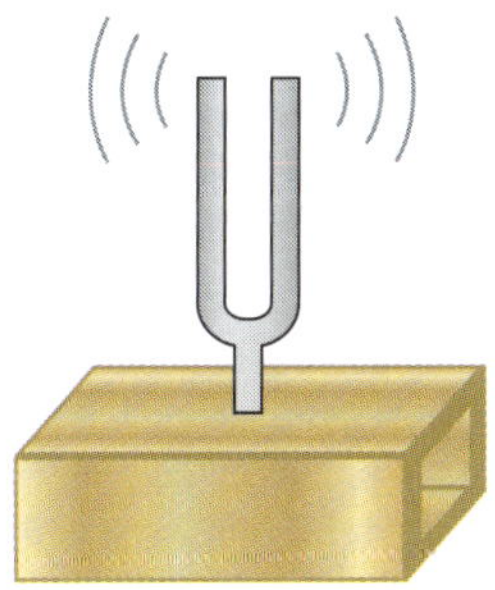

Figure 8.16 Tuning fork with resonance box

a **Explain why the resonance box must have a different length for different-frequency tuning forks.**

b **Find the shortest length box that would resonate with a 320 Hz tuning fork.**

Recall that resonance relates to driving a system at its natural frequency and the fundamental frequency is the lowest natural frequency of a system

Answer:

a To make the sound louder the tuning fork must drive one of the standing waves that can be set up in the box. The loudest sound will be produced when the fundamental frequency of the standing waves in the box is equal to the tuning fork frequency.

Figure 8.17 illustrates the fundamental frequency in the box.

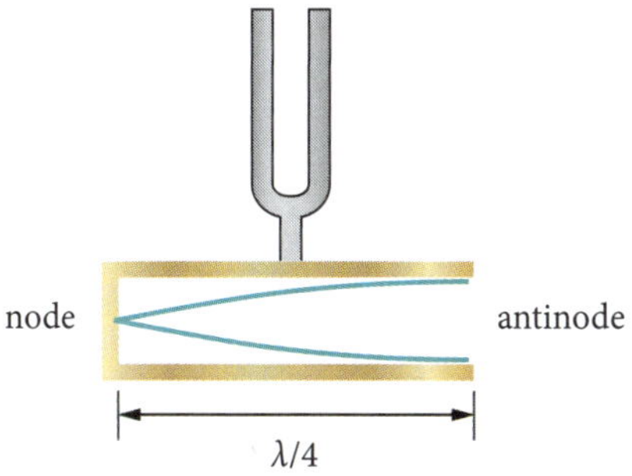

Figure 8.17 Fundamental mode of a tuning fork resonance box

As $f = v/\lambda$ and $\lambda = 4L$, we can write $f = v/4L$ or $L = v/4f$, which shows that the length of the resonance box will determine the frequency at which the box will resonate.

b As $L = v/4f = \dfrac{340}{4 \times 320} = 0.266$ m

➔ KEY QUESTIONS

7 **What standing wave harmonics can be set up in pipes open at both ends?**

8 **What standing wave harmonics can be set up in pipes open at one end?**

Answers ➲ p. 105

FIRSTHAND INVESTIGATION 7

Measuring resonance frequencies

Resonance in a pipe open at both ends can be demonstrated by placing a loudspeaker near one of the open ends of the pipe and using a signal generator to change the frequency of the sound produced by the loudspeaker until it coincides with one of the resonance frequencies of the pipe. When the pipe is driven at one of its resonance frequencies, energy will be coupled into the standing wave and the loudness of the sound increases greatly. The experimental set up is shown in Figure 8.18.

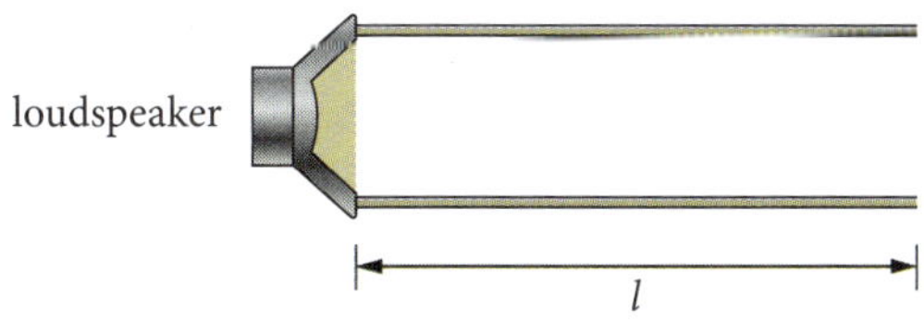

Figure 8.18 Investigating resonance in a tube open at both ends

By determining the lowest resonance frequency of the tube (the fundamental frequency) the wavelength can be determined and the speed of sound in the tube can be calculated.

The sound waves produced at resonance can actually be felt by placing your hand near the open end of the pipe.

The nodes and antinodes in the tube can be observed by sliding a microphone down the tube, as shown in Figure 8.19. When the nodes are reached the intensity of the sound will decrease to zero, while at the antinodes the amplitude will be maximum.

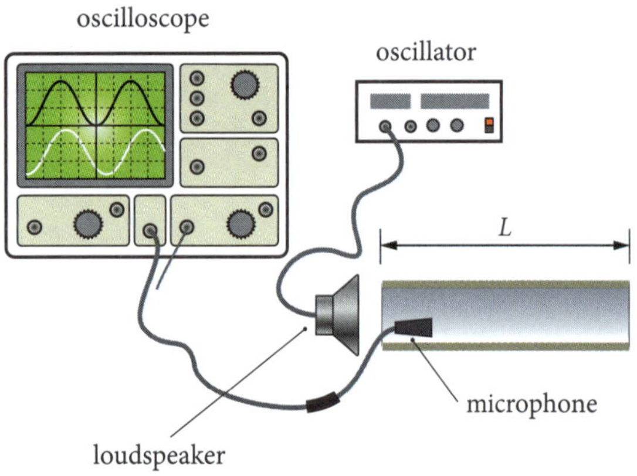

Figure 8.19 Using a microphone and CRO to observe the position of nodes and antinodes

6 Beats and the Doppler effect

» Students analyse qualitatively and quantitatively the relationships of the wave nature of sound to explain:

- beats ($f_{beat} = |f_2 - f_1|$)
- the Doppler effect $\left(f' = f\dfrac{(v_{wave} + v_{observer})}{(v_{wave} - v_{source})}\right)$.

➔ Recall that when we investigated the superposition of two sound waves with slightly different frequencies, a wave formed that changed its amplitude periodically (see Figure 8.10). We described this change in amplitude as a beat. The frequency of the beats produced is related to the frequency of the two waves that produced the beats by the equation:

$$f_{beat} = |f_2 - f_1|$$

➔ The straight brackets in the equation indicate the beat frequency is equal to the absolute value of the difference of the frequencies of the waves that produced the beat.

➔ The **Doppler effect** refers to the effect of a moving source or observer on the observed frequency of sound. An observer moving towards a source of sound would hear a higher frequency than an observer moving away from the sound. Similarly, if the source of sound moved towards the observer, the observer would hear a higher frequency than if the source of the sound was moving away from them. The sound wave produced by a moving source is shown in Figure 8.20.

Doppler effect: the change in frequency that an observer hears when there is relative motion between the source of sound and the observer

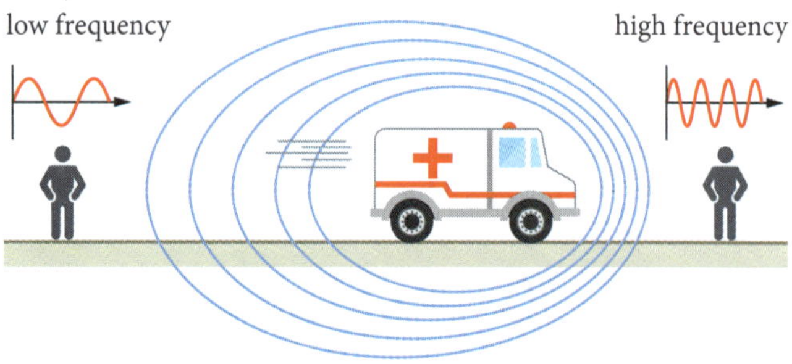

Figure 8.20 The Doppler effect produced by a moving source of sound

➔ Analysis of the change in frequency produced by the Doppler effect shows that:

$$f' = f\frac{(v_{wave} + v_{observer})}{(v_{wave} - v_{source})}$$

➔ This Doppler effect equation tells us that the observed frequency (f') is related to emitted frequency (f), the velocity of sound (v_{wave}), the velocity of the observer ($v_{observer}$) and the velocity of the source (v_{source}).

➔ The prime symbol (′) is used to show there is a relationship between similar measures.

➔ Note that the plus and minus signs in the equation are for an observer and/or source moving towards each other. Because velocity is a vector we need to reverse the sign of the observer's velocity and/or the source velocity when the source and/or observer are moving away from each other.

➔ If the source moves faster than the wave velocity, a conic shock wave is set up behind the source, as shown in Figure 8.21. A similar effect occurs when a boat travels faster than water waves and it makes a bow wave. When an aircraft travels faster than the speed of sound, observers on the ground do not hear it approach but are subjected to a loud sonic boom when the high-pressure shock wave behind the plane reaches them.

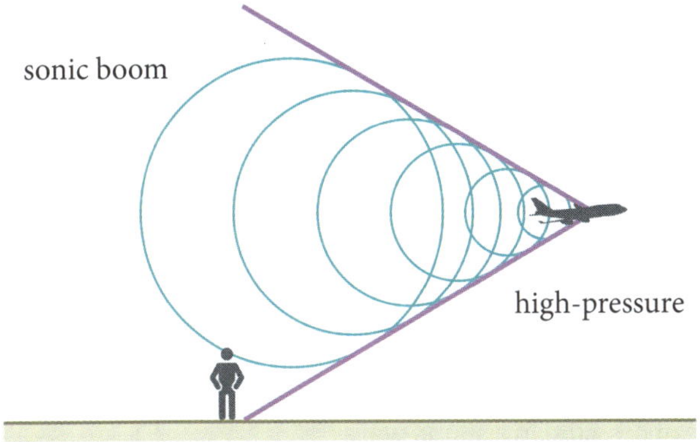

Figure 8.21 Shock wave produced by a jet travelling faster than sound

→ KEY QUESTIONS

9 What causes the phenomenon of beats in sound waves?

10 How does the relative motion of a source of sound and an observer affect the sound heard by the observer?

Answers ➲ p. 105

EXAMPLE 6

A piano tuner sounds a middle C tuning fork that oscillates at exactly 256 Hz and hits the middle C key on the piano at the same time. The piano tuner observes beats with a frequency of 4 Hz from the combined sounds.

a With what frequency is the middle C string on the piano oscillating?

b Suggest a method the piano tuner could use to adjust the piano string to the correct frequency.

Answer:

a As $f_{beat} = |f_2 - f_1|$ and $f_{beat} = 4$ Hz, the piano string may be oscillating at 260 Hz or 252 Hz, as either would produce beats of 4 Hz is sounded with a frequency of 256 Hz.

b The piano tuner could change the tension of the string, as this would change the speed of the waves on the string and the frequency at which the string oscillated (as the wavelength is fixed). He could tighten or loosen the string and listen for an increase in the periods of the beat. When the beat period increased until it could no longer be heard he would know that both sounds had the same frequency.

EXAMPLE 7

A police car travelling at 30 ms^{-1} north is chasing a second car that is travelling at 45 ms^{-1} north. Given that the police siren operates at a frequency of 350 Hz, what frequency would the second car hear?

Answer:

In this example the observer is receding from the source of sound but the source is moving towards the observer. As the second car is moving faster than the police car we would expect the second car to hear a lower frequency. We can calculate this frequency using the Doppler effect equation:

> The direction of motion of the source and observer are important when choosing what signs to use in the Doppler effect equation; choose the sign that will change the frequency in the direction you would expect from the relative motion

$$f' = f\frac{(v_{wave} + v_{observer})}{(v_{wave} - v_{source})}$$

As this equation is for objects approaching from each other, we have to change the sign of the observer's velocity as the observer is moving away from the source:

$$f' = f\frac{(v_{wave} + v_{observer})}{(v_{wave} - v_{source})}$$

$$f' = 350\frac{(340 - 45)}{(340 - 30)}$$

$$= 333 \text{ Hz}$$

EXAMPLE 8

An observant passenger on a train notices that when approaching a level crossing she hears the bells on the crossing sounding at 325 Hz rather than the 300 Hz that the observer had noticed when the train was at rest. Find the speed at which the train was approaching the level crossing.

Answer:

As the source is at rest we can write the Doppler effect equation as:

$$f' = f\frac{(v_{wave} + v_{observer})}{(v_{wave})}$$

> Take care to determine whether the observer, source or both are moving before substituting into the Doppler effect equation

Or $v_{observer} = (f'/f - 1)v_{wave}$

$$= \left(\frac{325}{300} - 1\right)340$$

$= 28.3$ ms^{-1} towards the level crossing

CHAPTER SYLLABUS CHECKLIST

Are you able to answer these questions from the syllabus for this chapter? Tick each question as you go through the checklist if you are able to answer it. If you cannot answer a question, turn to the relevant page in the study guide to find the answer. For NESA key word meanings, go to www.educationstandards.nsw.edu.au and search 'key words'.

FOR A COMPLETE UNDERSTANDING OF THIS TOPIC:		PAGE NO.	✓
1	Can I describe an experiment that would prove that sound is a mechanical wave?	72	
2	Can I relate the pitch and loudness of a sound wave to the frequency and amplitude of the wave?	92	
3	Can I relate the displacement of molecules to the change in pressure in a sound wave?	93	
4	Can I solve problems using the inverse square law of intensity?	94	
5	Can I describe and explain the results of investigations to examine reflection, diffraction, resonance and superposition of sound waves?	95–96	
6	Can I explain how standing waves of sound can be set up in pipes?	98	
7	Can I describe how resonance can be demonstrated using standing waves in pipes?	98	
8	Can I draw the standing waves that can be set up in pipes open at both ends and pipes open at only one end?	98	
9	Can I calculate the wavelength and frequency of the fundamental standing wave and other standing waves in pipes?	98–99	
10	Can I calculate the beat frequency of superimposed sound waves?	100	
11	Can I explain and solve problems involving the Doppler effect?	100–101	

YEAR 11 EXAM-TYPE QUESTIONS

Objective-response questions

(1 mark each)

1 What inference can be made from the results of the experiment shown in Figure 8.22?

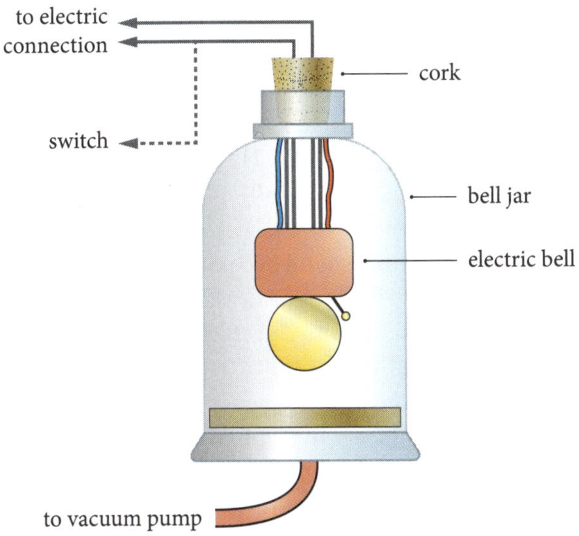

Figure 8.22 Sound experiment

A That sound is a longitudinal wave.

B That sound can be diffracted.

C That sound travels faster in a vacuum.

D That sound is a mechanical wave.

2 A sound wave causes air molecules to oscillate in air. Which of the following is most correct?

A The maximum displacement of the molecules from their equilibrium position occurs midway between compressions and rarefactions.

B The maximum displacement of the molecules from their equilibrium position occurs at the compressions and rarefactions.

C Only molecules at the compressions are displaced from their equilibrium positions.

D The minimum displacement of the molecules from their equilibrium position occurs at the compressions and rarefactions.

3 What is the fundamental frequency of a standing wave in a pipe of length 1.2 m, which is open at both ends?

A 283 Hz

B 567 Hz

C 142 Hz

D 71 Hz

4 If a 310 Hz tuning fork was sounded at the same time as a 320 Hz tuning fork, what sort of sound would be heard?

A a pure note with a frequency of 315 Hz

B beats with a beat frequency of 315 Hz

C beats with a frequency of 10 Hz

D beats with a period of 10 s

5 A spectator sees a racing car go past. Which of the following best describes the sound waves heard by the spectator?

A an increased frequency as the car approaches and a decreased frequency as it moves away

B an increased period as the car approaches and a decreased period as it moves away

C an increased wavelength as the car approaches and a decreased wavelength as it moves away

D a decreasing amplitude as the car approaches and an increasing amplitude as it moves away

Extended-response questions

6 A fire alarm produces sound waves with 5 watts of energy. Calculate the intensity of the sound wave:

a 2 m from the source (1 mark)

b 6 m from the source. (1 mark)

7 Consider the standing wave set up in the pipe shown in Figure 8.23. Note the length of the pipe, $L = 1.5$ m.

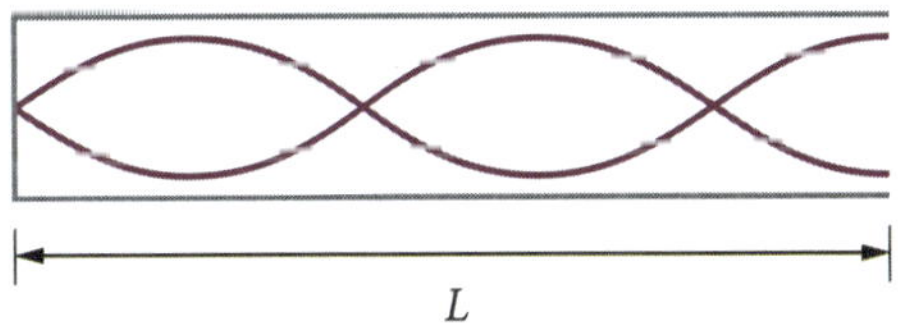

Figure 8.23 Standing waves in a pipe open at one end

a Redraw the standing wave and mark the nodes on your diagram. (2 marks)

b Given that the speed of sound is 340 ms^{-1}, find the wavelength and frequency of the sound produced by this standing wave. (2 marks)

c Many other standing waves could be set up in a pipe of this length. Find the fundamental frequency for this pipe and the wavelength of the fundamental standing wave. (2 marks)

8 A particular helicopter is known to emit sound waves with a frequency of 300 Hz, but when observing the helicopter, a stationary observer hears a frequency of 420 Hz.

a Is the helicopter approaching or receding from the observer? Justify your answer. (2 marks)

b Calculate the speed of the helicopter. (1 mark)

9 An oscilloscope is used to observe the sound produced when two loudspeakers emit slightly different frequencies. The oscilloscope trace is shown in Figure 8.24. This graph plots air pressure against time, and the distance between the vertical lines represents a time of 50 ms.

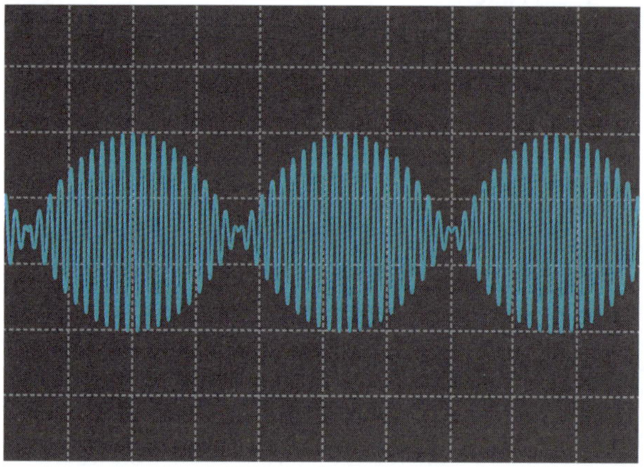

Figure 8.24 Oscilloscope trace of sound from two loudspeakers

a Explain the shape of the oscilloscope trace. (3 marks)

b Determine the beat frequency from the oscilloscope trace. (1 mark)

c If one of the loudspeakers was producing sound waves of 500 Hz, what frequencies could be emitted from the other loudspeaker to produce the beats shown on the oscilloscope? (2 marks)

10 A student conducts an experiment in which the frequency of sound produced by a loudspeaker placed in front of a tube open at both ends is gradually increased. The student observes that the volume remains constant until the loudspeaker passes a frequency of 170 Hz. At 170 Hz the volume emitted from the pipe increases greatly and the student observes another increase in intensity at 340 Hz.

a Explain why the sound becomes louder at 170 Hz and 340 Hz. (3 marks)

b Assuming that sound travels at 340 ms^{-1}, find the length of the pipe. (1 mark)

c Determine the next frequency that would produce a loud sound from the pipe and sketch the standing wave this frequency would produce in the pipe. (2 marks)

ANSWERS

KEY QUESTIONS

Key questions ➲ p. 94

1 The pitch of a sound is determined by the frequency of the sound wave, and the loudness is determined by the amplitude (maximum displacement of molecules or change in pressure) of the sound wave.

2 When groups of adjacent air molecules oscillate in opposite directions to one another they will come closer together and hence produce an increase in pressure. Figure A8.1 shows how the movement of air molecules towards one another produces a region of high pressure.

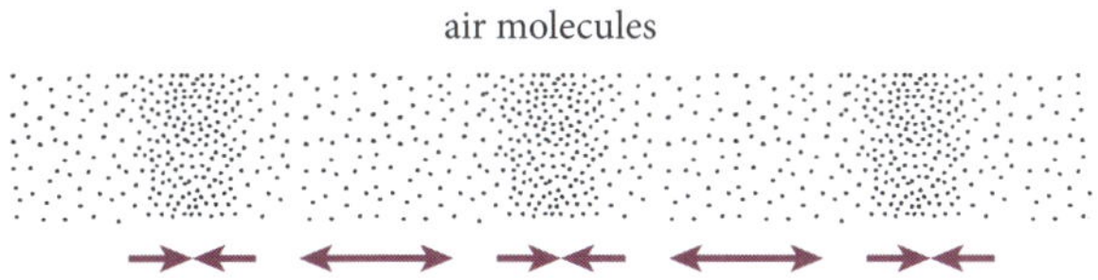

Figure A8.1 Motion of air molecules in sound waves

3 Because molecules or atoms are closer together in dense material, the motion of one molecule affects its neighbour more quickly than molecules would do in less-dense material. This faster transmission of motion causes the waves to move faster in the denser material.

Key questions ➲ p. 96

4 Intensity decreases as the inverse square of the distance from the source. That is, intensity $I \alpha 1/r^2$.

5 Yes, sound waves obey the same laws of reflection as other waves.

6 Lower frequency (longer wavelength) sound waves are diffracted more readily than higher frequency (shorter wavelength) sound waves.

Key questions ➲ p. 99

7 All harmonics can be set up in a pipe open at both ends. That is, all multiples of the fundamental frequency can form standing waves in a pipe open at both ends.

8 Only odd harmonics can form in a pipe closed at one end. That is, only odd multiples of the fundamental frequency will form standing waves in a pipe closed at one end.

Key questions ➲ p. 101

9 A beat (a wave that displays a periodic variation in amplitude) is produced when two sound waves with slightly different wavelengths are superimposed.

10 The observed frequency will be higher than the emitted frequency when the source/observer are moving closer together, and a lower frequency will be heard by the observer when the source/observer are moving apart from each other.

YEAR 11 EXAM-TYPE QUESTIONS

Objective-response questions

1 **D.** The sound from the bell will not pass through the evacuated bell jar. **A** is incorrect because this experiment cannot be used to distinguish between transverse and longitudinal waves. **B** is incorrect because diffraction is not involved in this experiment. **C** is incorrect because sound cannot pass through a vacuum.

2 **B.** Moving molecules produce the compressions and rarefactions. The displacement of the molecules is greatest (i.e. amplitude maximum) at the compressions and rarefactions, as shown in Figure A8.1. **A** is incorrect because the molecules will not be displaced at all between a compression and rarefaction. **C** and **D** are incorrect because molecules must be brought closer together or further apart to produce compressions and rarefactions.

3 **C.** The fundamental frequency occurs in an open pipe when $\lambda = 2L$ and as $f = v/\lambda = v/2L = \dfrac{340}{2 \times 1.2} = 142$ Hz. The other answers are all incorrect as they refer to different harmonics (i.e. not the fundamental, which is the longest-wavelength standing wave that can be set up in the pipe).

4 **C.** This is the correct answer as $f_{beat} = |f_2 - f_1| = |320 - 310| = 10$ Hz. **A** and **B** are incorrect because they are the wrong numerical answer. **D** is incorrect because the difference in frequencies is the beat frequency, not the period of the beats.

5 **A.** The wave fronts will come closer together in front of the car and spread out behind the car (see Figure 8.20), resulting in the observer hearing a higher pitch as the car approaches and a lower pitch as the car moves away. **B** is incorrect because the period would decrease as the car approached. **C** is incorrect because the observed wavelength would decrease as the car approached. **D** is incorrect because the sound intensity and hence the amplitude would increase as the car approached.

Extended-response questions

6 **EM** This question tests students' understanding of the inverse square law of wave intensity.

a As wave intensity is the energy that passes through a unit area per second, the intensity at 2 m would be given by the $\dfrac{\text{power}}{\text{surface area}}$ of a sphere with a radius of 2 m:

$$\text{Intensity} = \frac{\text{power}}{\text{area}} = \frac{5}{4\pi r^2} = 0.1 \text{ Wm}^{-2} \checkmark$$

b Recall that $I \alpha 1/r^2$. As 6 m is three times further from the source than 2 m, the intensity at 6 m must be the intensity at 2 m divided by 3^2:

$$I = \frac{0.1}{3^2} = 0.011 \text{ Wm}^{-2} \checkmark$$

7 **EM** This question tests students' understanding of nodes and antinodes, standing waves in pipes, applying the wave equation to standing waves, and the difference between fundamental frequency and overtones.

a The standing wave in Figure A8.2 has the two nodes labelled.

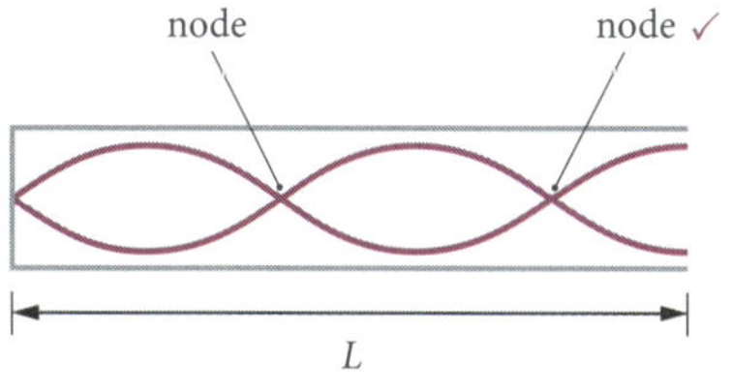

Figure A8.2 Standing waves in a pipe open at one end

b The distance between nodes is $\lambda/2$ and hence the wavelength is:

$\lambda = 4L/5 = \frac{4 \times 1.5}{5} = 1.2$ m ✓

The wave frequency can be determined from the wave equation:

$v = f\lambda$ and hence $f = v/\lambda = \frac{340}{1.2} = 283$ Hz ✓

c The wavelength of the fundamental frequency for a pipe closed at one end is four times the length of the pipe (see Figure 8.15), which gives:

$\lambda = 4L = 4 \times 1.5 = 6$ m ✓

The fundamental frequency is therefore:

$f = v/\lambda = \frac{340}{6} = 56.7$ Hz ✓

8 EM This question tests students' qualitative and quantitative understanding of the Doppler effect and requires students to rearrange the Doppler effect equation.

a As the observed frequency is greater than the emitted frequency, the helicopter must be approaching the observer. ✓ This is because the helicopter moves in the direction of the wave velocity and hence the distance between when it emits one wave front and the next (the wavelength) will be less than if it had been at rest. Observers will therefore measure a shorter wavelength and they will hear an increased pitch (frequency). This is because the speed of the waves remains constant and, as $f = v/\lambda$, decreasing the wavelength will increase the frequency. ✓

b First rearrange the wave equation to make the velocity of the source the subject:

$$f' = f\frac{(v_{wave} + v_{observer})}{(v_{wave} - v_{source})},\ v_{observer} = 0$$

$$v_{source} = v_{wave}\left(1 - \frac{f}{f'}\right)$$

$$= 340\left(1 - \frac{300}{420}\right)$$

$$v_{source} = 97.1\ \text{ms}^{-1}$$ ✓

9 EM This question requires students to use a scale to measure the beat period, to convert the period to a frequency, and to apply the equation for beats.

a The oscilloscope is displaying a superposition of the two sounds ✓ and, because the two waves go in and out of phase as time passes, the waves will sometimes add up to make a bigger wave (interfere constructively) and sometimes cancel out each other (interfere destructively). ✓ The superimposed sound consequently changes its amplitude periodically, as shown on the oscilloscope. We say that a wave which changes its amplitude in this way exhibits beats. ✓

b There are approximately 3.25 vertical divisions between the beats and hence the beat period (T) is:

3.25×50 ms = 175 ms = 0.165 s

Hence frequency = $f = 1/T = \frac{1}{0.165} = 6$ Hz ✓

c As $f_{beat} = |f_2 - f_1|$, $f_1 = f_2 \pm f_{beat} = 500 \pm 6 = 506$ Hz ✓ or 494 Hz ✓

10 EM This question requires students to understand resonance and standing waves in tubes open at both ends.

a The sound becomes louder at these frequencies because the tube has natural frequencies of oscillation at these frequencies. ✓ That is, standing waves can be set up in the pipe at these frequencies. As energy is very effectively coupled into an oscillating system when it is driven at its natural frequency of oscillation, ✓ a lot more energy is transferred to the sound waves in the pipe at these frequencies, resulting in a louder sound. ✓

b Applying the wave equation:

$v = f\lambda$ and hence $\lambda = v/f = \frac{340}{170} = 2$ m

As 170 Hz is the lowest frequency that causes the tube to resonate, it must be the fundamental frequency and the length of the tube (see Figure 8.15):

$L = \lambda/2 = \frac{2}{2} = 1$ m ✓

c As a pipe open at both ends can support standing waves of all integer multiples of the fundamental frequency, the next frequency that would produce resonance in the pipe would be:

$f = 3 \times 170 = 510$ Hz ✓

The third harmonic for this pipe is shown in Figure A8.3.

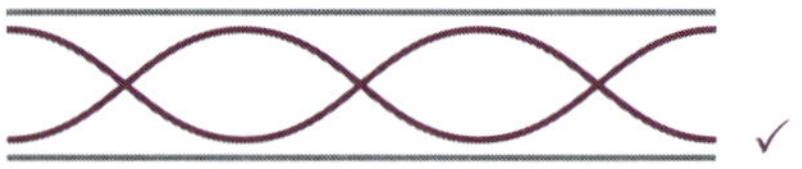

Figure A8.3 Third harmonic in a pipe open at both ends

CHAPTER 9 RAY MODEL OF LIGHT

MODULE 3
WAVES AND THERMODYNAMICS

INQUIRY QUESTION:

What properties can be demonstrated when using the ray model of light?

The ray model of light allows us to describe many of the observed behaviours of light, including: image formation by lenses and mirrors, refraction, reflection, critical angle and dispersion.

1 Geometric optics and image formation

» Students conduct a practical investigation to analyse the formation of images in mirrors and lenses via reflection and refraction using the ray model of light.

Image formation using lenses

- Geometric optics uses the ray model of light to analyse image formation from mirrors and lenses.
- Rays parallel to the principal axis are refracted to a **focal point** (F). The distance between the focal point and the lens is called the **focal length** (f).

focal point: point at which waves meet after reflection or refraction
focal length: the distance between the lens and the point where the rays that are approaching the lens parallel to the principal axis converge

- An *image* is that point where the rays that reach eye appear to emanate. We can think of this point as where the rays in a ray diagram converge after being refracted by a lens or reflected from a mirror.
- When the rays actually converge to a point, we say the image is a *real image*. A real image will appear on a screen if it was placed at the point of convergence. This is similar to the image projected onto a screen in a movie theatre. Figure 9.1 shows a real image formed by a lens.
- The image may be larger or smaller than the object. We define **magnification** (M) as:

M = height of image/height of object

magnification: the ratio of the height of the image to the height of the object

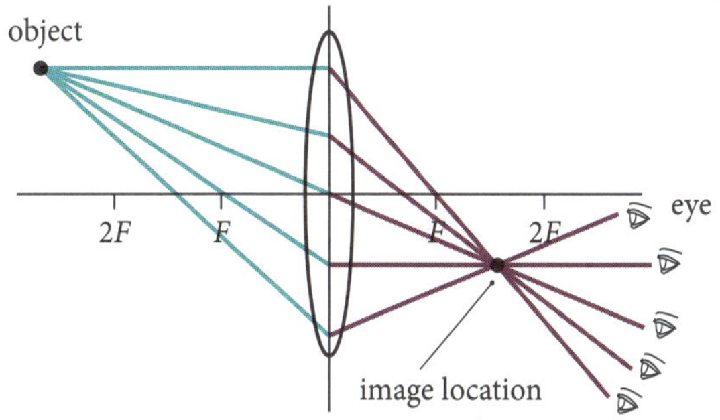

Figure 9.1 Real image formed by a convex lens

- In some ray diagrams the rays appear to converge to a point that rays cannot actually reach; for example, from behind a mirror. We call this type of image a *virtual image*. A virtual image cannot be projected onto a screen as no rays actually reach or come from the image.
- We can find the position and type of image formed by lenses and mirrors by using the principles of geometric optics to draw ray diagrams.

Image formation using curved mirrors

- Figure 9.2 shows that for light ray reflection from curved mirrors:
  - To a good approximation, the focal length (f) is half the distance between the mirror and the *centre of curvature* (C).
  - This relationship is approximate because rays parallel to the optic axis reflected from a spherical mirror do not quite meet at the same point. A parabolic mirror would be required to bring all the reflected rays to exactly the same focal point. To a first approximation, however, we can ignore this spherical aberration (spreading out of the focal length) and assume that the reflected rays from a spherical mirror meet at the one point.
  - Hence we assume that for a spherical concave mirror, an incident ray parallel to the **principal axis** (PA) will be reflected through the focal point of the mirror.

principal axis: the line drawn perpendicular to the plane of a curved mirror or lens that passes through the centre of the mirror or lens

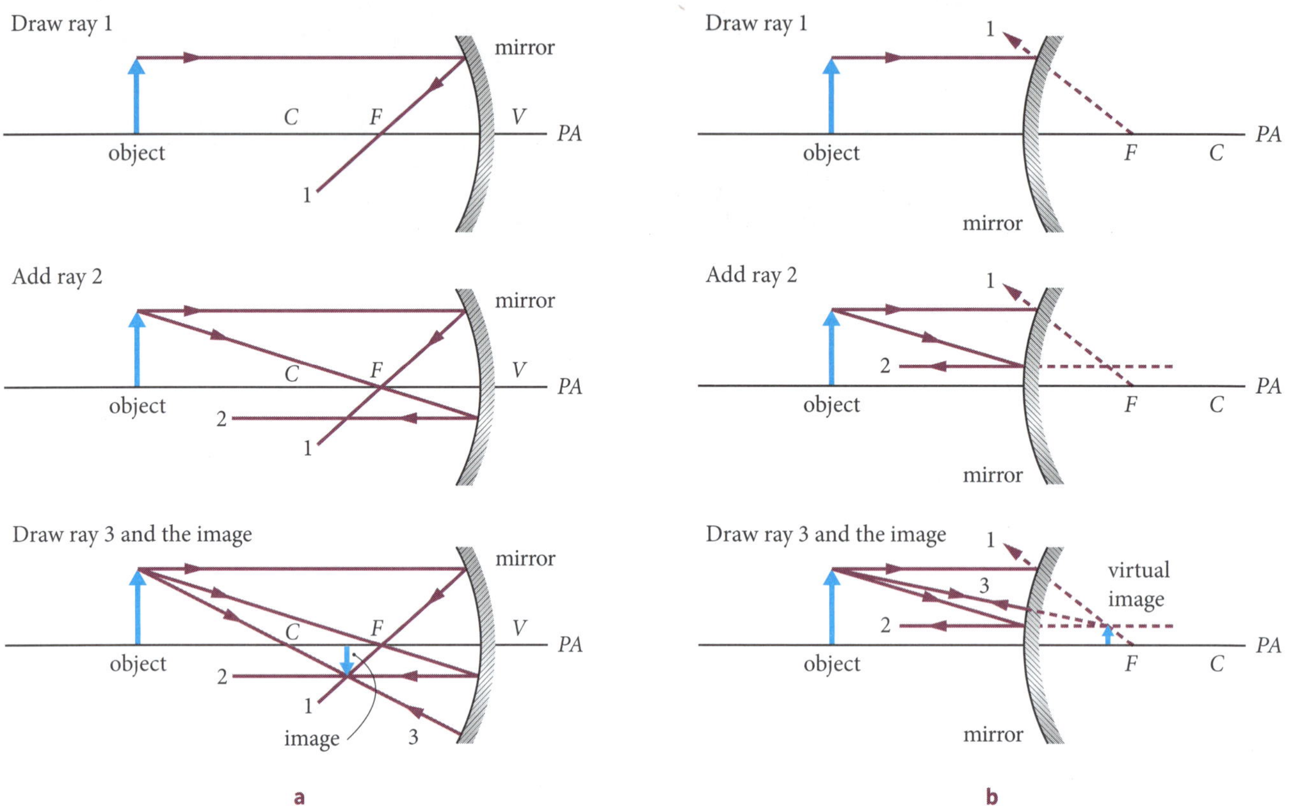

Figure 9.2 Using the rules of geometric optics to draw **a** a real image formed by a converging spherical mirror, and **b** a virtual image formed by a diverging spherical mirror

- For a convex mirror, the focus is determined by tracing back the reflected ray to the virtual point of convergence behind the mirror, as shown in Figure 9.2(b).
- A ray that passes through the focus will be reflected back along the principal axis.
- A ray that is directed through the centre of curvature of the mirror will be reflected back through the centre of curvature.

Image formation using lenses

→ Figure 9.3 illustrates how rules of geometric optics can be used to find the image produced by light refracted by a lens. For rays refracted by a lens:
- A ray drawn parallel to the principal axis will be refracted through the focal point of the lens.
- A ray drawn through the centre of the lens will pass straight through the lens without being refracted.
- A ray drawn through the focal point will be refracted to run parallel to the principal axis of the lens.

→ If the image appears on the other side of the principal axis to the object we say the image is *inverted*, and if the image is larger than the object we say the image is *enlarged*. If the image is smaller than the object the image is said to be *diminished*.

→ Figure 9.4 shows the formation of an upright, diminished virtual image by a concave lens.

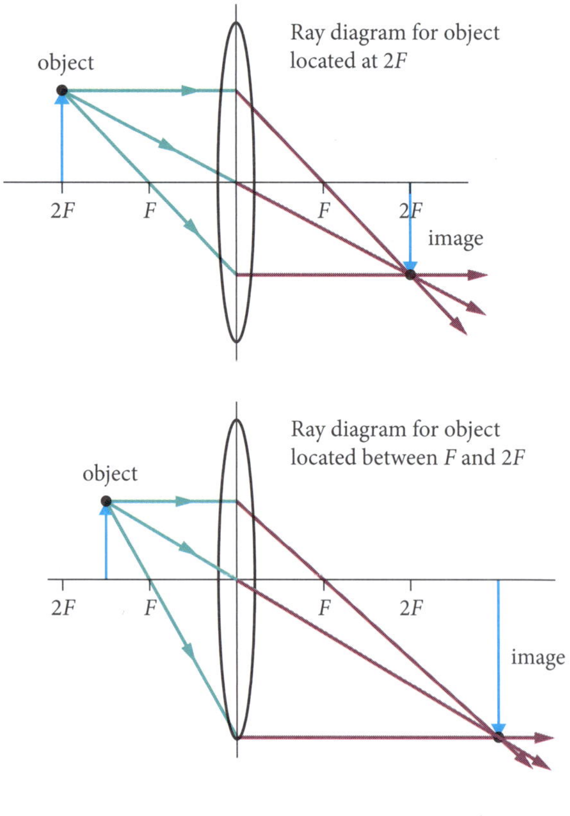

Figure 9.3 Ray diagram showing the formation of a magnified, inverted real image by a convex lens

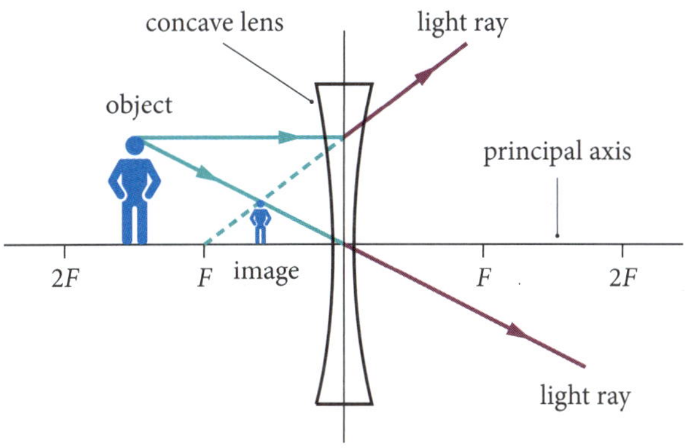

Figure 9.4 Formation of a diminished virtual image by a convex lens

- Because a ray drawn parallel to the principal axis of a convex lens diverges, the focal point of a concave lens is determined by tracing back to the point where all the emerging rays appear to converge (see Figure 9.4).
- An observer looking through a concave lens at an object placed between *F* and 2*F* would see a diminished, upright image of the object that would appear to be between the lens and *F*, as shown in Figure 9.4.
- Although it is not required by the syllabus, students may wish to investigate the lens equation, which relates the object distance (u), the image distance (v) and the focal length f):

  $1/u + 1/v = 1/f$

  Students may also wish to investigate the equation for magnification (M):

  $$M = -\frac{v}{u}$$

KEY QUESTIONS

1. What is the difference between a real and a virtual image?
2. What are the rules of geometric optics for rays reflected from a spherical curved mirror?
3. What are the rules of geometric optics for rays refracted by a lens?

Answers ➲ p. 121

EXAMPLE 1

Use a ray diagram to find the position and type of image that would be formed of an object placed 30 cm from a convex lens with a focal length of 20 cm.

Answer:

Rays that follow the rules of geometric optics converge at the image

Draw a diagram to scale and trace rays from the object according to the principles of geometric optics to produce an image, as shown in Figure 9.5.

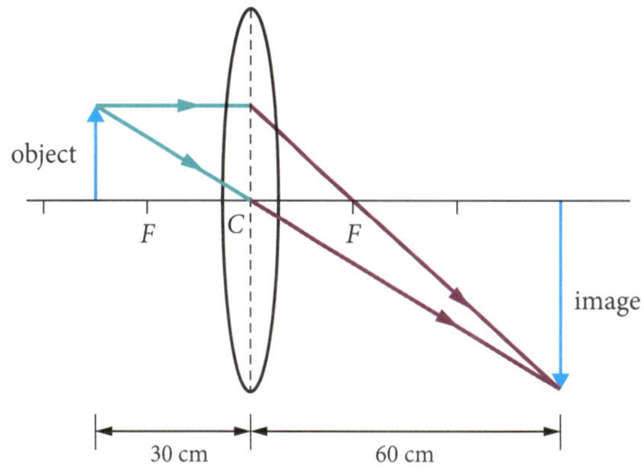

Figure 9.5 Ray diagram to find position and type of image formed by a convex lens

By measuring from the scaled diagram we note the image is real, inverted, magnified and will appear 60 cm from the lens.

FIRSTHAND INVESTIGATION 1

Image formation by a spherical mirror

Students can be given a spherical mirror of known focal length and asked to predict where images will form and what type of images will form when objects are placed at different distances from the mirror. If the focal length (f) of the mirror is not given it can be measured, as shown in Figure 9.6, by using the mirror to bring parallel light rays (e.g. from the Sun) to a focus and measuring the distance from the lens to the focal point. Care should be exercised if the Sun is used as the light intensity at the focus might be very high.

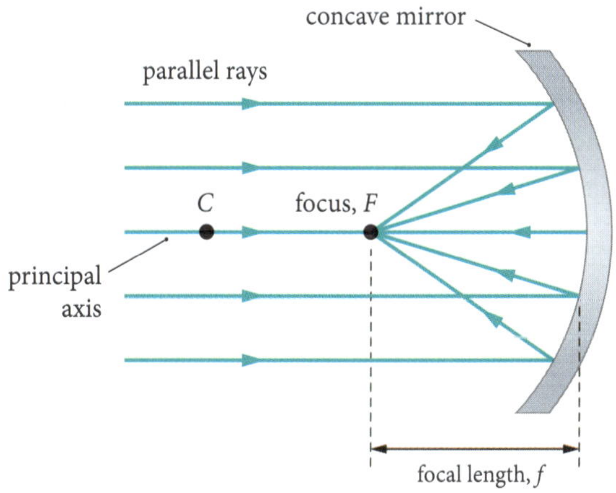

Figure 9.6 Measuring the focal length of a concave mirror

Students draw a number of ray diagrams to scale (similar to the one shown in Figure 9.7) to predict the position and type of image that will be formed.

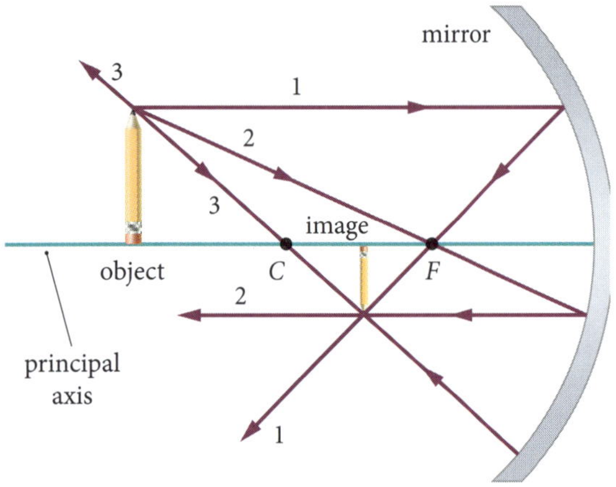

Figure 9.7 Using a ray diagram to predict the position and type of image formed by a convex mirror. In this case the image is diminished, real and inverted

Students can then test their predictions of the type and position of image formed from the mirror, using a candle as an object.

FIRSTHAND INVESTIGATION 2

Image formation by a convex lens

In this investigation students predict the position of an image and its magnification by drawing a ray diagram. They can then test the prediction by using the lens to project an image on a screen. If the focal length (f) of the lens is not given it can be measured (as shown in Figure 9.8) by using the lens to bring parallel light rays (e.g. from the Sun) to a focus and measuring the distance from the lens to the focal point.

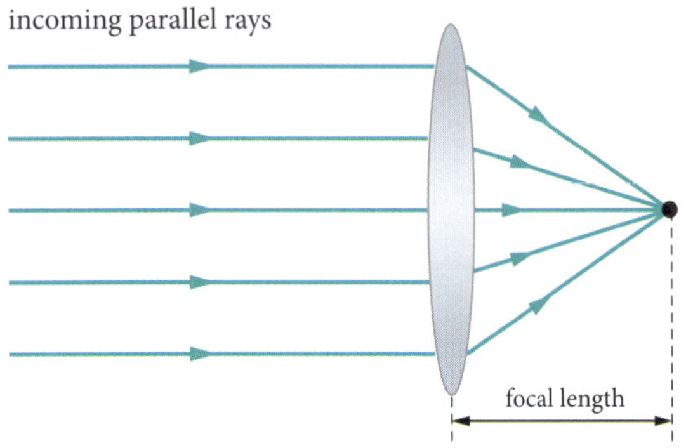

Figure 9.8 Measuring focal length of a convex lens

Once the focal length is known, students can draw ray diagrams to scale, as shown in Figure 9.9, to determine the position and magnification of the image for several different object distances.

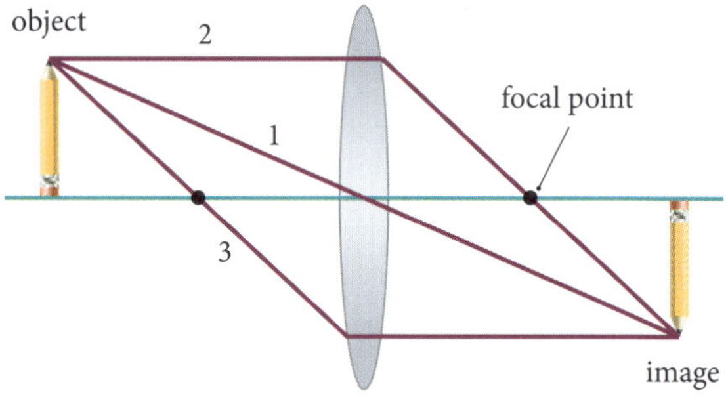

Figure 9.9 Example of a ray diagram drawn to find the position and magnification of the object

Students then test their predictions for the position and magnification of real images by placing a screen at the calculated image distance and seeing if a sharp image of the object is projected on the screen.

2 Refraction

» Students conduct investigations to examine qualitatively and quantitatively the refraction and total internal reflection of light.

» Students predict quantitatively, using Snell's law, the refraction and total internal reflection of light in a variety of situations.

→ Snell's law relates the angle of incidence (θ_1) and refraction (θ_2) to the ratio of the velocity (v_1) in the medium the ray came from, and the velocity (v_2) in the medium the wave moved into:

$$\sin \theta_1 / \sin \theta_2 = v_1 / v_2$$

→ We can derive Snell's law by considering a wavefront moving from one medium (1) to another (2), as shown in Figure 9.10. Note that the line drawn between P and P' represents the boundary between the two mediums.

→ Referring to Figure 9.10, if we call the time (t) for the wave to travel from B to C, then the distance $BC = v_1 t$ and, during this time, the other end of the wave front has moved a distance $AE = v_2 t$. For the triangles ABC and AEC we can write:

For triangle ABC we have $\sin \theta_1 = v_1 t / AC$

For triangle AEC we have $\sin \theta_2 = v_2 t / AC$

Taking the ratio of these two equations gives us Snell's law:

$$\sin \theta_1 / \sin \theta_2 = v_1 / v_2$$

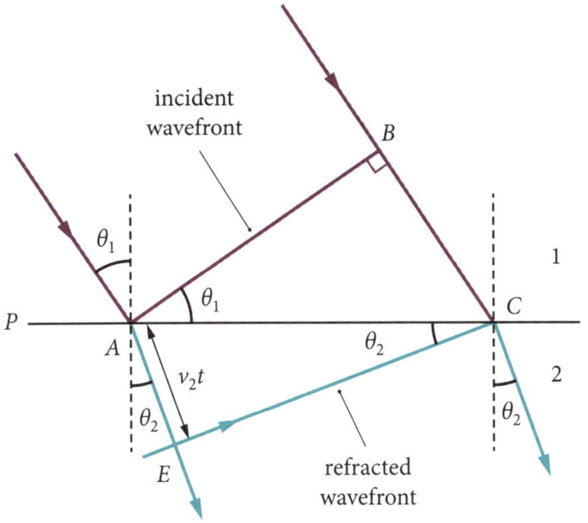

Figure 9.10 Refracted wavefronts

- When a ray moves into a medium in which it moves faster, the ray is refracted away from the normal. If the angle of incidence (θ_1) continues to be increased, a **critical angle** (θ_C) is reached where the angle of refraction (θ_2) will be 90°.
- For rays with an angle of incidence greater than the critical angle, the incident ray will be **totally internally reflected**. Figure 9.11 shows the path of a ray with an angle of incidence (a) below the critical angle, (b) at the critical angle, and (c) greater than the critical angle.

critical angle: the angle of incidence that results in an angle of refraction of 90°; the critical angle for light always refers to rays moving from the denser medium towards a less dense medium

total internal reflection: occurs when the angle of incidence is greater than the critical angle and the ray is reflected back into the first medium

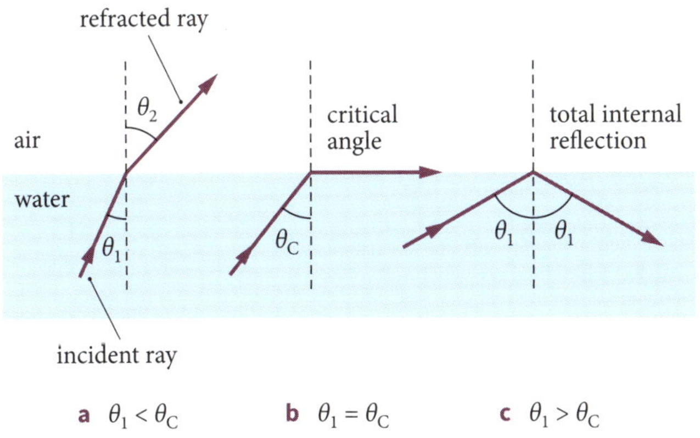

Figure 9.11 Refraction and critical angle for a ray passing from water to air

- Applying Snell's law to the critical angle gives:

  $\sin\theta_C = v_1/v_2$
- As shown in Figure 9.11, rays with an angle of incidence greater than the critical angle are totally internally reflected. These rays obey the laws of reflection and because total internal reflection occurs without loss of light intensity it is used in many important applications, including optic fibres and optical instruments.
- Because a wave must be continuous across a boundary, the frequency of a wave must remain constant when it moves from one medium to another. Because the velocity changes but the frequency remains constant, the wavelength of the wave must change when it enters the new medium and, as $v_1 = f\lambda_1$ and $v_2 = f\lambda_2$, we can divide one equation by the other to obtain:

$$v_1/v_2 = \lambda_1/\lambda_2$$

where v = velocity

λ = wavelength

- When dealing with light rays we sometimes use **refractive index** (n), which is defined as the ratio of the speed of light in a vacuum (c) to the speed of light in the material (v):

$$\text{Refractive index} = n = c/v$$

where c is speed of light in a vacuum

v is the speed of light in the material

refractive index: the ratio of the speed of light in a vacuum to the speed of light in the material

- The refractive index of a vacuum and for air is $n = 1$. For all other materials the refractive index $n > 1$, because nothing can travel faster than light.
- We can write Snell's law and the critical angle equation in terms of refractive index and wavelength as follows:

$$\sin\theta_1/\sin\theta_2 = v_1/v_2 = \lambda_1/\lambda_2 = n_2/n_1$$

and

$$\sin\theta_C = v_1/v_2 = \lambda_1/\lambda_2 = n_2/n_1$$

- The refractive index and speed of light in a transparent material can be determined by measuring the angle of refraction as a function of the angle of incidence for a ray moving from air to the material. A graph of the sine of the angle of incidence against sine of the angle of refraction would produce a straight line with a gradient equal to the refractive index of the material (see Figure 9.12).

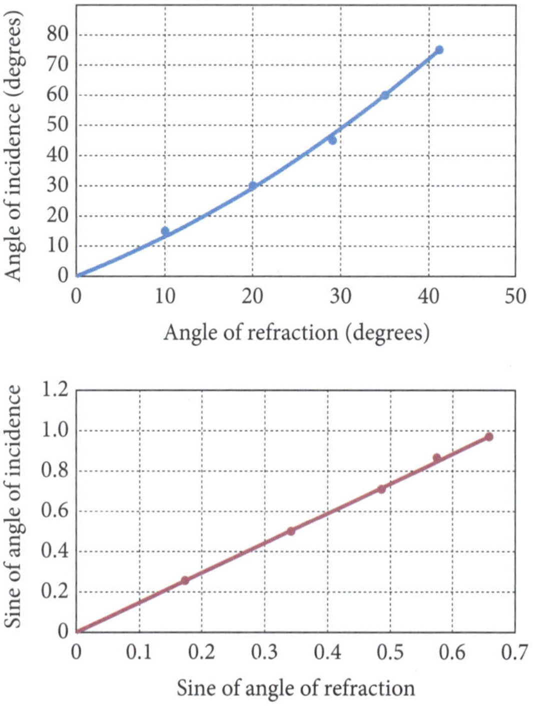

Figure 9.12 Angle of incidence graphed against the angle of reflection and sine of the angle of incidence against the sine of the angle of refraction

→ KEY QUESTIONS

4 **What wave properties change and what wave properties remain unchanged when a ray enters a new medium?**

5 **When a wave enters a new medium and increases its velocity, is the ray refracted towards or away from the normal?**

Answers ➲ p. 121

FIRSTHAND INVESTIGATION 3

Verifying Snell's Law

To investigate Snell's law students can measure the angle of refraction for various angles of incidence for a ray of light passing from air into a block of Perspex (or glass), as shown in Figure 9.13. Snell's law states that the ratio of sin θ_1 to sin θ_2 is a constant and therefore a graph of sin θ_1 against sin θ_2 should be produce a straight line (a linear graph).

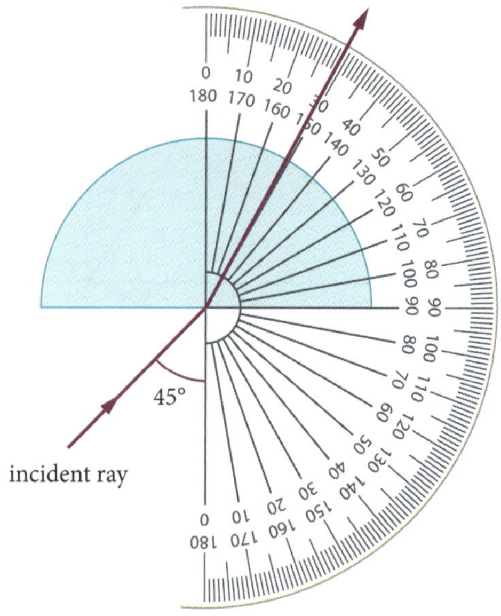

Figure 9.13 Measuring the angle of refraction for an angle of incidence of 45°

After measuring the angle of refraction for various angles of incidence, students can plot a graph of sin θ_2 against sin θ_1. Typical results are shown in Figure 9.14.

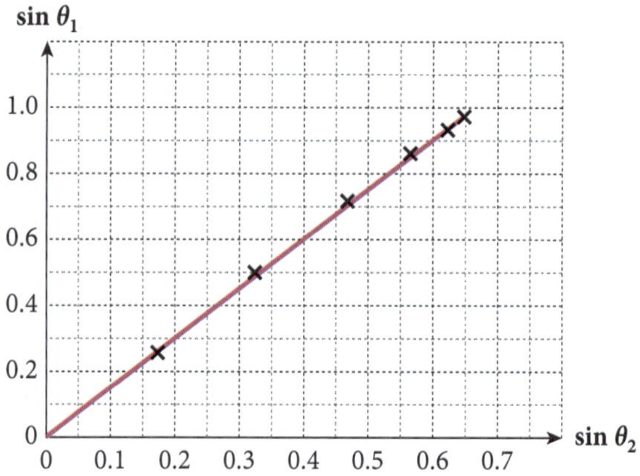

Figure 9.14 Plot of the sine of the angle of incidence against the sine of the angle of refraction

Snell's law states:

$$\sin \theta_1 / \sin \theta_2 = v_1/v_2 = n_2/n_1$$

But as the first medium is air, $n_1 = 1$ and Snell's law becomes:

$$\sin \theta_1 / \sin \theta_2 = c/v_2 = n_2$$

This relationship shows that the gradient of the graph (sin θ_1/sin θ_2) shown in Figure 9.14 is equal to the refractive index of the material. Once the refractive index is determined from the gradient, the speed of light in the material can be calculated using $n = c/v$.

EXAMPLE 2

Given that the refractive index of diamond is $n_D = 2.4$ and the refractive index of water is $n_W = 1.3$, find:

a the speed of light in diamond
b the critical angle for a ray passing from diamond into air
c the critical angle for a ray passing from diamond into water
d the angle of refraction for a ray passing from air to diamond with an angle of incidence $\theta_1 = 65°$
e the wavelength of red light in diamond, given that red light has a wavelength 620 nm in air.

Don't forget that Snell's law uses n_2/n_1 where n_1 is the medium the ray is coming from

Answer:

a Refractive index $n = c/v$ and hence:

$$v = c/n = \frac{3\times10^8}{2.4} = 1.25 \times 10^8 \text{ ms}^{-1}$$

b At the critical angle $\theta_2 = 90°$ and Snell's law becomes $\sin\theta_C = n_2/n_1$, or:

$$\theta_C = \sin^{-1}(n_2/n_1) = \sin^{-1}\left(\frac{1}{2.4}\right) = 24.6°$$

c Using $n_2 = 1.3$ in the previous equation gives for water:

$$\theta_C = \sin^{-1}(n_2/n_1) = \sin^{-1}\left(\frac{1.3}{2.4}\right) = 32.8°$$

d Applying Snell's law:

$\sin\theta_1/\sin\theta_2 = n_2/n_1$ and hence

$$\theta_2 = \sin^{-1}(n_1\sin\theta_1/n_2) = \sin^{-1}\left(1 \times \frac{\sin 65°}{2.4}\right) = 22.2°$$

e As the frequency is constant in the two mediums of diamond and air, the ratio of the wavelengths is equal to the ratio of the velocities:

$v_D/v_A = \lambda_D/\lambda_A$ and hence $\lambda_D = \lambda_A v_D/v_A$

$$= 620 \times 10^{-9}\left(\frac{1.25\times10^8}{3\times10^8}\right) = 258 \text{ nm}$$

3 Dispersion

» Students conduct a practical investigation to demonstrate and explain the phenomenon of the dispersion of light.

→ Because different frequencies of light travel at slightly different velocities in transparent materials, colours are refracted by slightly different angles (see Figure 9.15). Blue light is slower than red light in glass, which is why a prism can split white light into a spectrum of colours. This splitting up of a wave into its constituent wavelengths is called **dispersion**.

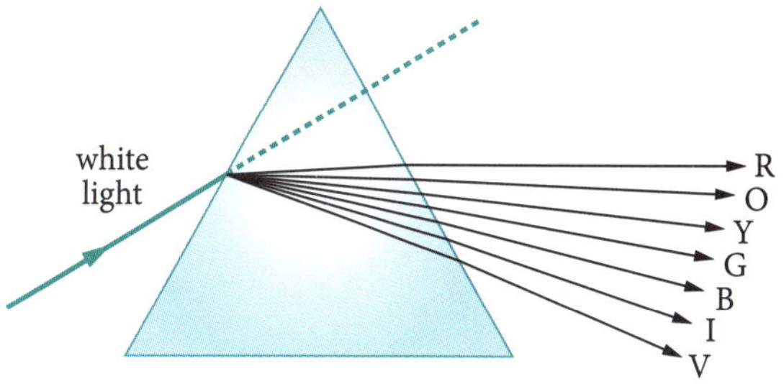

Figure 9.15 Dispersion of white light by a glass prism

dispersion: occurs when the velocity of a wave in a material is a function of wave frequency; dispersion causes different colours of light to be refracted by different angles

FIRSTHAND INVESTIGATION 4

Dispersion of light rays

The critical angle for red light and blue light for Perspex in air can be used to illustrate dispersion and to find the velocity of red light and blue light in Perspex. Students set up a lightbox to produce a white light ray and direct the ray into a semicircular block of Perspex, as shown in Figure 9.16.

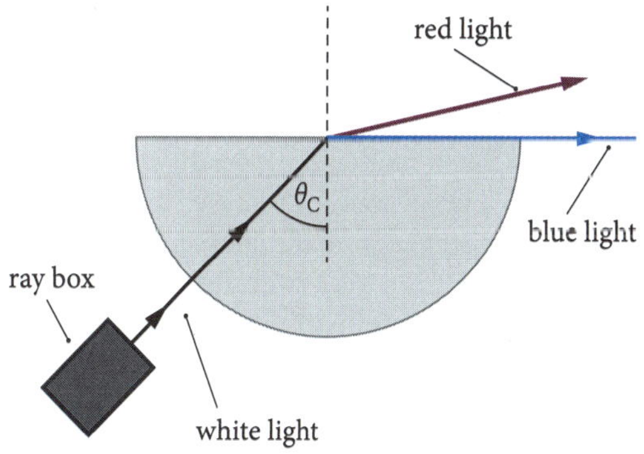

Figure 9.16 Finding the critical angles for blue light and red light at a Perspex–air boundary

The incident white light is dispersed as it is refracted at the Perspex–air boundary. By changing the angle of incidence of the white light ray until the angle of refection for the blue light is 90°, the critical angle for blue light can be determined (this is the angle marked θ_C in Figure 9.16. The angle of incidence can then be increased until the red ray is refracted by 90° and the critical angle for red light is determined.

The speed of light and refractive index for blue light and for red light in Perspex can then be calculated using Snell's law. As the light is moving into air $n_2 = 1$ and hence:

$$\sin \theta_C = v/c = 1/n$$

where θ_C is the critical angle

v is the speed of the light in Perspex

n is the refractive index in Perspex

➔ KEY QUESTIONS

6 **Why is blue light refracted by a greater angle than red light at an air–glass boundary?**

7 **Do light waves with different frequencies travel at different velocities in air?**

Answers ➲ p. 121

4 Intensity and distance

» Students conduct investigations to demonstrate the relationship between inverse square law, the intensity of light and the transfer of energy.

➔ The intensity of light is measured by the energy that passes through a unit area perpendicular to the wave velocity each second. The SI units of light intensity are therefore Wm^{-2}.

➔ Recall from Chapter 8 that the **intensity** of a wave emitted from a point source decreased as the inverse square of the distance from the source. This occurs because the power is spread over the surface area of a sphere surrounding the source and, as the area of the sphere increases with the square of the distance from the source, the intensity of the wave must decease by the inverse square of the distance. Figure 9.17 illustrates this concept.

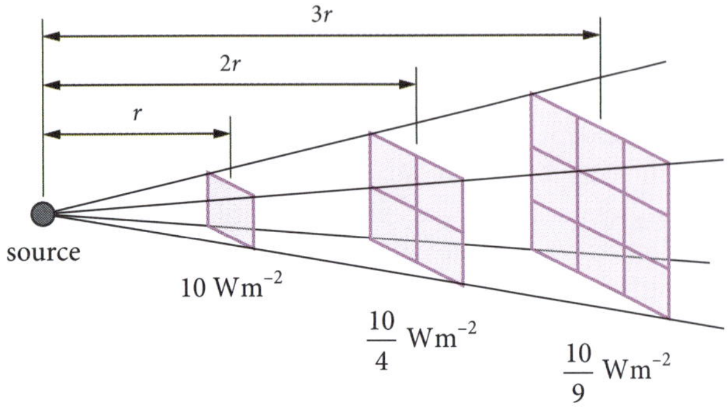

Figure 9.17 Inverse square law of wave intensity with distance (r) from the source

intensity: a measure of the energy carried by a wave in power per unit-area perpendicular to the wave velocity; intensity uses the SI units Wm^{-2}

➔ We can represent the **inverse square law** as:

$$I = P/4\pi r^2$$

where P is the power emitted by the source
I is the intensity at distance (r) from the source

inverse square law: a law that equates a quantity to the inverse square of another quantity (i.e. $I \alpha 1/r^2$)

➔ By taking ratios of the previous equation, the intensity at two distances from the source (r_1 and r_2) can be related to the intensity (I_1) at r_1 and the intensity (I_2) at r_2 by:

$$I_1 r_1^2 = I_2 r_2^2$$

FIRSTHAND INVESTIGATION 5

Inverse square law for light intensity

In this investigation students use a light meter or a light-intensity mobile-phone application to measure the intensity of light at different distances (r) from a light globe. The experiment should be conducted on a black tabletop, as shown in Figure 9.18 to prevent reflection from the tabletop interfering with the results.

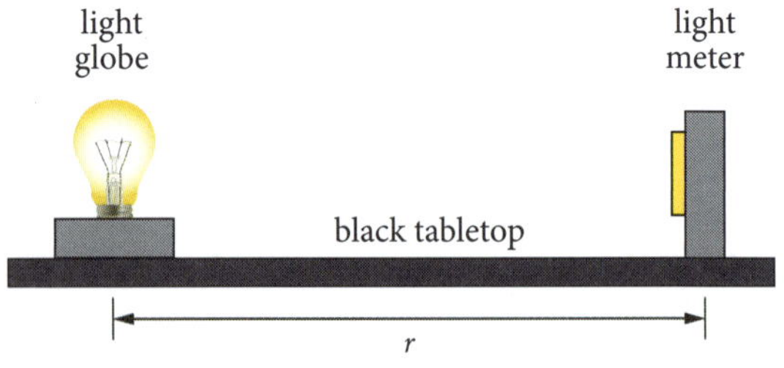

Figure 9.18 Experimental set up for measuring light intensity as a function of distance from the source

Plotting the measured light intensity against the inverse square of the distance ($1/r^2$) will yield a straight line (as shown in Figure 9.19), which confirms the intensity (I) of light is proportional to the inverse square of the distance ($1/r^2$) from the light source.

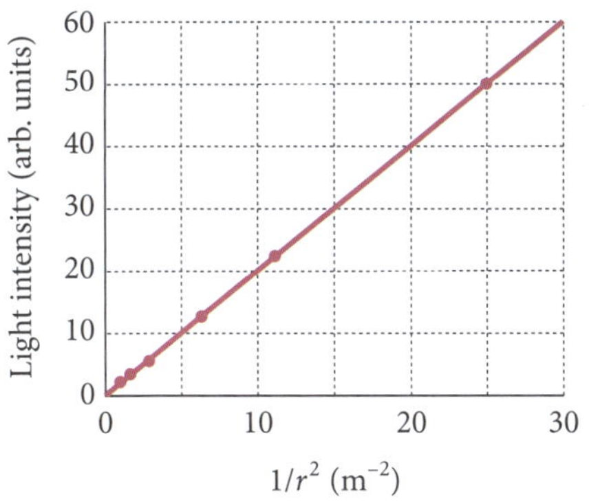

Figure 9.19 Graph of light intensity against the inverse square of the distance from the source

FIRSTHAND INVESTIGATION 6

Area over which light falls as a function of distance

Students can investigate how light energy spreads over a greater area as it moves away from a source by setting up an experiment like the one shown in Figure 9.20.

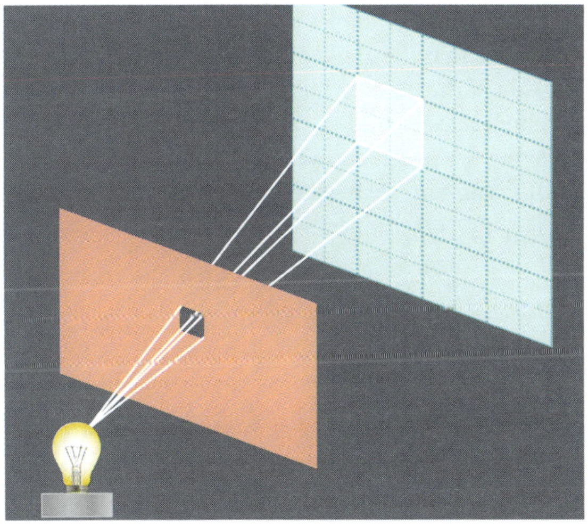

Figure 9.20 Experiment to investigate the area over which light spreads as a function of distance from the source

In this experiment the light that passes through a 1 cm^2 hole in a mask placed 10 cm from a light source is permitted to fall on a sheet of graph paper. By moving the graph paper from the 1 cm^2 hole and recording the lightened area on the graph paper as a function of distance, students deduce that the lightened area increases as the square of the distance from the source and hence the intensity of light falling on the graph paper decreases as the inverse square of the distance from the source.

➔ KEY QUESTION

8 **How is the intensity of light emitted from a point source related to the distance between the source and the observer?**

Answers ➲ p. 121

5 Solving problems using the ray model

» Students solve problems or make quantitative predictions in a variety of situations by applying the following relationships to:

- $n_x = \frac{c}{v_x}$ (for the refractive index of medium x, v_x is the speed of light in the medium)
- $n_1 \sin\theta_1 = n_2 \sin\theta_2$ (Snell's law)
- $\sin\theta_C = \frac{n_2}{n_1}$
- $I_1 {r_1}^2 = I_2 {r_2}^2$ (to compare the intensity of light at two points, r_1 and r_2).

➔ The examples that follow show how the ray model of light and the quantitative relationships we have examined earlier in the chapter can be used to solve problems and make predictions about the behaviour of light in a variety of situations.

EXAMPLE 3

A student conducts a refraction experiment in which the angle of refraction (θ_2) is measured as a function of the angle of incidence (θ_1) for a ray of red light moving from a glass block to air. The student then plots $\sin\theta_2$ against $\sin\theta_1$, as shown in Figure 9.21.

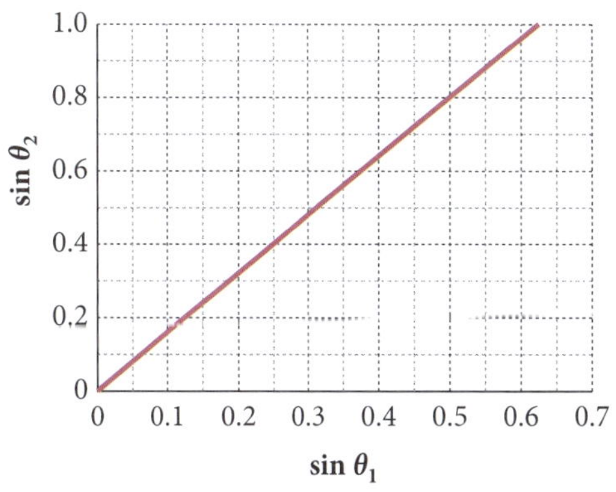

Figure 9.21 Sine of the angle of refraction (θ_2) plotted against the sine of the angle of incidence (θ_1)

a Explain why the student plotted the sine of the angles rather than the angles themselves?

b Use the graph to find the refractive index of the glass used in the experiment and the speed of light in glass.

c Explain how the results obtained by the student would have differed if the student had used blue light rather than red light for the experiment.

Answer:

If the ray moves from glass to air the velocity will increase; remember also that $n = 1$ for air

a Snell's law states that the ratio of the sines of the angles of incidence and refraction equal a constant, and hence plotting $\sin\theta_1$ against $\sin\theta_2$ will produce a straight-line (linear) graph. As the second medium is air, in this case $n_2 = 1$, and the gradient of the graph will be equal to the refractive index of the glass.

b Applying Snell's law with $n_2 = 1$ yields:

$\sin\theta_1/\sin\theta_2 = 1/n_1$

Hence the gradient $= \sin\theta_2/\sin\theta_1 = n_1 = \dfrac{1}{0.65} = 1.54$

Once the refractive index is known we can calculate the velocity of light:

$n = c/v$ and hence $v = c/n = \dfrac{3\times10^8}{1.54} = 1.95\times10^8\ \text{ms}^{-1}$

c We know from our study of dispersion that blue light travels slower in glass than red light and therefore if blue light was used rather than red light, the gradient of the graph would have been steeper, the refractive index greater, and the velocity of light in glass lower for the blue light.

EXAMPLE 4

Glass has a refractive index of 1.52 and water has a refractive index of 1.33.

a **Find the critical angle for glass in air.**

b **Find the critical angle for glass in water.**

Answer:

The angle of refraction is 90° when the angle of incidence is equal to the critical angle and $n = 1$ for air

a For glass in air, we apply Snell's law with $\theta_2 = 90°$ and $n_2 = 1$:

$\sin\theta_C = 1/n_1$ and hence $\theta_C = \sin^{-1}(1/n_1)$

$= \sin^{-1}\left(\dfrac{1}{1.52}\right) = 41.1°$

b For glass in water, we have $n_2 = 1.33$ and hence:

$\theta_C = \sin^{-1}(n_2/n_1) = \sin^{-1}\left(\dfrac{1.33}{1.52}\right) = 61°$

EXAMPLE 5

Find the angle (θ) for the ray shown in Figure 9.22. Note that the refractive index of the glass prism is $n = 1.5$.

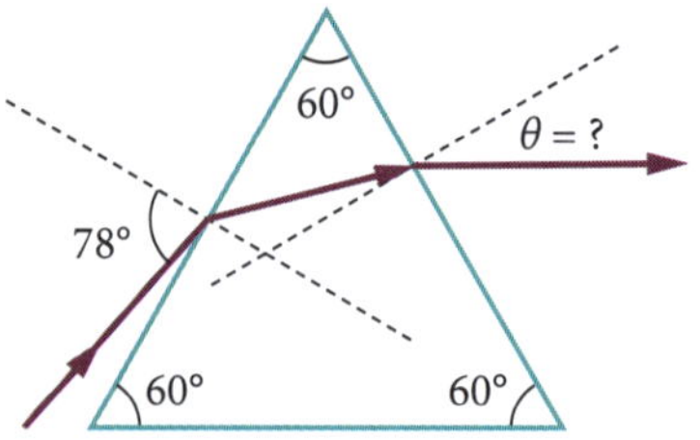

Figure 9.22 Ray of light passing through a glass prism in air

Answer:

Remember that n_1 in Snell's law refers to the medium the light ray is coming from and that a light ray will be refracted towards the normal when it enters a denser medium

The angle of refraction for the ray entering the prism would be given by Snell's law and, as the first medium is air, we have $n_1 = 1$:

$\sin\theta_1/\sin\theta_2 = n_2/1$

Therefore $\theta_2 = \sin^{-1}(\sin\theta_1/n_2) = \sin^{-1}\left(\dfrac{\sin 78°}{1.5}\right)$

$= 40.7°$

As the three interior angles of a triangle must add to 180°, we see that the angle of incidence of the ray on the other side of the prism will be $\theta_1 = 19.3°$. We can now apply Snell's law to find the angle of refraction of the ray leaving the prism. However, this time the second medium is air and we have $n_2 = 1$. Hence:

$\sin\theta_1/\sin\theta_2 = 1/n_1$

Hence $\theta_2 = \sin^{-1}(n_2\sin\theta_1) = \sin^{-1}(1.5\times(\sin 19.3°))$

$= 29.7°$

EXAMPLE 6

The Sun continually radiates 3.828×10^{26} W of power in the form of electromagnetic waves. What is the intensity of sunlight that reaches the Earth, given that the Earth is 150 million kilometres from the Sun?

Answer:

Remember to change to SI units and recall that light will spread equally in all directions from the Sun

Intensity is given by:

$I = P/4\pi r^2 = \dfrac{3.828\times10^{26}}{4\pi(150\times10^9)^2}$

$= 1354\ \text{Wm}^{-2}$

This is the intensity of the Sun's radiation that passes through a square metre perpendicular to the Sun's rays at the top of the atmosphere.

EXAMPLE 7

The planet Neptune is 30 times further from the Sun than the Earth. Given that the intensity of light from the Sun is 1354 Wm^{-2} when it reaches the Earth, find the intensity of the light reaching Neptune from the Sun.

Answer:

> Recall that we can use ratios of intensity and the inverse square law of distance to solve problems involving the intensity at two different distances from a source

We can express the ratio of intensities at two distances from a source as:

$$I_1 r_1^2 = I_2 r_2^2$$

$$\text{Or } I_2 = I_1 r_1^2 / r_2^2 = \frac{1354 \times (150 \times 10^9)^2}{(30 \times 150 \times 10^9)^2}$$

$$= \frac{1354}{30^2}$$

$$= 1.5\ Wm^{-2}$$

CHAPTER SYLLABUS CHECKLIST

Are you able to answer these questions from the syllabus for this chapter? Tick each question as you go through the checklist if you are able to answer it. If you cannot answer a question, turn to the relevant page in the study guide to find the answer. For NESA key word meanings, go to www.educationstandards.nsw.edu.au and search 'key words'.

	FOR A COMPLETE UNDERSTANDING OF THIS TOPIC:	PAGE NO.	✓
1	Can I explain the difference between a real image and a virtual image?	107	
2	Can I outline the laws of geometric optics for rays refracted by lenses and reflected from curved mirrors?	107–108	
3	Can I use the laws of geometric optics to draw ray diagrams to show where images are formed and determine the type of image formed from lenses and curved mirrors?	108–109	
4	Can I define refractive index and relate it to the speed of light in a material?	111	
5	Can I relate total internal reflection at boundaries to critical angle?	111	
6	Can I conduct an investigation to verify Snell's law experimentally and relate the gradient of a graph of sin *i* against sin *r* to the refractive index?	112	
7	Can I use Snell's law to calculate the critical angle for a light beam incident on a boundary between different mediums?	113	
8	Can I conduct an experiment to demonstrate dispersion?	113	
9	Can I explain the dispersion of light in terms of the relationship between frequency and refractive index?	113	
10	Can I conduct an investigation to verify the inverse square law for the decrease of light intensity with distance?	114	
11	Can I solve problems involving the inverse square law for the decrease of light intensity with distance?	115	
12	Can I apply Snell's law to solve a wide range of refraction problems?	115–117	

YEAR 11 EXAM-TYPE QUESTIONS

Objective-response questions

(1 mark each)

1 Which of the flowing conditions are necessary for a light ray to undergo total internal reflection when it encounters a boundary?

Answer	Angle of incidence (i)	Refractive index in mediums (n)
A	$\theta_1 > \theta_C$	$n_1 < n_2$
B	$\theta_1 > \theta_C$	$n_1 > n_2$
C	$\theta_1 < \theta_C$	$n_1 < n_2$
D	$\theta_1 < \theta_C$	$n_1 > n_2$

2 Two stars viewed from the Earth have the following intensities:

Star	Light intensity (Wm^{-2})
Sirius	20
Polaris	80

Given that Sirius is twice as far from the Earth as Polaris, what can we deduce about the absolute energy radiated (the luminosity) from each star?

A The luminosity of Sirius is equal to the luminosity of Polaris.

B Polaris has twice the luminosity of Sirius.

C Polaris has four times the luminosity of Sirius.

D Polaris has half the luminosity of Sirius.

3 What can we conclude about the refracted light ray in Figure 9.23?

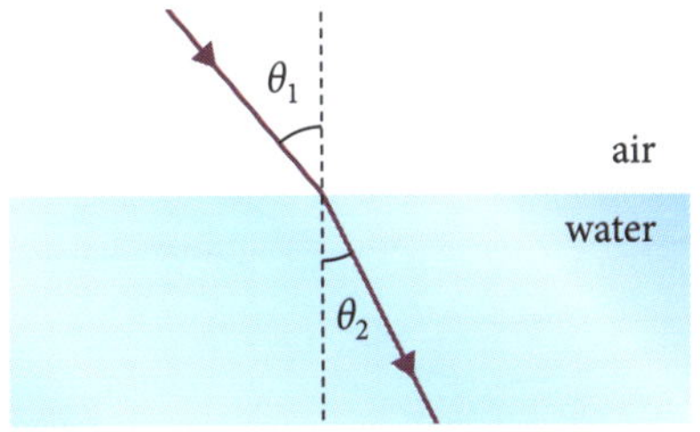

Figure 9.23 Ray moving from air to water

A The frequency of the refracted ray is less than the frequency of the incidence ray.

B The wavelength of the refracted ray is less than the wavelength of the incident ray.

C The wavelength of the refracted ray is greater than the wavelength of the incident ray.

D The speed of light in the air is less than the speed of light in the water.

4 Which of the following options describe the image formed in Figure 9.24?

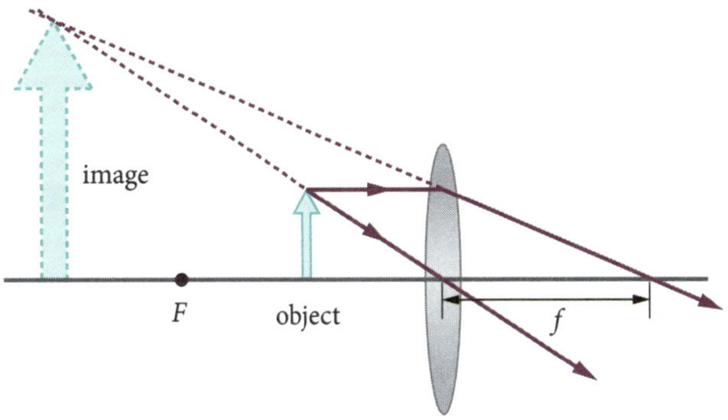

Figure 9.24 Ray diagram showing image formed by a convex lens

A Real, magnified, upright

B Virtual, diminished, upright

C Virtual, magnified, inverted

D Virtual, magnified, upright

5 Figure 9.25 shows how light rays from the Sun are refracted by the atmosphere. Which of the following best describes what the observer would see if the Sun's rays were not refracted by the atmosphere in this way?

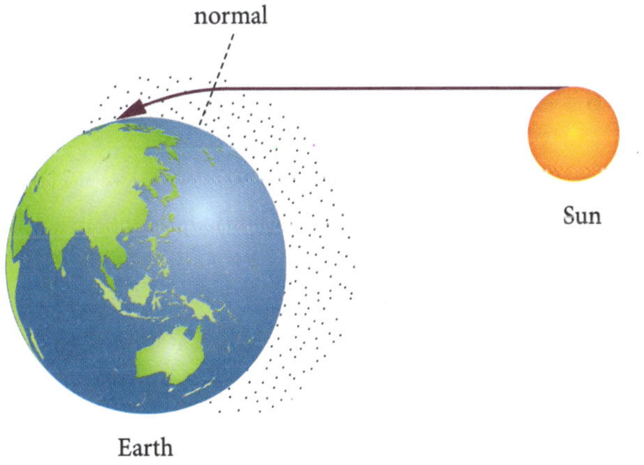

Figure 9.25 Refraction of sunlight by the atmosphere

A The observer would see the Sun rise later and the Sun set later.

B The observer would see the Sun rise later and set earlier.

C The observer would see the Sun rise earlier and the Sun set earlier.

D The observer would not notice any difference in the length of the day.

Extended-response questions

6 Figure 9.26 illustrates an experiment designed to measure the intensity of light at different distances from a light source. The results obtained are shown in Table 9.1.

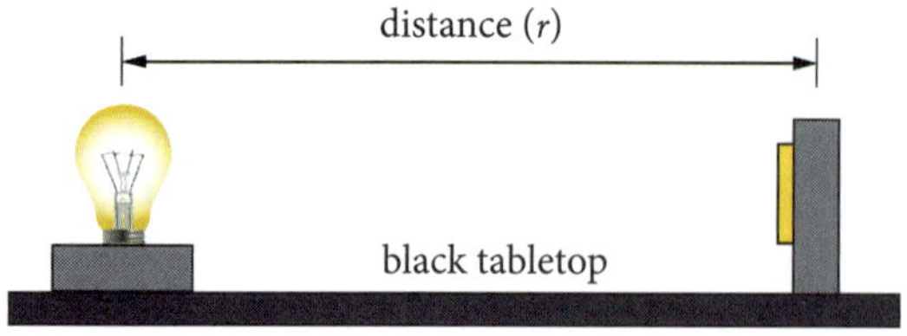

Figure 9.26 Experimental setup to measure light intensity at different distances (r) from the source

Table 9.1 Results

Intensity of light (Wm^{-2})	Distance (r) from light source (cm)	
41	7	
32	8	
20	10	
14	12	
9	15	
5	20	

a Label the third column as 'Inverse square of the distance ($1/r^2$)' with the correct units, and calculate the missing values. (2 marks)

b Plot a graph of the inverse square of the distance ($1/r^2$) against the intensity of the light. (Place the dependent variable on the vertical axis.) (3 marks)

c Find the gradient of the graph and its units. (1 mark)

d Explain why the student used a black table top rather than a white table top. (1 mark)

7 A spherical concave mirror is illustrated in Figure 9.27. The focal point is marked as F and the centre of curvature of the mirror is marked as C.

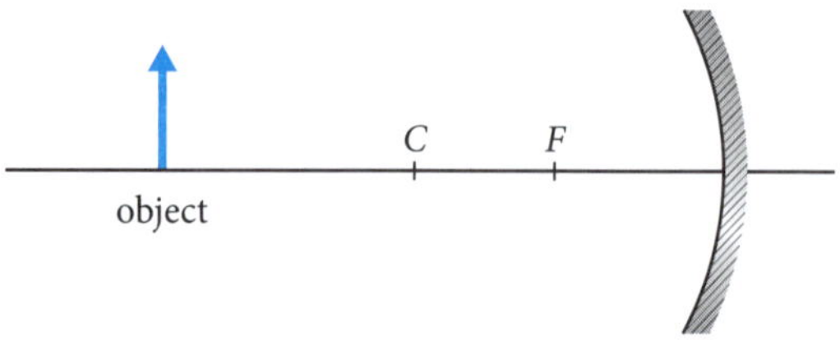

Figure 9.27 Concave mirror and object

a Draw a ray diagram to show where the image of the object would form. (2 marks)

b Is the image is real or virtual? Justify your answer. (2 marks)

c Draw a second ray diagram to show the path the rays would take if a small light globe was placed at the focal point (F). (1 mark)

8 Given that water has a refractive index of 1.33, find:

a the angle of refraction for a laser beam that moves from air into water with an angle of incidence of 35° (1 mark)

b the critical angle for a ray passing from water to air (1 mark)

c the speed *and* the wavelength of the laser beam in water, given that it has a wavelength of 630 nm and travels at 3×10^8 ms^{-1} in air. (2 marks)

9 a Draw a ray diagram to illustrate the type of image that would result from an object placed at the focal point and the concave lens shown in Figure 9.28. (2 marks)

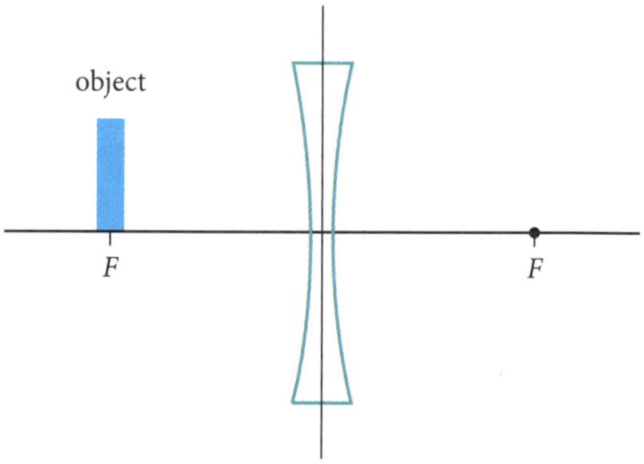

Figure 9.28 Concave lens with an object placed at the focal point

b Describe the type of image formed by the lens when the object is placed at the focal point. (3 marks)

10 An optic fibre (shown in Figure 9.29) consists of an inner glass core with a refractive index 1.7 and an outer glass core of refractive index 1.5.

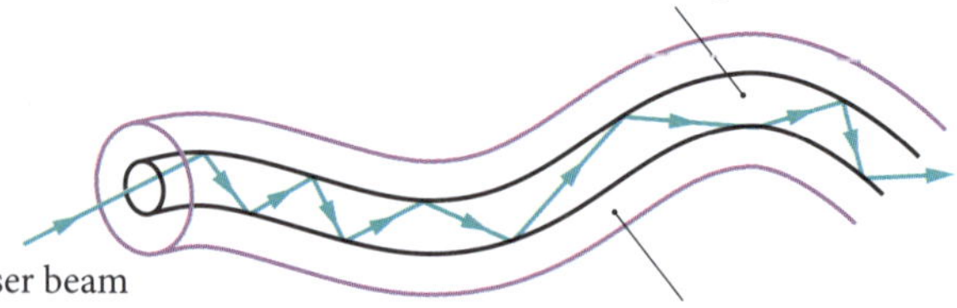

Figure 9.29 Laser beam passing through an optic fibre

a Calculate the critical angle for the laser beam at the boundary of the inner and outer glass cores. (1 mark)

b Explain why the laser beam stays within the inner core. (1 mark)

c Explain why the inner core must have a higher refractive index than the outer core in an optic fibre. (1 mark)

ANSWERS

KEY QUESTIONS

Key questions ⊃ p. 109

1 An image is produced when the rays from an object converge. A real image can be projected onto a screen, while a virtual image cannot be projected onto a screen.

2 There are three rules for spherical curved mirrors:
- An incident ray parallel to the optic axis will be reflected through the focus of the mirror. For a convex mirror the focus is determined by tracing back the reflected ray to its virtual point of convergence behind the mirror.
- A ray that passes through the focus will be reflected parallel the principal axis.
- A ray that is directed through the centre of curvature of the mirror will be reflected back through the centre of curvature.

3 There are three rules for lenses:
- A ray drawn parallel to the principal axis will pass through the focal point of the lens.
- A ray drawn through the centre of the lens will pass straight through the lens without being refracted.
- A ray drawn through the focus will be refracted to run parallel to the principal axis of the lens.

Key questions ⊃ p.112

4 When a ray enters a new medium it changes velocity, wavelength and, if it is not an incident ray perpendicular to the surface, it also changes direction. Note that the frequency does not change.

5 If the velocity increases, the ray will be refracted away from the normal. If the velocity decreases, the ray will be refracted towards the normal.

Key questions ⊃ p. 114

6 Blue light travels slower in glass than red light and hence blue light is refracted more than red light at the air–glass boundary (i.e. $r_{blue} > r_{red}$).

7 No, light of all frequencies travels at the same speed in air.

Key questions ⊃ p. 115

8 Intensity decreases as the inverse square of the distance from the source.

YEAR 11 EXAM-TYPE QUESTIONS

Objective-response questions

1 **B.** The angle of incidence must be greater than the critical angle or the ray would pass into the second medium rather than being reflected. The velocity in the second medium must be also be greater and hence the refractive index of the second medium must be less than the refractive index of the first medium. **A** is incorrect because if the second medium had a greater refractive index the ray would not be internally reflected. **C** and **D** are incorrect because if the angle of incidence was less than the critical angle the ray would pass into the second medium rather than be internally reflected.

2 **A.** As Sirius is twice as far away as Polaris it will have $\frac{1}{2^2} = \frac{1}{4}$ of the intensity of Polaris or $\frac{80}{4} = 20$ Wm^{-2}. **B, C** and **D** are all incorrect as they do not obey the inverse square law of intensity with distance.

3 **B.** The frequency of the wave remains constant as it moves into the second medium, but its velocity decreases and, as $v = f\lambda$, the wavelength must also decrease when it enters the second medium. We can also see this from the refraction equation $v_1/v_2 = \lambda_1/\lambda_2$. **A** is incorrect because frequency does not change when a ray is refracted. **C** is incorrect because if the velocity decreases the wavelength must also decrease if the frequency remains constant. **D** is incorrect because the velocity of light decreases with increasing density and, as glass is much denser than air, the velocity is glass must be less than in air.

4 **D.** The rays do not actually reach the image and it is therefore a virtual image. The image is on the same side of the principal axis as the object and is therefore upright and, as the image is bigger than the object, it is magnified. **A** is incorrect because this image could not be projected on a screen and hence is not a real image. **B** is incorrect because the image is bigger than the object and is therefore magnified rather than diminished. **C** is incorrect because the image is not on the opposite side of the principal axis to the object and hence is upright rather than inverted.

5 **B.** Because the rays are refracted by the atmosphere the rays would reach the observer even when the Sun was below the horizon. Removing the atmosphere would mean the observer would not see the Sun until it was above the horizon and would see it set as soon as it went below the horizon, making the days shorter. **A** is incorrect because the observer would see the Sun set earlier if the atmosphere was removed. **C** is incorrect because the observer would see a later sunrise. **D** is incorrect because the length of day would increase.

Extended-response questions

6 EM This question tests students' understanding of the inverse square law of light intensity, SI units and graphing.

a The right-hand column is completed in Table A9.1.

Table A9.1 Inverse square calculations

Intensity of light (Wm^{-2})	Distance (r) from light source (cm)	Inverse square of distance $1/r^2$ (m^{-2}) ✓
41	7	204
32	8	156
20	10	100
14	12	69
9	15	44
5	20	25 ✓

b A graph of the results is shown in Figure A9.1.

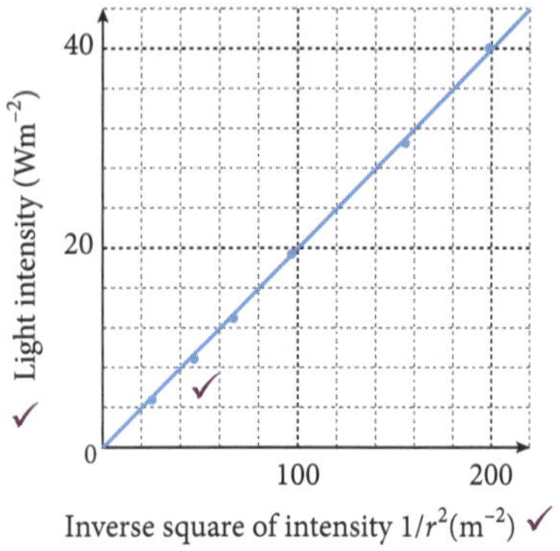

Figure A9.1 Graph of experimental results

c Gradient = rise/run = $\dfrac{41 - 0}{204 - 0}$ = 0.2 watts ✓

d A white bench top would have reflected light into the detector, which would mean the intensity would not decrease as the inverse square of the distance from the source. ✓

7 **EM** This question tests students' understanding of ray tracing and image formation when light is reflected by a concave mirror.

a The ray diagram required is shown in Figure A9.2.

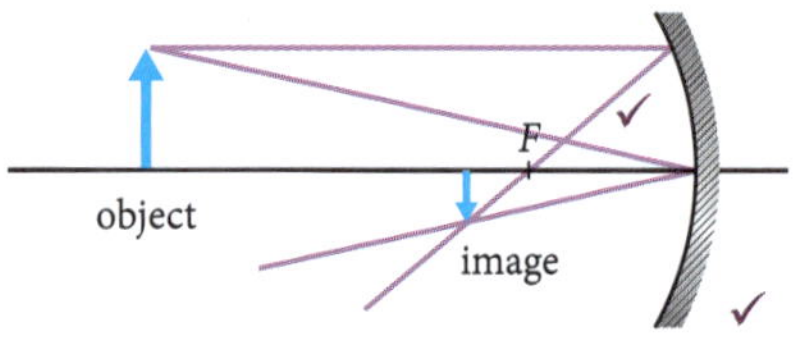

Figure A9.2 Ray diagram for image formation from a concave mirror

b The image is a real image ✓ as light rays actually converge at the image and it could be projected on a screen. ✓

c The required ray diagram is shown in Figure A9.3.

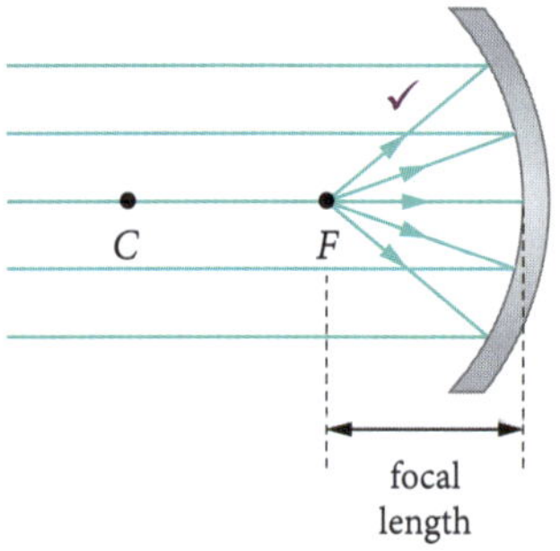

Figure A9.3 Ray diagram showing reflection from a concave mirror with a light source at the focus

8 **EM** This question tests students' understanding of Snell's law, refraction, critical angle and the effect of refraction on wavelength.

a Here $\sin\theta_1/\sin\theta_2 = n_2/1$, and as the first medium is air, $n_1 = 1$.

Hence $\theta_2 = \sin^{-1}(1/n_2(\sin\theta_1)) = \sin^{-1}\left(\dfrac{\sin 35°}{1.33}\right)$

$= 25.5°$ ✓

b In this case $\theta_2 = 90°$ and, as the second medium is air, $n_2 = 1$. Snell's law becomes $\sin\theta_C = 1/n_1$ and hence:

$\theta_C = \sin^{-1}(1/n_1) = \sin^{-1}\left(\dfrac{1}{1.33}\right) = 48.8°$ ✓

c $n = c/v$ and hence $v = c/n = \dfrac{3\times10^8}{1.33} = 2.26 \times 10^8\ \text{ms}^{-1}$ ✓

$v_1/v_2 = \lambda_1/\lambda_2$ and hence $\lambda_2 = v_2\lambda_1/v_1 = \dfrac{(2.26\times10^8)(630\times10^{-9})}{3\times10^8}$

$= 474$ nm ✓

9 **EM** This question tests students' ability to use ray tracing to find the type and position of an image formed when light is refracted by a concave lens.

a The required ray diagram is shown in Figure A9.4.

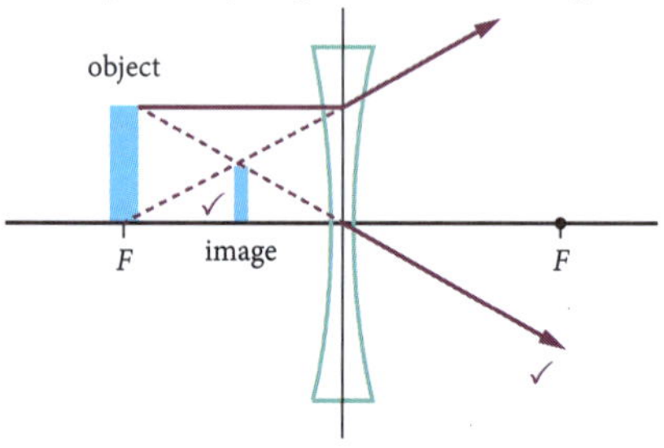

Figure A9.4 Ray diagram showing image formation for a concave lens with the object at the focal point

b The image formed is virtual, ✓ diminished ✓ and upright. ✓

10 **EM** This question tests student's understanding of Snell's law, critical angle, total internal reflection and refractive index.

a The critical angle occurs when the angle of refraction is 90° and hence:

$\sin\theta_C = n_2/n_1$ or $\theta_C = \sin^{-1}(n_2/n_1) = \sin^{-1}\left(\dfrac{1.5}{1.7}\right) = 61.9°$ ✓

b The laser beam hits the boundary between the inner and outer cores at an angle of incidence (θ_1) that is always greater than the critical angle (θ_C) ✓ and hence the beam is totally internally reflected at the boundary.

c Total internal reflection can only occur for light rays that are incident on a boundary to a medium in which they will travel faster. In the optic fibre, the speed of light must therefore be lower in the inner core than the outer core. ✓ As the refractive index $n = c/v$, this means that the refractive index of the inner core of the fibre must be greater than the refractive index of the outer glass core.

CHAPTER 10 THERMODYNAMICS

MODULE 3
WAVES AND THERMODYNAMICS

INQUIRY QUESTION:

How are temperature, thermal energy and particle motion related?

Temperature is a measure of the hotness or coldness of a sample of matter, with reference to some standard. The temperature of a sample of matter is a measure of the average kinetic energy of the particles in the sample. Thermal energy is the internal heat energy contained in a sample of matter due to the motion of particles in the sample.

1 Temperature

» Students explain the relationship between the temperature of an object and the kinetic energy of the particles within it.

- **Temperature** measures hotness or coolness. You will recall from your previous studies that the Celsius temperature scale is based on the freezing point of water (0 °C) and the boiling point of water (100 °C).
- In physics we generally use the absolute temperature scale, which measures temperature in Kelvin. The zero of the **Kelvin** scale (0 K = –273.15 °C) is the point where the motion of molecules would approach zero. Studies of gasses show that, at constant pressure, the volume of a gas is proportional to its temperature. Thus by plotting the volume of a gas as a function of its temperature at constant pressure, as shown in Figure 10.1, we obtain a graph that can be extrapolated back to a point where the volume of the gas approaches zero. This is the point where the gas molecules would not be moving and corresponds to absolute zero on the Kevin scale.

temperature: a measure of hotness or coldness; temperature is proportional to the average kinetic energy of the atoms or molecules in the substance
Kelvin: a unit of absolute temperature; zero kelvin represents the temperature where the kinetic energy and hence the velocity of molecules would be zero

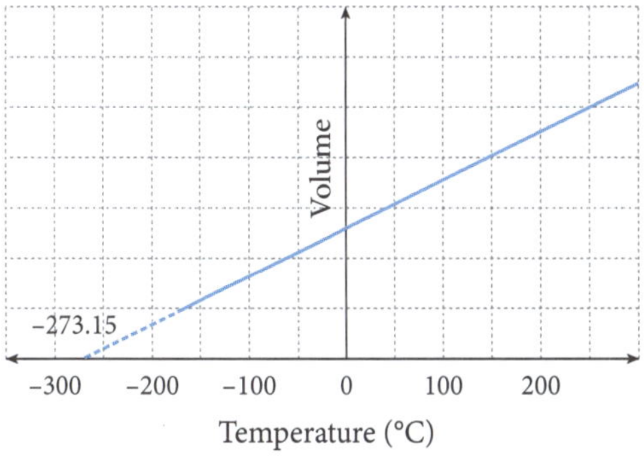

Figure 10.1 Volume of an ideal gas as a function of temperature at constant temperature

- The atoms and molecules in a sample of matter are in constant motion. Absolute temperature is a measure of the average kinetic energy of the particles in a sample. By treating gas molecules as point masses that collide elastically, it is possible to show that the average kinetic energy of gas atom/molecule and the absolute temperature of the gas are related by:

$$\text{Average } K = \langle mv^2/2 \rangle = 3kT/2$$

where $\langle mv^2/2 \rangle$ is the average kinetic energy of an atom/molecule
T is the absolute temperature
$k = 1.38 \times 10^{-23}\ \text{kgm}^2\text{s}^{-2}\text{K}^{-1}$ (which is a constant called Boltzmann's constant)

(Note that angular brackets < > around a quantity are used to designate the average value of the quantity.)

- We can think of temperature as a *measure* of the **internal** (thermal) **energy** of a body, and **heat** as the *transfer* of thermal energy from or to a body.

internal energy: the energy contained in the kinetic energy of the molecules in the body
heat: the transfer of thermal energy between bodies

EXAMPLE 1

Given that air is largely nitrogen (N_2) and that the mass of a nitrogen molecule is 4.65×10^{-26} kg, calculate the average speed of air molecules at 20 °C.

Answer:

First, convert the temperature to Kelvin:
20 °C = 20 + 273 = 293 K

Remember that the average kinetic energy of a gas molecule is related to the temperature of the gas

Now as the average kinetic energy of the air molecules is related to the absolute temperature by $<mv^2>/2 = 3/2kT$, the average speed will be given by:

$$<v> = \sqrt{3kT/m} = \sqrt{\frac{3 \times 1.38 \times 10^{-23} \times 293}{4.65 \times 10^{-26}}} = 511 \text{ ms}^{-1}$$

Surprisingly, the average speed of an air molecule is faster than the speed of sound.

➔ KEY QUESTIONS

1 How is the Kelvin temperature scale related to the Celsius scale?

2 What relationship does temperature have to the average kinetic energy of particles?

Answers ➲ p. 135

2 Thermal equilibrium

» Students explain the concept of thermal equilibrium.

➔ If two bodies are placed in thermal contact (i.e. heat can flow between them), heat will flow from the hotter body to the cooler body until both bodies reach the same temperature. At this point we say the bodies have reached **thermal equilibrium** and, provided no heat or energy is added or removed from either body, they will remain in thermal equilibrium with the same temperature.

thermal equilibrium: the condition where no energy flows between two bodies in thermal contact

➔ The concept of thermal equilibrium is expressed more formally by the zeroth law of thermodynamics, which states that if two bodies are in thermal equilibrium with a third body, then the two bodies will be in thermal equilibrium with each other.

➔ Thus two bodies are in thermal equilibrium with each other if they have the same temperature and no heat will be transferred between bodies in thermal equilibrium if they are placed in thermal contact with each other.

➔ The first law of thermodynamics uses the concept of conservation of energy to explain why the internal energy (temperature) of a body remains constant and why it changes. The first law of thermodynamics states that the internal energy of a body will remain constant unless heat is added or removed from the body or work is done on or by the body. The first law can be expressed as:

$$\Delta U = Q - W$$

where ΔU is the change in internal energy
Q is the amount of heat added to the system
W is the work done by the system

➔ Work done in this context refers to work done by the thermal energy contained in the body and hence, if the system did work, it would lose internal energy and its temperature would decrease.

EXAMPLE 2

A bowl of rice with a temperature of 50 °C is placed into a refrigerator that has an internal temperature of 5 °C, and is left overnight.

a Explain how the internal energy of the rice would change overnight.

b How could you prove that the internal energy of the rice had changed overnight?

c What caused the internal energy of the rice to change?

d How could you test to see if the rice and refrigerator were in thermal equilibrium in the morning?

Answer:

Heat moves from the hotter to the cooler body until both reach the same temperature

a The internal energy of the rice would decrease overnight because the rice would cool down (i.e. temperature decreases).

b Temperature is proportional to internal energy and hence measuring the temperature of the rice and showing it was lower in the morning would prove that its internal energy had decreased.

c Heat would flow from the hot rice to the cold refrigerator, reducing the internal energy of the rice.

d If the temperature of the rice and the temperature inside the refrigerator were equal, the rice and refrigerator would be in thermal equilibrium.

3 Specific heat capacity

» Students analyse the relationship between the change in temperature of an object and its specific heat capacity through the equation $Q = mc\Delta T$.

➔ The **specific heat capacity** (c) of a substance in a specific phase (i.e. solid, liquid or gas) is a measure of the amount of energy that must be added to a unit mass of the substance to increase its temperature by one degree Celsius. The relationship between heat added, specific heat and the change in temperature of a body can be expressed as:

$$Q = mc\Delta T$$

where Q is the heat added
m is the mass of the body
c is the specific heat capacity
ΔT is the increase in temperature that will result from the heat energy added

(Important note: This expression does not apply during changes of phase.)

specific heat capacity: the thermal energy required to increase the temperature of a unit mass of a specific substance by 1 °C

➔ The specific heat of a substance is different for different states of matter. Table 10.1 shows the specific heat of water in its solid, liquid and vapour phase. Note that in the special case of water, 1 L has a mass of 1 kg.

Table 10.1 Specific heat of water in different phases

State of matter (phase)	Specific heat ($Jkg^{-1}K^{-1}$)
Water vapour (gas)	2.02×10^3
Liquid water (liquid)	4.184×10^3
Ice (solid)	2.06×10^3

➔ KEY QUESTIONS

3 Which direction does heat flow when two bodies at different temperatures are brought into contact?

4 When will heat stop flowing between two bodies in thermal contact?

5 What factors determine the temperature change of a body when it is heated?

Answers ➲ p. 135

EXAMPLE 3

A 2 kW kettle holds 1.5 L of water. How long would it take to boil 1.5 L of water if the initial temperature of the water was 20 °C? Note that the specific heat of water is 4.184×10^3 $Jkg^{-1}K^{-1}$.

Answer:

We first must calculate the energy (heat) required to raise 1.5 L of water from 20 °C to 100 °C.

Remember that power is the rate at which energy is transferred to the fluid

The change in temperature $\Delta T = T_{final} - T_{initial}$ $= 100 - 20 = 80$ °C and the mass of water is 1.5 kg:

$Q = mc\Delta T = 1.5 \times 4.184 \times 10^3 \times 80 = 5.02 \times 10^5$ J

Now as power $P = E/t$, $t = E/P = \dfrac{5.02\times10^5}{2000} = 251$ s

EXAMPLE 4

If the energy required to increase the temperature of 1 L of water from 20 °C to 100 °C was converted to work, how high could it lift a 50 kg student?

Answer:

First calculate the energy required to boil 1 L of water from 20 °C:

Recall that work done to lift an object is equal to potential energy gained

$Q = mc\Delta T = 1 \times 4.184 \times 10^3 \times 80 = 3.34 \times 10^5$ J

Now $U = mgh$ and hence $h = U/mg = \dfrac{3.34\times10^5}{50\times9.8}$
$= 683$ m

FIRSTHAND INVESTIGATION 1

Finding the specific heat capacity of water

The specific heat of water can be found by timing how long it takes an electric kettle of a known power rating to raise the temperature of a measured quantity of water by a specified amount. Typical results are described below.

It takes a 1 kW kettle 102 s to raise the temperature of 1.2 L of water by 20 °C.

First find the energy (heat) input:

Power $= P = U/t$ and hence $U = Pt$
$= 1000 \times 102$
$= 1.02 \times 10^5$ J

$Q = mc\Delta T$ and hence $c = Q/m\Delta T$
$= \dfrac{1.02\times10^5}{1.2\times20}$
$= 4.25 \times 10^3$ $Jkg^{-1}K^{-1}$

Note that the measured value of specific heat in this experiment is a little larger than the accepted value of 4.184×10^3. This difference could be due to some heat being used to heat the plastic body of the kettle and some heat escaping to the environment. To measure specific heat accurately we would have to use a thermally insulated vessel (calorimeter) and take into account the heat lost to the device.

4 Heat transfer

» Students investigate energy transfer by the process of:
- conduction
- convection
- radiation.

➔ Thermal energy can be transferred by conduction, convection or radiation.

➔ **Conduction** transfers thermal energy via direct atomic or molecular collisions. Faster moving molecules in hotter regions will transfer energy to slower molecules in collisions. In this way heat is conducted from hotter regions to cooler regions. The rate at which heat is transferred by convection (Q/t) between two points with different temperatures is described by the relationship:

$$Q/t = kA\Delta T/d$$

where k is the thermal conductivity of the material
A is the cross-sectional area
ΔT is the temperature difference between the two points in Kelvin
d the distance between the points

conduction: the transfer of heat via collisions between atoms and molecules

➔ The conduction of heat between two points at different temperatures is shown schematically in Figure 10.2.

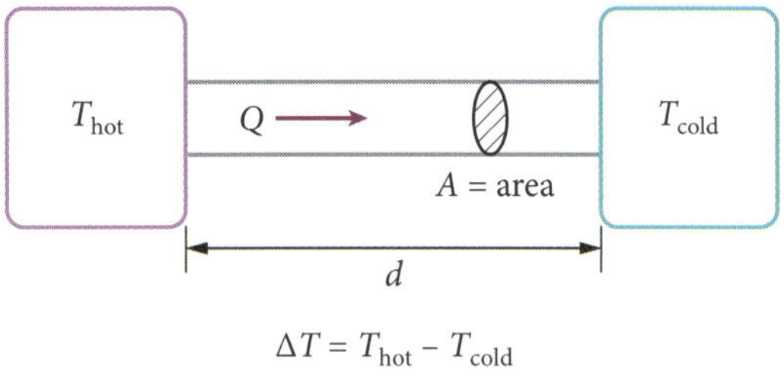

Figure 10.2 Illustration of the variables that determine the rate of heat conduction

EXAMPLE 5

A polystyrene esky contains ice water at 0 °C. Because the esky was left in the sun, the outside temperature of the esky is 40 °C. Given that the thermal conductivity of polystyrene is $k = 0.033$ Wm^{-1}K^{-1}, the sides of the esky are 3 cm thick, and the surface area of the esky is 1.2 m^2, find the rate at which thermal energy would pass into the esky.

Answer:

Applying the conductivity equation for the rate of heat transfer:

We can apply the thermal conductivity equation to determine the rate of heat transfer through a material

$$Q/t = kA\Delta T/d = \frac{0.033 \times 1.2 \times 40}{0.03} = 52.8 \text{ Js}^{-1} = 52.8 \text{ W}$$

➔ **Convection** is the transfer of heat via the mass motion of particles in a fluid (i.e. a gas or liquid). We can understand convection by recalling that when heat is added to a fluid it expands and becomes less dense and hence more buoyant. This less-dense mass rises, carrying the heat it absorbed with it, and displaces cooler parts of the fluid downwards. These convection currents transfer heat through the fluid. The process is illustrated in Figure 10.3.

convection: the transfer of heat via the mass movement of particles in a gas or liquid

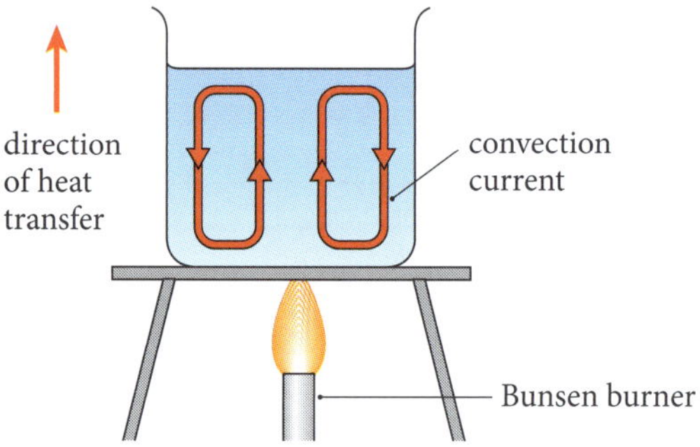

Figure 10.3 Heat transfer by convection

➔ **Radiation** transfers thermal energy via electromagnetic waves. Atoms are made up of charged particles, and accelerating charged particles emit electromagnetic waves. All bodies above absolute zero (0 K) contain moving particles and hence all bodies above absolute zero emit electromagnetic radiation. The loss of thermal energy through radiation reduces the kinetic energy of the particles in the body, causing the body to cool.

radiation: the transfer of heat via electromagnetic waves

➔ Unlike convection and conduction, which require a material medium, radiant heat can be transferred through a vacuum.

➔ Studies of radiant heat show that the power of the radiation emitted from a hot body increases with the fourth power of the temperature:

$$P \propto T^4$$

where P is the power emitted per unit area
T is the absolute temperature
α is the Greek lowercase letter *alpha*, representing 'proportional to'

Hence, if the absolute temperature of a body is doubled, the energy radiated from the body per second would be 16 times greater.

- The wavelength at which most heat is radiated decreases with temperature, as shown in Figure 10.4. Studies have shown that the wavelength of maximum emission is proportional to the inverse if the absolute temperature:

  $\lambda_{max} \propto 1/T$

  where λ_{max} is the wavelength of maximum emission (using Greek lowercase letter *lambda*)
  T is the absolute temperature

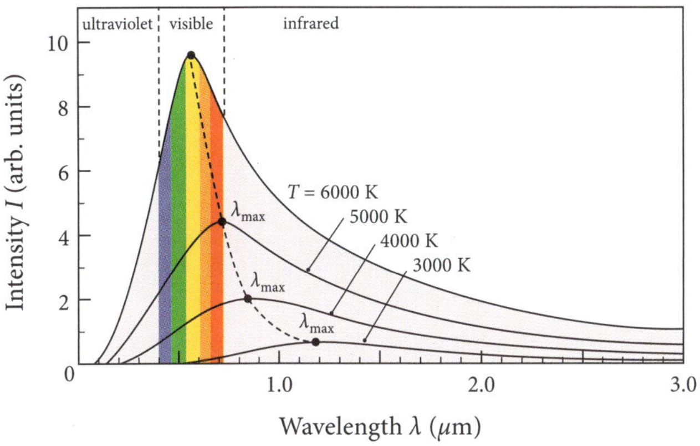

Figure 10.4 Radiation emitted at different wavelengths from a black body at different temperatures

- To simplify the study of the radiation emitted by hot bodies, physicists often look at the emission from perfectly black bodies. A **black body** is defined as a body that absorbs all the radiation (light) that falls on it and it reflects none. The radiant emission from a black body is therefore due to its temperature alone. The emission from a black body obeys the Stefan-Boltzmann law:

  $P/A = \sigma T^4$

  where A is the surface area of the black body
  $\sigma = 5.67 \times 10^{-8}$ Wm^{-2}K^{-4}, is a constant (using Greek lowercase letter *sigma*)

black body: a body that absorbs all the electromagnetic radiation that falls on it and reflects none

- For a black body the wavelength of maximum emission is given by Wien's law:

  $\lambda_{max} = b/T$

  where T is the absolute temperature
  $b = 2.898 \times 10^{-3}$ mK , is a constant

KEY QUESTIONS

6 How can thermal energy be transferred?

7 How can energy be transferred in a vacuum?

Answers ➲ p. 135

FIRSTHAND INVESTIGATION 2

Convection

Convection can be observed in the laboratory by placing some potassium permanganate crystals in a beaker of cold water and heating the edge of the beaker with a low Bunsen burner flame, as shown in Figure 10.5. The flame on the side of the beaker will produce convection currents that will be clearly visible in the beaker.

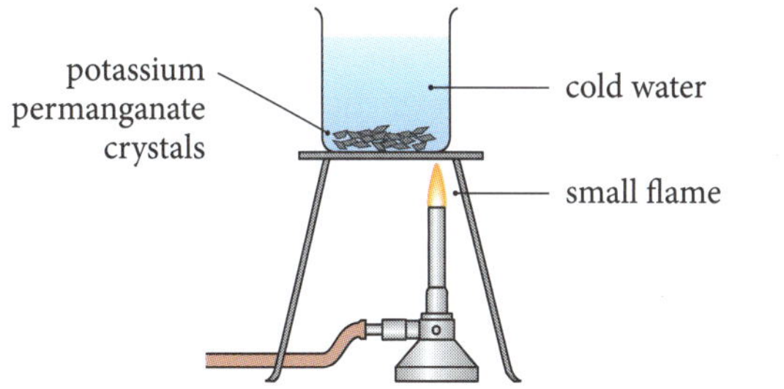

Figure 10.5 A simple experiment to demonstrate heat transfer by convection

FIRSTHAND INVESTIGATION 3

Conduction

In this investigation the rate of conduction for a range of materials is determined. A measured quantity of water is heated to a specific temperature (e.g. 60 °C) and then added to a cup. A cap is placed on the cup and the cup lowered into a large beaker containing ice and water, as shown in Figure 10.6. After 2 minutes the temperature of the water in the cup is measured. From the change in temperature of the water in the cup, the amount of heat added (Q) to the cup in 2 minutes can be calculated. Students can compare the rate of heat transfer (conduction) through polystyrene, plastic ceramic and metal cups.

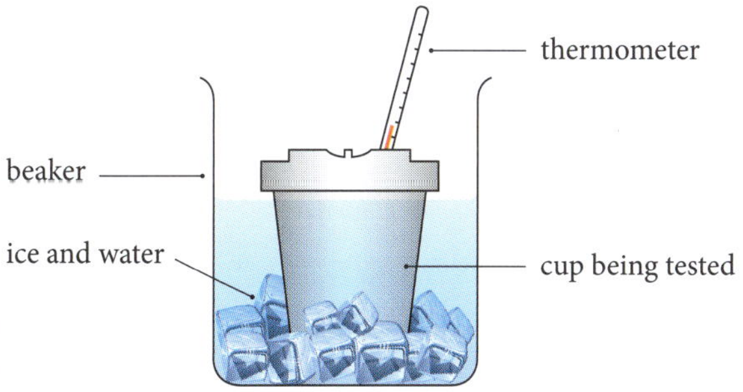

Figure 10.6 Apparatus for comparing rates of heat conduction in various materials

By measuring the area and thickness of the section of the cup in contact with the ice water, and the average temperature of the water in the cup during the test, students can apply the conduction equation to estimate the thermal conductivity (k) of the material used in the cups.

FIRSTHAND INVESTIGATION 4

Radiation

A handheld prism spectrometer can be used to investigate the electromagnetic radiation emitted from a hot body at different temperatures, as shown in Figure 10.7. The temperature of the filament in the light globe can be increased by increasing the applied voltage.

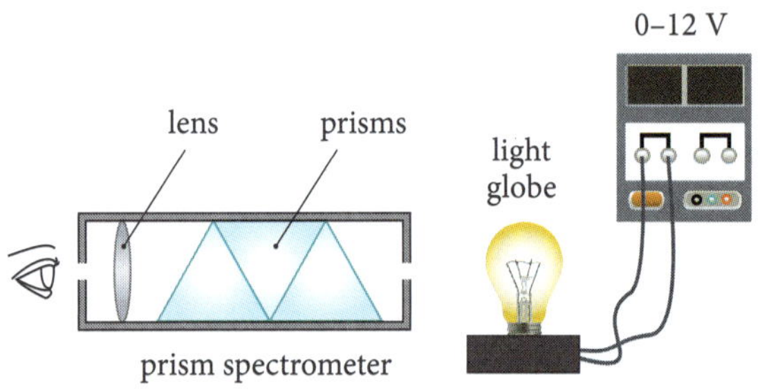

Figure 10.7 Examining the wavelengths of light emitted from a light globe filament at different temperatures

At low voltages the lightbulb becomes warm and emits infrared wavelengths. As the voltage and hence the temperature of the globe is increased, the globe begins to emit visible red light, then red/yellow and, at full power, the entire spectrum of visible light is emitted.

This experiment shows that, as the temperature of a body increases, the intensity of the radiation it emits increases, and the wavelength at which the radiation is emitted decreases.

5 Latent heat

» Students conduct an investigation to analyse qualitatively and quantitatively the latent heat involved in a change of state.

➔ **Latent heat** is the heat that is absorbed or emitted when a substance changes phase (i.e. its state of matter). Because it takes energy to break the bonds between molecules, changing a substance from a solid to a liquid or from a liquid to a gas requires thermal energy to be added to the substance. Latent heat is defined as the heat required to change the phase of a unit mass of a substance:

$$L = Q/m$$

where L is the latent heat
Q is the heat added
m is the mass

latent heat: the thermal energy that is absorbed or emitted when a substance changes phase (i.e. from one state of matter to another)

➔ The heat required to change a unit mass of a specific substance from one phase to another is therefore:

$$Q = mL$$

where L is the latent heat of the substance
Q is the heat required to change the phase of a mass m of the substance

➔ The latent heats for a range of substances are shown in Table 10.2. Note that the latent heat required to melt a substance is called the **latent heat of fusion**, and the latent heat to change from liquid to a gas is the **latent heat of vaporisation**.

latent heat of fusion: the thermal energy required to change a unit mass of a substance from a solid to a liquid
latent heat of vaporisation: the thermal energy required to change a unit mass of a substance from a liquid to a gas

Table 10.2 Latent heats of fusion and vaporisation

Substance	Melting point (°C)	Latent heat of fusion ($J\,kg^{-1}$)	Boiling point (°C)	Latent heat of vaporisation ($J\,kg^{-1}$)
Oxygen	−218.79	1.38×10^4	−182.97	2.13×10^5
Water	0.00	3.33×10^5	100.00	2.26×10^6
Lead	327.30	2.45×10^4	1750.00	8.70×10^5
Copper	1083.00	1.34×10^5	1187.00	5.06×10^6

→ When a pure substance is heated its temperature increases until it begins to change phase. As shown in Figure 10.8, the temperature remains constant during the phase change as the thermal energy going into the substance is used to change the phase rather than move the particles faster. When the phase change is complete, the temperature of the substance in the new phase begins to increase again. We call the temperatures at which a substance changes phase the *melting point* and *boiling point* of the substance.

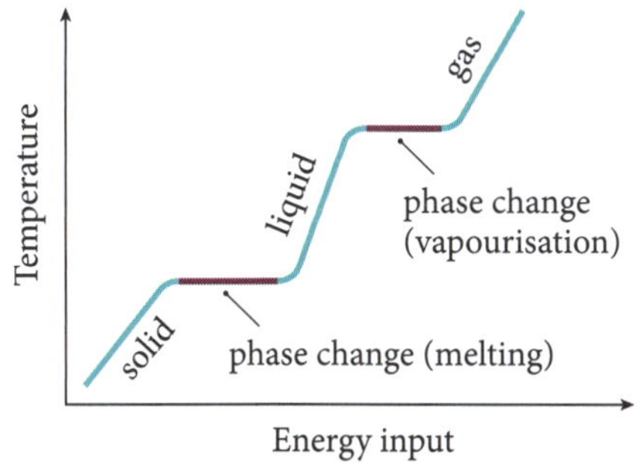

Figure 10.8 Change in temperature of a pure substance as heat is added to the substance

→ KEY QUESTIONS

8 **Why does perspiring make you cooler?**

9 **Why does the temperature of boiling water remain at 100 °C even when it continues to be heated?**

Answers ➲ p. 135

EXAMPLE 6

Use the data given in Tables 10.1 and 10.2 to find the energy required to change 200 mL of ice at –20 °C to liquid water at 25 °C.

Answer:

Remember: It takes energy to raise the temperature of an object and to change its state

We have to solve this problem in three parts. Find the heat required to raise the temperature of the ice to melting point, the heat to melt the ice, and the heat to raise the temperature of the water to 25 °C. Note also that 200 mL of water has a mass of 0.2 kg.

The heat to raise the temperature of the ice from 20 °C to 0 °C is:

$Q = mc\Delta T = 0.2 \times 2.06 \times 10^3 \times 20 = 8240$ J

The heat to melt the ice:

$Q = mL = 0.2 \times 3.33 \times 10^5 = 66\,600$ J

The heat to raise water temperature to 25 °C:

$Q = mc\Delta T = 0.2 \times 4184 \times 25 = 20\,920$ J

Hence the total thermal energy required is:

$8240 + 66\,600 + 20\,920 = 9.576 \times 10^4$ J

FIRSTHAND INVESTIGATION 5

Finding the latent heat of fusion for water

The latent heat of fusion for water can be determined by boiling water in an open-top electric kettle with a known power rating. An open-topped kettle is used to allow the steam to escape rather than recondensing on the lid of the kettle. We can calculate the latent heat by boiling the kettle for a measured time and measuring the amount of water that has been converted to steam.

For example, if a 2 kW kettle containing 1 L of boiling water was boiled for 2 minutes and 900 mL of water remained, 100 mL would have been converted to steam.

Note that 100 mL of water = 100 g and, as power = energy/time, the energy used by the kettle would go into heating the water:

$Q = Pt = 2000 \times (2 \times 60) = 2.4 \times 10^5$ J

Now as $Q = mL$, $L = Q/m = \dfrac{2.4\times10^5}{0.100} = 2.4 \times 10^6\ \mathrm{J\,kg^{-1}}$

This is a little higher than the accepted value of $2.26 \times 10^6\ \mathrm{J\,kg^{-1}}$ but this could be due to some of the heat being lost to the environment.

FIRSTHAND INVESTIGATION 6

Heating curve of a pure substance

Students can investigate heating curves, like the one shown in Figure 10.8, which can be produced by measuring the temperature of a pure substance like stearic acid ($C_{18}H_{36}O_2$) or water as they are heated.

6 Thermal conductivity

» Students model and predict quantitatively energy transfer from hot objects by the process of thermal conductivity.

→ We saw previously that the rate at which thermal energy is transferred through a material is related to the temperature gradient in the material ($\Delta T/d$). The rate of energy transfer also depends on the cross-sectional area (A) of the conductor and a constant called the **thermal conductivity**, which is a measure of how readily the substance conducts heat.

thermal conductivity: a measure of how readily a substance conducts heat

- Table 10.3 lists the thermal conductivity of a range of materials. Metals are good thermal conductors because they have mobile valence elections that can readily transfer heat (particle kinetic energy) between atoms in the metal. Solids are generally better heat conductors than liquids, and liquids are better heat conductors than gases. This is because heat is conducted by collisions between atoms/molecules and these particles are much more tightly bound together in solids than liquids and they are well separated and not bonded in gasses.

Table 10.3 Thermal conductivity of selected materials

Material	Thermal conductivity ($Wm^{-1}K^{-1}$)
Water	0.6
Air	0.024
Glass	0.9
Polystyrene	0.033
Copper	386.0
Aluminium	237.0
Concrete	1.0

→ KEY QUESTION

10 **What factors determine the rate of heat transfer by conduction through a material?**

Answers ⊃ p. 135

EXAMPLE 7

A block of ice at 0 °C is placed on top of a copper kettle containing boiling water. Given that the thermal conductivity of copper is 386 $Wm^{-1}K^{-1}$ and that the side of the kettle is 1 mm thick, calculate the heat lost per square centimetre per second ($cm^{-2}s^{-1}$) through the side of the kettle where the block of ice is placed.

Answer:

Note that 1 $cm^2 = 10^{-4}m^2$, and the temperature difference would be $\Delta T = (100 - 0) = 100$ °C.

> The thermal conductivity equation can be used to determine the rate of heat transfer through a material

Applying the conduction equation:

$$Q/t = kA\Delta T/d = \frac{386 \times 10^{-4} \times 100}{0.001}$$
$$= 3.86 \times 10^3 \text{ Js}^{-1}\text{cm}^{-2}$$
$$= 3.86 \times 10^3 \text{ Wcm}^{-2}$$

EXAMPLE 8

In cold climates, houses often use double-glazed windows, which comprise two panes with an air gap between them to reduce heat loss. Use the data in Table 10.3 to explain how double-glazing reduces heat loss through a window.

Answer:

From Table 10.3, we see that the thermal conductivity of glass is 0.9 $Wm^{-1}K^{-1}$, while the conductivity of air is 0.024 $Wm^{-1}K^{-1}$. Taking the ratio of these two thermal conductivities we see that $\frac{0.024}{0.9} = 0.027$. This means that the air will conduct only 2.7% of the heat that the glass of the same thickness would conduct. Hence an air gap between two glass windowpanes would conduct significantly less heat than a single glass windowpane alone.

> The rate of heat flow through a material is proportional to the thermal conductivity of the material

7 Heat problems

» Students apply the following relationships to solve problems and make quantitative predictions in a variety of situations:

- $Q = mc\Delta T$, where c is the specific heat capacity of a substance
- $\frac{Q}{t} = \frac{kA\Delta T}{d}$, where k is the thermal conductivity of a material.

- The specific heat equation ($Q = mc\Delta T$) enables us to make predictions and solve problems that involve temperature change and the heat emitted or absorbed by an object. It should be noted that the specific heat equation can only be applied up to the temperature where the state of matter changes. At this point we must use the latent heat equation that relates the change of phase to the heat energy emitted or absorbed.
- The thermal conduction equation ($Q/t = kA\Delta T/d$) can be used to make predictions and solve problems involving the rate at which thermal energy (heat) is conducted through a material.
- The rate of heat loss by conduction and convection is related to the temperature difference between the object and its surroundings, and the rate at which heat is radiated depends on the absolute temperature of the body. This means that a hotter body will lose heat and cool faster than a cooler body. Because the rate of heat loss decreases as the body cools, the rate at which the temperature falls slows as an object cools.

This is illustrated by the curve shown in Figure 10.9. The gradient of the curve between phase changes represents the rate at which heat is lost.

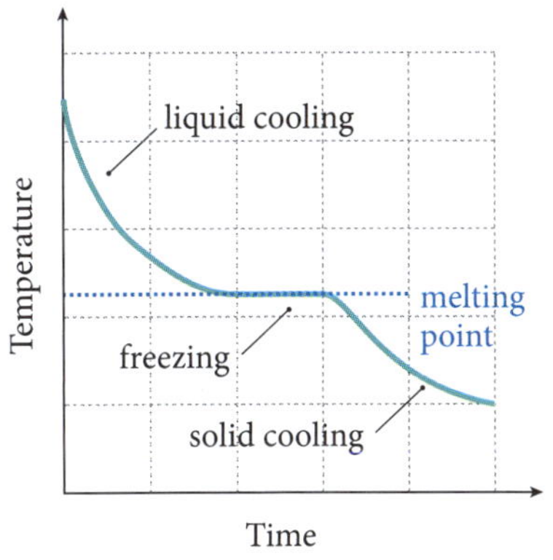

Figure 10.9 Cooling curve for a pure substance showing that the rate of temperature change decreases as the substance cools

EXAMPLE 9

In an experiment to find the specific heat capacity of gold, a 100 g gold coin is heated to 300 °C and dropped into 500 mL of water. The water had an initial temperature of 20 °C, but after the gold coin was dropped into the water, the temperature of the water was found to be 21.7 °C. Note that the specific heat capacity of water is 4184 Jkg^{-1}K^{-1}. Use this information to calculate the specific heat capacity of gold.

Answer:

> The change in temperature of a body is related to the heat gained or lost by the body

Recall that 500 mL of water has a mass of 0.5 kg. Now we can use the change in water temperature to determine the heat added to the water from the gold coin:

Heat added to water = $Q = mc\Delta T = 0.5 \times 4184 \times 1.7$
$= 3556.4$ J

This must be equal to the heat lost from the gold coin when its temperature changes from 300 °C to 21.7 °C:

For the gold coin, $Q = mc\Delta T$, hence

$$c = Q/(m\Delta T) = \frac{3556.4}{0.1 \times 278.3} = 128 \text{ Jkg}^{-1}\text{K}^{-1}$$

EXAMPLE 10

The thermal conductivity of PET plastic is 0.2 Wm^{-1}K^{-1} and of glass is 0.9 Wm^{-1}K^{-1}. A plastic PET bottle and a glass bottle both have a surface area of 400 cm^2 and contain water at 20 °C. The plastic bottle has walls of thickness 1 mm, while the glass bottle has a wall thickness of 3 mm. If both bottles were placed in a large container of ice water at 0 °C, calculate the initial rate of heat loss from each bottle and determine which bottle would cool faster.

Answer:

> The rate at which heat passes through a material is related to the thermal conductivity of the material

First note that 400 cm^2

$$= \frac{400}{100 \times 100} = 0.04 \text{ m}^2$$

Applying the conduction equation:

For the glass bottle, $Q/t = kA\Delta T/d = \dfrac{0.9 \times 0.04 \times 20}{0.003}$
$= 240$ W

For the plastic bottle, $Q/t = kA\Delta T/d = \dfrac{0.2 \times 0.04 \times 20}{0.001}$
$= 160$ W

Because heat is lost faster from the glass bottle, it will cool faster in the ice water than the plastic bottle. However, this also means that when the cool glass bottle is removed from the ice water it will warm faster than the plastic bottle.

CHAPTER SYLLABUS CHECKLIST

Are you able to answer these questions from the syllabus for this chapter? Tick each question as you go through the checklist if you are able to answer it. If you cannot answer a question, turn to the relevant page in the study guide to find the answer. For NESA key word meanings, go to www.educationstandards.nsw.edu.au and search 'key words'.

FOR A COMPLETE UNDERSTANDING OF THIS TOPIC:		PAGE NO.	✓
1	Can I explain the relationship between the temperature of an object and the average kinetic energy of the particles in the object?	123	
2	Can I explain the difference between the absolute temperature scale and a scale based on arbitrary reference points?	123	
3	Can I relate the Kelvin temperature scale to the Celsius temperature scale?	123	
4	Can I explain the direction of heat flow between objects?	124	
5	Can I explain the concept of thermal equilibrium?	124	
6	Can I explain heat transfer in term of conduction, convection and radiation?	126	
7	Can I make predictions and solve problems involving heat transfer by conduction, convection and radiation?	126–127	
8	Can I explain temperature curves in terms of specific heat capacity and latent heat?	129–131	
9	Can I make predictions and solve quantitative problems involving specific and latent heat?	129–131	
10	Can I explain why different materials have different thermal conductivities?	130	
11	Can I solve quantitative problems involving heat conduction in materials?	130–131	

YEAR 11 EXAM-TYPE QUESTIONS

Objective-response questions

(1 mark each)

1 **Object A has a higher temperature than object B. If the objects are placed in thermal contact with each other, which of the following statements best describes the particle motion and heat transfer between the bodies?**

A Heat will flow from object B to A until the average velocity of the particles in both objects is equal.

B Heat will flow from object A to B until the average velocity of the particles in both objects is equal.

C Heat will flow from object B to A until the objects are in thermal equilibrium.

D The average kinetic energy of the particles in both objects will decrease as heat flows between the objects.

2 **How would the temperature of a solid object change if it was heated at a constant rate?**

A The temperature would increase at an increasing rate until the object began to melt.

B The temperature would increase at a decreasing rate until the object began to melt.

C The temperature would increase at a constant rate until the object had entirely melted.

D The temperature would increase at a constant rate until the object started to melt.

3 **Why is less heat lost when we wear multiple layers of clothes than just one thick layer?**

A Using multiple layers with different thermal conductivities reduces conductive heat loss.

B Using multiple layers with an air gap between them increases heat loss by convection.

C Using multiple layers creates pockets of air between the layers thereby decreasing heat loss by conduction as air has a low thermal conductivity.

D Using multiple layers increases heat loss by radiation.

4 **Two objects A and B have the same mass but object A is hotter and has a higher specific heat capacity than object B. If the objects were brought into thermal contact, which statement best describes what would happen?**

A Heat would flow from object A to object B until the temperature of both objects was the average of their initial temperatures.

B Heat would flow from object A to object B until both objects reached the initial temperature of object A.

C Heat would flow from object A to object B until both objects reached the same temperature, which would be higher than the average of their initial temperatures.

D Heat would flow from object A to object B until both objects reached the same temperature, which would be lower than the average of their initial temperatures.

5 **Why do humans perspire when they are hot?**

A Because the perspiration lubricates the cells in the skin, which causes them to cool.

B Because perspiration evaporates and the latent heat of vaporisation takes heat from the body.

C Because perspiration will increase convection currents near the skin, which will cool the body.

D Because perspiration lowers the thermal conductivity of the skin, allowing more heat to escape from the body.

Extended-response questions

6 **A thermos flask is designed to keep hot liquids hot and cold liquids cold. A typical thermos flask is shown in Figure 10.10.**

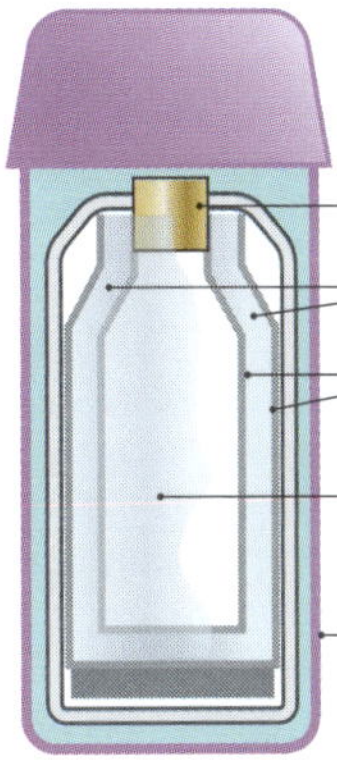

Figure 10.10 Glass vacuum thermos flask

a **Explain why the thermos flask uses a double-glass vessel to contain the liquid rather than a double-stainless-steel vessel.** (2 marks)

b **Explain why the flask uses a vacuum between the inner and outer vessels.** (2 marks)

c **Explain why the inner surface is silver rather than black.** (2 marks)

7 **A 2 kW electric kettle contains 400 mL of ice at −25 °C. Use the data in the Table 10.4 to answer the following questions.**

Table 10.4 Heat parameters for water in different phases

Specific heat capacity of ice ($Jkg^{-1}K^{-1}$)	2.06×10^3
Melting point (°C)	0.00
Specific heat of water ($Jkg^{-1}K^{-1}$)	4184
Latent heat of fusion (Jkg^{-1})	3.33×10^5
Boiling point (°C)	100.00
Latent heat of vaporisation (Jkg^{-1})	2.26×10^6

a. How long will it take the kettle to melt all the ice? (2 marks)

b. How long will it take after the ice is melted to boil the kettle? (1 mark)

c. How long would the kettle take to boil dry from the time it started to boil? (1 mark)

d. What assumption have you made in your calculations? (1 mark)

8 A stainless-steel saucepan with a wall thickness of 1.2 mm contains water at 20 °C and is placed on a hotplate with a temperature of 700 °C. Note that stainless steel has a thermal conductivity of 16 $Wm^{-1}K^{-1}$.

a. Find the rate at which heat will initially pass through each cm^2 of the saucepan base. (1 mark)

b. Explain in terms of molecular and bulk movement how this heat will spread through the water. (2 marks)

c. Explain why the cooktop element appears black when it is at 200 °C but glows bright red when its temperature is increased to 700 °C. (2 marks)

9 Disposable coffee cups are often made from polystyrene and use a plastic lid with a small drinking hole.

a. Using your knowledge of heat transfer, explain why coffee cups are made this way. (2 marks)

b. Suggest one structural change that could be made to the cup to enable it to keep coffee warm even longer. Explain how this change would reduce heat loss. (2 marks)

10 A 0.5 kg block of copper at 400 °C is dropped into a 10 L bucket of water initially at 20 °C. The specific heat capacity of water 4184 $Jkg^{-1}K^{-1}$ and the specific heat capacity of copper is 386 $Jkg^{-1}K^{-1}$. You may assume no heat is lost to the environment.

a. Find the temperature at which the copper block and water will reach thermal equilibrium. (2 marks)

b. Explain how the time to reach thermal equilibrium and the equilibrium temperature would change if the copper block was replaced in the experiment with a metal with a lower heat capacity and lower thermal conductivity than copper. (2 marks)

ANSWERS

KEY QUESTIONS

Key questions ➲ p. 124

1 The divisions in the Kelvin scale are the same as the Celsius scale but they start at a different point. Zero on the Celsius scale is the freezing point of water, while zero on the Kelvin scale is the point where molecules would have zero kinetic energy. Hence 0 K = –273.15 °C.

2 Absolute temperature (measured in Kelvin) is proportional to the average kinetic energy of the molecules in the sample being measured. For an ideal gas the average kinetic energy of the molecules = $\langle mv^2/2 \rangle = 3kT/2$.

Key questions ➲ p. 125

3 Heat always flows from the hotter body to the cooler body. This is because heat is conducted from one body to another by the faster molecules in the hotter body transferring energy to the slower molecules in the cooler body.

4 Heat will stop flowing between thermally connected bodies when the bodies reach the same temperature; that is, when the bodies are in thermal equilibrium with each other.

5 Provided the phase of the body is not changing, the temperature change depends on the amount of heat added, the mass of the body, and its specific heat capacity. Specific heat capacity is a measure of the energy required to raise the temperature of a unit mass of a substance by one degree Kelvin.

Key questions ➲ p. 127

6 Thermal energy can be transferred by conduction, convection or radiation.

7 The conduction of heat requires a material medium and the convection of heat requires a fluid material medium, hence only radiation can transfer heat in a vacuum.

Key questions ➲ p. 129

8 Perspiring puts water on the surface of your skin that evaporates, taking heat from the body (i.e. latent heat of vaporisation).

9 At 100 °C water undergoes a phase change from liquid water to steam, and heating the water when it is at 100 °C will simply put energy into breaking the bonds between molecules as the liquid changes to a molecular gas (steam). Heating will only increase the temperature when all the liquid has been changed into a gas.

Key questions ➲ p. 130

10 The rate at which thermal energy moves through a material is proportional to the temperature gradient ($\Delta T/d$), the cross-sectional area (A) and the thermal conductivity of the material (k).

YEAR 11 EXAM-TYPE QUESTIONS

Objective-response questions

1 B. Heat always flows from the hotter body to the cooler body and bodies in thermal contact will come to thermal equilibrium when their temperatures and the average kinetic energy of their molecules are equal. **A** and **C** are incorrect because heat cannot flow from the cooler body to the hotter body. **D** is incorrect because heat transfer must increase the kinetic energy of the molecules in the body absorbing the heat. In addition, energy would not be conserved in an isolated system if the internal energy of both objects decreased.

2 D. Because $Q = mc\Delta T$, $\Delta T = Q/mc$ and hence if heat is put in at a constant rate (Q = a constant) the temperature will increase at a constant rate. However, when the object starts to melt the temperature will stop increasing as the heat goes into changing the phase. **A** and **B** are incorrect because the increase in temperature must be constant. **C** is incorrect because the temperature will stop increasing as soon as the solid begins to melt.

3 C. Air is a poor conductor of heat and hence multiple air layers will reduce heat loss through conduction. **B** and **D** are incorrect because to keep warm heat loss must be reduced not increased. **A** is incorrect because the rate of heat loss through conduction is not related to the number of different materials but to the thermal conductivity of the materials.

4 C. The temperature will change until both bodies have the same temperature but, because body A has a higher specific heat, its temperature will change less than body B because body B requires less energy to change its temperature by 1 °C than body A. **A** is incorrect because the final temperature would only become the average if the initial temperatures of both bodies had the same specific heat capacity. **B** is incorrect because if heat flows between the bodies, both of their temperatures must change. **D** is incorrect because the temperature change of body A would have to be greater than the temperature change of body B. Body A loses the same amount of energy that body B gains, but body B must change its temperature by a greater amount as it has a lower specific heat capacity than body A.

5 B. When perspiration evaporates the latent heat of vaporisation comes from the body, reducing the temperature of the body. **A** is incorrect because lubrication would not remove heat. **C** is incorrect because convection currents are related to the temperature of the surface, and having a liquid on the surface will not increase the surface temperature. **D** is incorrect because a layer of water will not change the thermal conductivity of the skin; it will just put another layer on top that will reduce conductivity.

Extended-response questions

6 EM This question tests students' understanding of the ways in which heat can be transferred and the necessary requirements for heat transfer via conduction, convection and radiation.

a Double-glass vessels are used to reduce heat loss by conduction. ✓ Glass has a much lower thermal conductivity than stainless steel and hence the rate at which heat is conducted through the glass layers is much slower than the rate of heat transfer would be if stainless steel was used. ✓

b A vacuum jacket is used to reduce heat loss by conduction and convection. ✓ Heat transfer via conduction and convection require a material medium and hence the vacuum jacket prevents this type of heat transfer from or to the inner vessel. ✓

c A silver inner surface reduces heat loss by radiation. ✓ A black surface would absorb radiation, while a silver surface would reflect radiation back to the inner vessel. ✓

7 EM This question tests students' understanding of specific heat, latent heat, phase change and power. The question also requires students to understand the limitations of their calculations.

a Note that power = energy/time = $\Delta Q/t$ and hence $t = \Delta Q$/power. Note also that 400 mL of water has a mass of 0.4 kg.

The energy required to melt the ice will be the sum of the heat required to raise the temperature of the ice to the melting point (0 °C) plus the heat to melt the ice:

$$\text{Heat} = Q = mc\Delta T + mL$$
$$= 0.4 \times 2.06 \times 10^3 \times 25 + 0.4 \times 3.33 \times 10^5$$
$$= 1.538 \times 10^5 \text{ J} \checkmark$$

$$\text{Time} = \text{heat/power} = \frac{1.538\times10^5}{2000} = 76.9 \text{ s} \checkmark$$

b The energy to raise the temperature of water in liquid phase from 0 °C to 100 °C is:

$$Q = mc\Delta T = 0.4 \times 4184 \times 100 = 1.674 \times 10^5 \text{ J}$$

$$\text{Time} = Q/t = \frac{1.6774\times10^5}{2000} = 83.7 \text{ s} \checkmark$$

c The heat required to change 0.4 kg of water to steam is:

$$Q = mL = 0.4 \times 2.26 \times 10^6 = 9.04 \times 10^5 \text{ J}$$

$$\text{Time} = Q/t = \frac{9.04\times10^5}{2000} = 452 \text{ s} \checkmark$$

d All the heat produced by the kettle goes into the water and none is lost to the environment or heating the kettle. ✓

8 **EM** This question tests students' ability to apply the thermal conductivity equation, the underlying processes that drive heat transfer by convection and conduction, and how the temperature of a body affects the wavelength of the radiation emitted from the body.

a Applying the heat conduction equation and recalling that

$$1 \text{ cm}^2 = \frac{1}{100 \times 100} = 10^{-4}\text{m}^2\text{:}$$

$$Q/t = kA\Delta T/d = \frac{16\times10^{-4}\times(700-20)}{0.0012} = 906.7 \text{ Wcm}^{-2} \checkmark$$

b The fast-moving (hot) molecules in the metal saucepan transfer kinetic energy to the adjacent water molecules. These molecules then transfer kinetic energy by collisions to other water molecules, transferring heat via conduction through the water. ✓ Because the warm water at the bottom of the saucepan expands and rises, it creates convection currents that transfer heat through the liquid via bulk movement and circulation of the liquid. ✓

c The wavelength of maximum emission of radiation decreases when the temperature of the emitting body is increased. ✓ At 200 °C the element is emitting radiation in the infrared region, which has a wavelength that is too long to be visible. But when the element has reached a temperature of 700 °C, some of emitted radiation has a wavelength short enough to reach the red end (the long-wavelength end) of the visible spectrum. ✓

9 **EM** This question tests students' ability to apply their knowledge of heat transfer to a real-life application.

a Polystyrene is used because it has a low thermal conductivity, which reduces heat loss via conduction. ✓ The cap is made of plastic because it also has a low thermal conductivity to reduce heat loss by conduction. The warm air gap between the cap and the coffee also reduces heat loss by conduction as air has a low thermal conductivity. More importantly, the cap prevents the hot coffee coming into contact with cold air, which stops heat loss through convection. ✓

b There are many possible answers. For example, the cup could be made from thicker polystyrene, ✓ which would reduce the rate of conductive heat loss. ✓ Or a cap could be used with a flap that closed the drinking hole when it was not in use, preventing heat loss via convection and conduction through the hole.

10 **EM** This question tests students' understanding of specific heat and thermal equilibrium. The question also tests students' knowledge of the rate of heat conduction.

a Heat will move from the copper block to the water until the water and the copper block reach the same temperature (i.e. come into thermal equilibrium). As the heat lost from the copper block is equal to the heat gained by the water we can write:

$$\Delta Q_{copper} = \Delta Q_{water} \checkmark$$

If we call the final temperature T, we can write:

$$(mc\Delta T)_{copper} = (mc\Delta T)_{water}$$

Thus $(0.5 \times 386 \times (400 - T)) = 10 \times 4184(T - 20)$

$(10 \times 4186 + 0.5 \times 386)T = (0.5 \times 400 \times 386 + 200 \times 4184)$

Hence thermal equilibrium would occur when $T = 21.7$ °C. ✓

b If the thermal conductivity of the metal was lower, it would take longer for the thermal energy to move from the metal to the water and hence it would take a little longer to reach thermal equilibrium. ✓ If the specific heat capacity of the metal was lower, it would transfer less energy to the water than copper for the same temperature change and hence the water would not increase its temperature as much as it did when copper was used. Therefore thermal equilibrium would be reached at a lower temperature. ✓

CHAPTER 11 ELECTROSTATICS

MODULE 4
ELECTRICITY AND MAGNETISM

INQUIRY QUESTION:

How do charged objects interact with other charged objects and with neutral objects?

Charged objects exert a non-contact electrostatic force on one another. This force causes like charges to repel one another, and unlike charges to attract one another. Charged objects can induce charge separation in nearby neutral objects. This charge separation produces an electrostatic force of attraction between charged objects and neutral objects.

1 Static charge and electrostatic force

» Students conduct investigations to describe and analyse qualitatively and quantitatively:
- processes by which objects become electrically charged
- the forces produced by other objects as a result of their interactions with charged objects
- variables that affect electrostatic forces between those objects.

- This chapter investigates electrostatics, which is the study of stationary charges.
- You will recall from previous studies that atoms are made up of charged particles and that a charge can be positive or negative. Charged particles exert a non-contact force on other charged particles. Like charges repel one another, and unlike charges attract one another.
- Atoms are generally neutral because the positive charge of the nucleus is balanced by the negative charge on the electrons that surround the nucleus. However, it is possible to add or remove electrons from atoms to produce a charged atom called an ion.
- The SI unit of charge is the **coulomb** (C). The elementary charge in electrostatic interactions is the charge carried by a single *proton* ($+1.602 \times 10^{-19}$ C), or the magnitude of the charge carried by a single *electron* (-1.602×10^{-19} C).
- Charge is conserved in any closed system. This means that net charge, like energy and momentum, always remains the same in a closed system.
- Some atoms hold their electrons more strongly than other atoms and, when two different materials with different electron affinities are rubbed together, electrons are transferred from one material to the other, leaving both materials with a net charge. For example, ebonite (very hard rubber) holds its electrons strongly, while wool holds its electrons less strongly. Rubbing an ebonite rod with wool will therefore transfer electrons from the wool to the ebonite, leaving the wool with a net positive charge and the ebonite with a net negative charge. This process is called static or **triboelectric charging**. The triboelectric series shown in Table 11.1 allows us to determine which materials are likely to gain electrons or lose electrons when they are rubbed together.

coulomb: the SI unit of charge
triboelectric charging: occurs when charges move from one material to another when they are rubbed together

Table 11.1 Triboelectric series

Holds electrons less tightly ↑	Air	↑ More likely to be left positively charged
	Human skin	
	Glass	
	Human Hair	
	Nylon	
	Wool	
	Fur	
	Silk	
Holds electrons more tightly ↓	Hard rubber	↓ More likely to be left negatively charged
	Polyester	
	Styrene (Styrofoam)	
	Polyethylene (sticky tape)	
	Polypropylene vinyl (PVC)	
	Ebonite	

➔ Charged objects can cause the charges in neutral objects to move in such a way that the neutral object and charged object will attract one another. The charge movement causes the opposite charges to be closer to the charged object than the like charges, which produces a greater force on the unlike charges than the like charges and hence a net force of attraction. In conductors this charge separation is called **induction**, while in insulators the effect is called **polarisation of charge**. Both types of charge separation are illustrated in Figure 11.1.

induction: the charge movement in a conductor that results from the presence of a nearby charged object
polarisation of charge: the slight separation of opposite charges in the molecules of an insulator produced by a nearby charged object

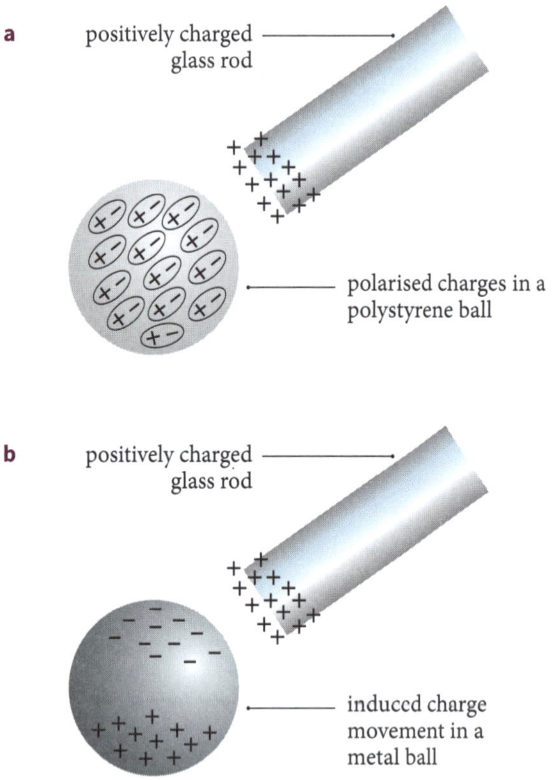

Figure 11.1 Polarisation of charge in an insulator **a** and induced charge movement in a conductor **b**. In both cases the uncharged object is attracted to the charged object

➔ KEY QUESTIONS

1 **How can an electrostatic charge be imparted to an object?**

2 **Why are some uncharged objects attracted to charged objects?**

Answers ➲ p. 151

FIRSTHAND INVESTIGATION

Static electricity

Triboelectric charging

We can investigate triboelectric charging by rubbing different materials together. For example, by rubbing a glass rod with a silk cloth the rod can be charged positively and the cloth negatively, as shown in Figure 11.2.

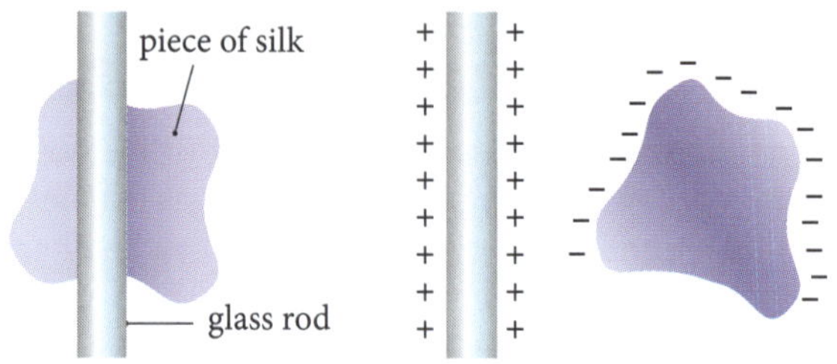

Figure 11.2 Transferring charge by rubbing a glass rod with a silk cloth

Simple experiments like the one shown in Figure 11.3 can be used to show how objects acquire the charges predicted by the triboelectric series (Table 11.1).

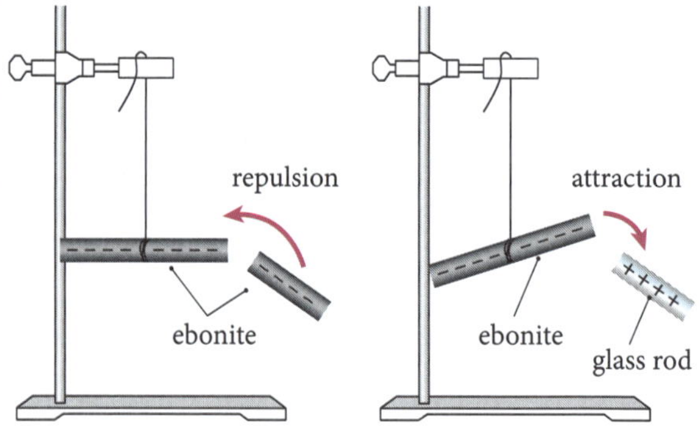

Figure 11.3 A simple experiment to investigate repulsion by like charges and attraction by unlike charges

Static charge

Static charge can be measured using an **electroscope** like the one shown in Figure 11.4. If a charged object is placed in contact with the top disk on the electroscope, the charge will spread down the electroscope (because like charges repel) and the vane and post will repel one another as they will both have the same charge.

electroscope: a device for measuring static charge

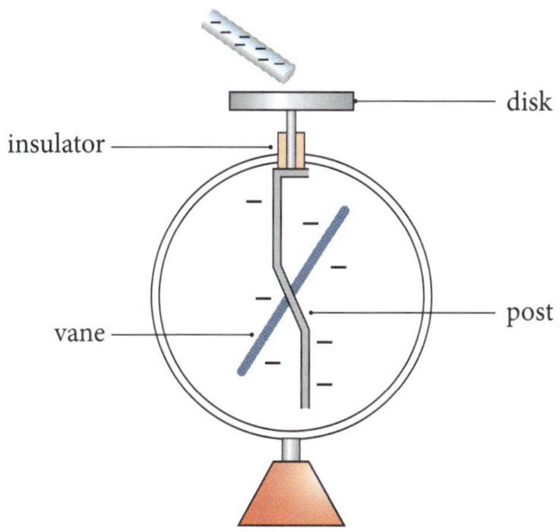

Figure 11.4 Electroscope measuring the charge on a rod

The vane on the electroscope will also move apart when a charged object is brought near the electroscope due to induced charge movement, as shown in Figure 11.5. This induced charge movement causes separation of some of the charges in the electroscope. When the charged rod is removed the vane returns to its rest position adjacent to the post as the separated charges on the electroscope come back together. Because no charge is transferred to or from the electroscope, the electroscope remains neutral (i.e. has no net charge).

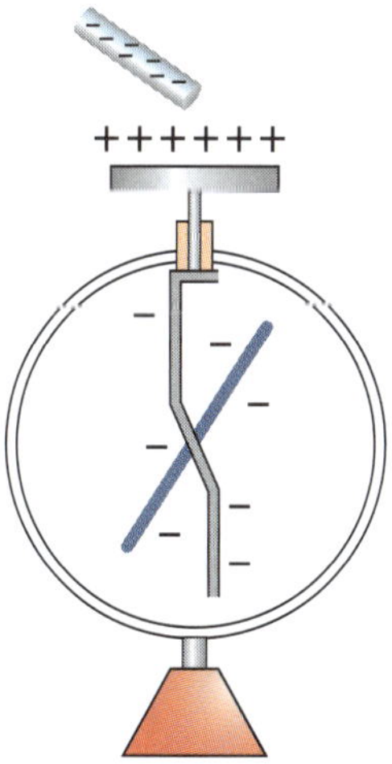

Figure 11.5 A nearby charged object inducing the opposite charge to move to the vane of an electroscope

Induction

As shown in Figure 11.6(a), an electroscope charged by contact with a charged rod is left with the same charge as the rod. In contrast, charging by induction (shown in Figure 11.6(b)), which involves momentarily earthing the electroscope when a charged object is nearby, leaves the electroscope with the opposite charge to the rod.

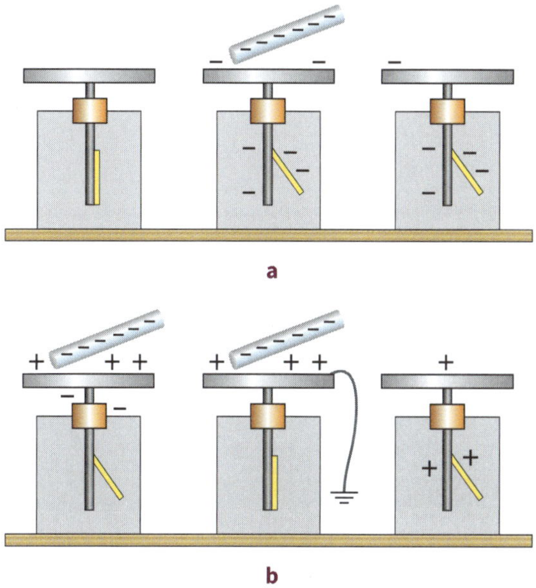

Figure 11.6 Charging by contact **a** and charging by induction **b**

Polarisation of charge

The attraction of an uncharged insulator can be illustrated by using a charged ebonite rod or glass rod to attract small pieces of paper. The effect can also be demonstrated by holding a charged rod near a stream of water (water molecules are naturally polarised), as shown in Figure 11.7. The electrostatic force on the water molecules causes them to align with the charged rod, resulting in the negative side of the water molecule being closer to the positively charged rod than the positive side. Thus the force of attraction on each molecule is greater than the force of repulsion, and it is this difference that causes a net force of attraction between the water and the charged rod.

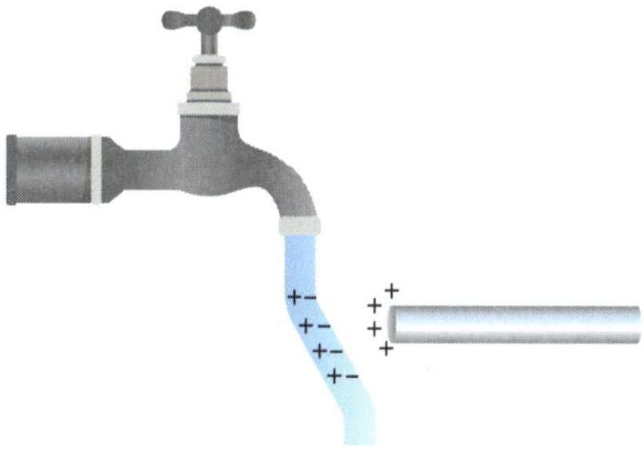

Figure 11.7 Charged rod attracting a stream of water due to alignment of polarised molecules

Van der Graaf generator

Many electrostatic phenomena that can be investigated with charged rods can also be demonstrated with a Van de Graaff generator. A Van der Graaff generator is a device that uses an electric motor to drive a rubber belt to transfer static charge to a large metal sphere, as shown in Figure 11.8.

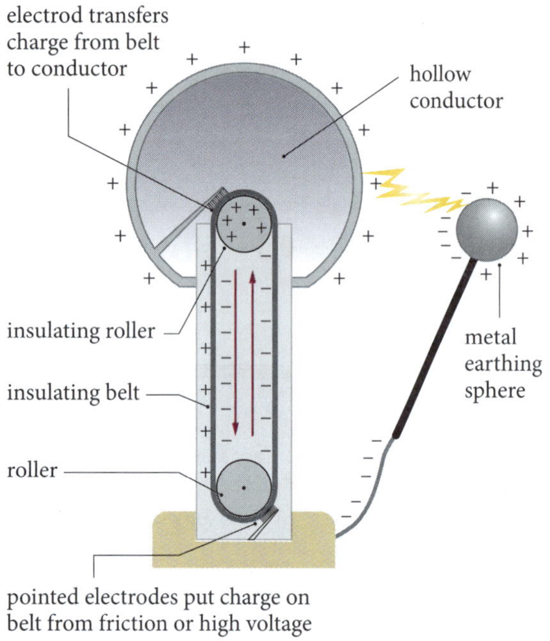

Figure 11.8 Schematic diagram of a Van der Graaff generator

The metal sphere on the top of the Van der Graaff generator can become highly charged and will produce sparks several centimetres' long if an earthed object is brought towards it. A force of attraction due to charge-separation in the uncharged, metal earthing ball shown in Figure 11.8 can be felt between the generator and the earthing ball when they are moved towards each other. Polarisation of charge in nearby small pieces of paper will cause the paper to be attracted to the charged generator.

EXAMPLE 1

Explain what would happen to the vane of an electroscope in the following situations:

a **A positively charged object is brought into contact and then removed from an uncharged electroscope.**

b **A positively charged object is moved towards, but does not touch, a negatively charged electroscope.**

c **A positively charged object is moved towards, but does not touch, a positively charged electroscope.**

d **A negatively charged object is brought into contact with a positively charged electroscope.**

Answer:

A nearby charged object will induce charge movement in a conductor

a Some negative charge would leave the electroscope, causing the vane to move away from the post and remain there when the charged object was removed.

b Charge movement would be induced in the electroscope. The nearby positively charged body would cause negative charges to move to the top of the electroscope. This would neutralise the negatively charged vane and post, causing the vane to collapse. When the charged object was removed, the net negative charge on the electroscope would again cause the vane to move away from the post. Hence as the charged object was brought nearer, the vane would collapse, but when the charged object was moved away, the vane would again move away from the post.

c In this case, induced charge movement would increase the positive charge on the vane and cause it to move further from the post as the positively charged object was brought nearer. The vane would return to its original position when the positively charged object was removed.

d When the negatively charged rod touched the electroscope, negative charges would flow onto the electroscope, neutralising the positive charge on the electroscope. The vane would therefore move back to the post.

EXAMPLE 2

Three small conducting spheres are suspended from insulating threads, as shown in Figure 11.9.

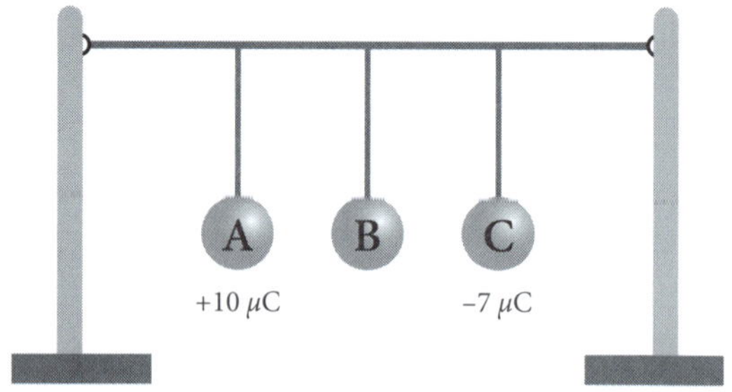

Figure 11.9 Three identical conducting spheres

Sphere A has a charge of +10 μC, sphere B has no net charge and sphere C has a charge of -7 μC.

a **Would there be a force on sphere B? Justify your answer.**

b **If sphere A is momentarily brought into contact sphere B, and then sphere C is momentarily brought into contact with sphere B, what will the final charge on each sphere be?**

Answer:

a Induced-charge movement on sphere B will move positive charges to the left and negative charges to the right of the sphere. Provided the spheres are separated by some distance, this will result in a force of attraction between sphere A and B and between sphere B and C. But because sphere A has the larger charge, the net force on sphere B will be to the left.

Charge will spread evenly between identical conductors when they come in contact with each other

b When spheres A and B are brought together the charge will spread evenly over both, leaving both spheres with a charge of $\frac{10}{2} = 5\ \mu$C. Now if sphere B was brought into contact with sphere C, the charge in both would combine and then spread evenly, leaving each sphere with $\frac{+5-7}{2} = -1\ \mu$C on each sphere.

→ KEY QUESTIONS

3 **How can a charged rod be used to give an electroscope the same type of charge as the rod?**

4 **How can a charged rod be used to give an electroscope the opposite charge as the rod?**

Answers ⊃ p. 151

2 The electric field model

» Students, using the electric field lines representation, model qualitatively the direction and strength of electric fields produced by:
- simple point charges
- pairs of charges
- dipoles
- parallel charged plates.

→ Field models provide a powerful tool for describing and analysing forces that act at a distance. We say an **electric field** exists at a point if a charged particle experiences a force at the point. The electric field at any point in space is defined as the force that a unit positive charge (i.e. +1 coulomb) would experience if it was placed at the point. The electric field strength is therefore given by:

$$E = \frac{F}{q}$$

where F is the force that a charge q would experience in an electric field of strength E

The SI units for electric field strength are therefore NC^{-1}.

electric field: a region in space where a charge would experience a force; the magnitude and strength of the field at a point is equal to the force that a unit positive charge would experience at the point

→ Electric fields are vector fields because the field has a specific magnitude and direction at each point in space. We can picture the electric field by drawing electric **lines of force** that represent the direction a positive test charge would experience a force if it was placed at each point in the region of interest. Electric lines of force are also called **electric field lines**. For example, the electric field lines around a positive point charge and around a negative point charge are shown in Figure 11.10.

lines of force or electric field lines: directed lines that represent the force that a unit positive test charge would experience at the point and hence they represent the electric field

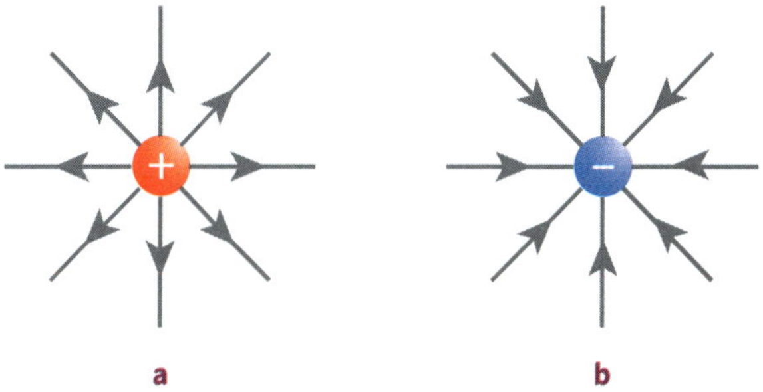

Figure 11.10 Electric field lines around a positive point charge **a** and a negative point charge **b**

→ If two charges are close to each other, the electric field lines must be determined by considering the vector sum of the forces produced by the charges. We call a pair of nearby opposite charges a **dipole**. The electric field lines around a dipole are illustrated in Figure 11.11.

dipole: a pair of equal and opposite charges separated by a small distance

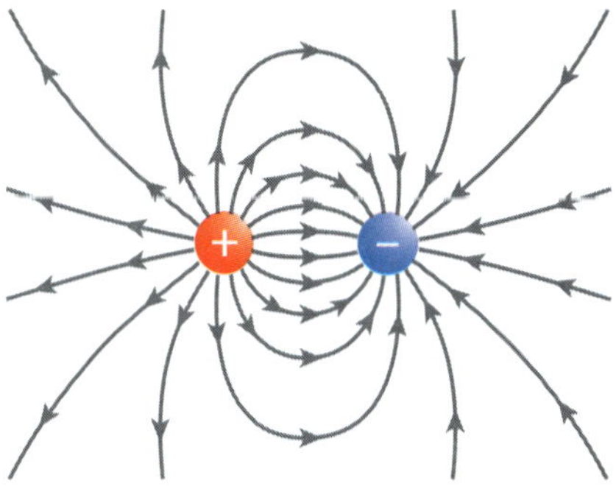

Figure 11.11 Electric field lines surrounding a dipole

- The *direction* of the electric field lines shows the direction of the electric field, and the *density* of the lines (i.e. how close they are together) reflects the strength of the electric field. In Figure 11.11 we see the electric field lines are close together and hence the field is stronger closer to the charge.
- Parallel charged plates produce a uniform electric field, as shown in Figure 11.12. Note the field lines spread a little near the edges, but the lines are parallel and hence the field is uniform across most of the region between the plates.

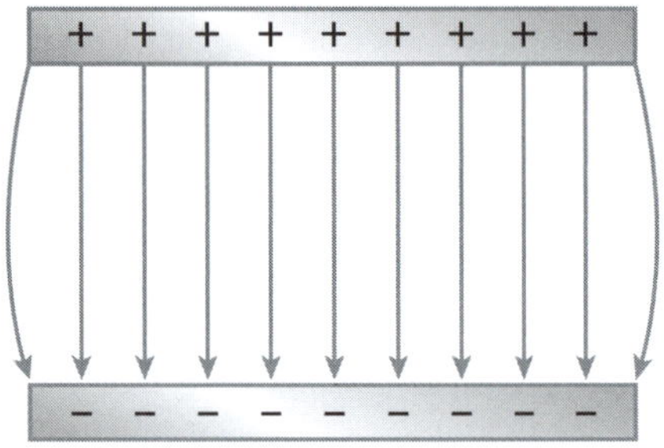

Figure 11.12 Electric field lines between a pair of charged parallel pates

- As the electric field is defined as the net force on a unit positive charge at the point, the field can only have one value and one direction at each point. This means that when we draw electric field lines they can never cross one another as this would imply the electric field was operating in two directions at the point of intersection.

EXAMPLE 3

Sketch the electric field lines around two adjacent positive point charges.

Answer:

> Think about the force that would be on a small positive test charge at each point

By thinking about the net force that would result at each point due to the two nearby positive charges, we would obtain the electric field lines shown in Figure 11.13.

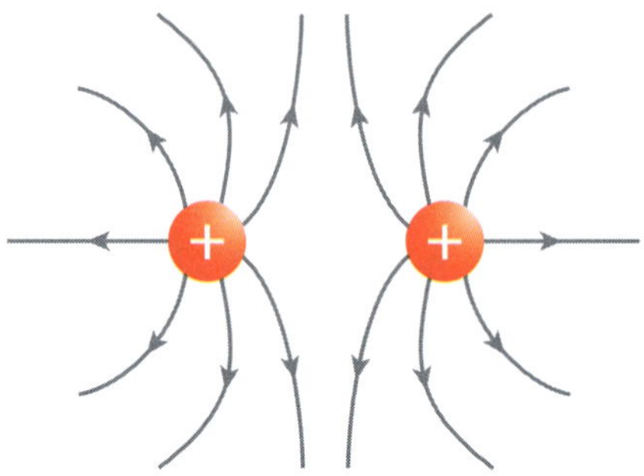

Figure 11.13 Electric field lines around two adjacent positive point charges

EXAMPLE 4

An electron of mass 9.1×10^{-31} kg and charge -1.6×10^{-19} C is placed in a uniform electric field of strength 25 NC^{-1} west. Find:

a the magnitude and direction of the force on the charge

b the acceleration of the charge.

Answer:

> Recall that the electric field strength at a point is the force on a charge of +1 coulomb at the point

a From the definition of the electric field, $E = F/q$ and therefore:

$$F = qE = -1.6 \times 10^{-19} \times 25$$
$$= -4 \times 10^{-18} \text{ N or } 4 \times 10^{-18} \text{ N east}$$

As the charge is negative it experiences a force in the opposite direction to the electric field because, by definition, the electric field direction is the direction that a positive charge would experience a force.

b $F = ma \therefore a = F/m = \dfrac{4 \times 10^{-18}}{9.1 \times 10^{-31}} = 4.4 \times 10^{12}\ ms^{-2}$ east

→ KEY QUESTIONS

5 How can we visualise the electric fields around charge objects?

6 What do electric field lines tell us about the electric field?

Answers ➲ p. 151

3 Electric fields and forces

» Students apply the electric field model to account for and quantitatively analyse interactions between charged objects using:

- $\vec{F} = q\vec{E}$
- $E = \dfrac{V}{d}$
- $F = \dfrac{1}{4\pi\varepsilon_0}\dfrac{q_1 q_2}{r^2}$.

- The force between two charges (F) is given by **Coulomb's law** (illustrated in Figure 11.14):

$$F = \frac{1}{4\pi\varepsilon_0}\frac{q_1 q_2}{r^2}$$

where $1/4\pi\varepsilon_0 = 9 \times 10^9\ NC^{-2}m^2$ is a constant

q_1 and q_2 is charge in coulombs

r is the separation between the charges

(The lowercase Greek letter *epsilon* ε is a constant called the *electrical permittivity*. For forces between charges in a vacuum we use the constant ε_0, which is called the *permittivity of free space*. This constant (ε) is related to the ability of the space between the charges to support electric field lines.)

Coulomb's law: the electrostatic force between two charged objects is proportional to the product of their charges and inversely proportional to the distance between the charges

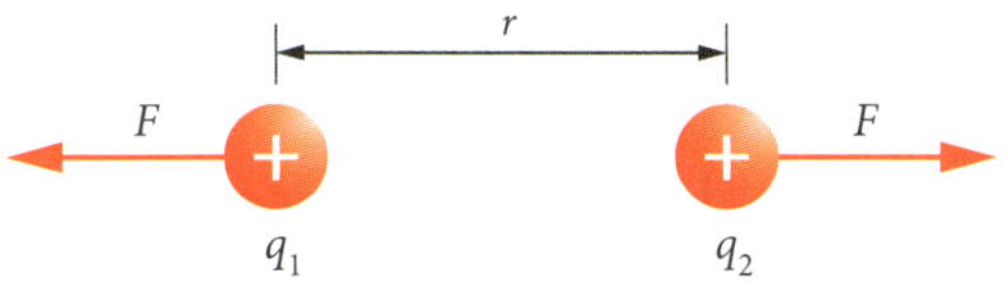

Figure 11.14 Electrostatic force between two positive charges

- ➔ If the charges are not in a vacuum or air, we have to replace the constant ε_0 with a different constant (ε) that is related to the electrical properties of the material between the charges. In accordance with Newton's third law, the force on each of the charges is always equal and opposite.
- ➔ If more than two charges are involved, Coulomb's law can be used to find the magnitude and direction of the force on a specific charge due to each of the other charges, and these individual forces can be added to find the net force on the charge.

➔ KEY QUESTION

7 **What factors determine the electrostatic force between charged particles?**

Answers ➲ p. 151

EXAMPLE 5

Figure 11.15 shows three charges fixed in the positions shown. Find the magnitude and direction of the force on the charge q_1 in Figure 11.15.

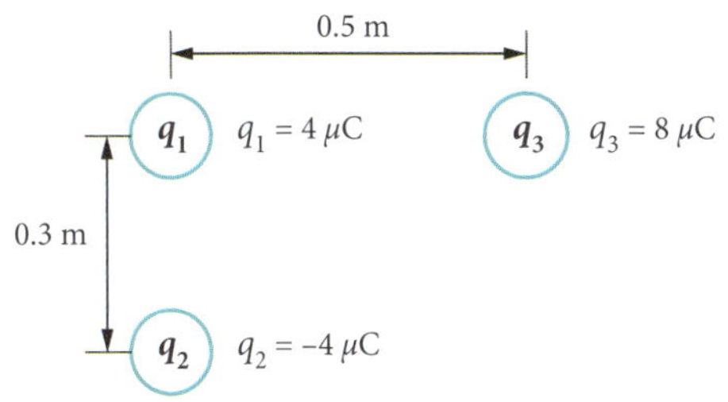

Figure 11.15 Three charges placed at 90°

Answer:

The net force is the vector sum of the individual forces acting on an object

Using Coulomb's law to find the force of attraction between q_1 and q_2:

$$F_{12} = (1/4\pi\varepsilon_0)(q_1q_2/r^2) = \frac{9\times10^9\times4\times10^{-6}\times4\times10^{-6}}{0.3^2}$$
$$= 1.6\ \text{N}$$

Hence there is a downwards force of 1.6 N on q_1 due to q_2.

Similarly, to find the force of repulsion between q_1 and q_3:

$$F_{13} = (1/4\pi\varepsilon_0)(q_1q_2/r^2) = \frac{9\times10^9\times4\times10^{-6}\times8\times10^{-6}}{0.5^2}$$
$$= 1.15\ \text{N}$$

Hence there is a force to the left of 1.15 N on q_1 due to q_3.

Adding the two forces as vectors (as shown in Figure 11.16) enables us to calculate the net force on the charge.

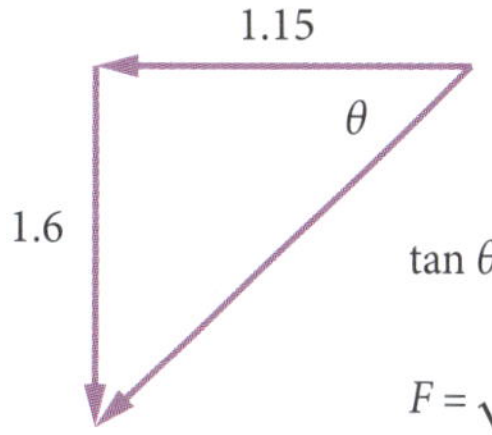

$$\tan\theta = \frac{1.6}{1.153} \text{ and hence } \theta = 54°$$

$$F = \sqrt{1.6^2 + 1.15^2} = 1.97\ \text{N}$$

Figure 11.16 Vector diagram required to find net force on the charge

The net force on q_1 is therefore $F = 1.97$ N at N216°.

EXAMPLE 6

Use Coulomb's law to write an expression for the electric field strength as a function of distance (r) from a point charge (q).

Answer:

Remember the electric field strength is defined as the force on a +1 coulomb test charge placed at the point

The electric field at any point in space is defined as the force on a unit positive test charge placed at the point. We can use Coulomb's law to find the force on a unit positive charge as follows:

$F = (1/4\pi\varepsilon_0)(q_1q_2/r^2)$ and, by setting $q_1 = q$ and $q_2 = 1$ C, we obtain:

$$E = (1/4\pi\varepsilon_0)(q/r^2)$$

This expression enables us to calculate the electric field strength (E) at a distance r from a point charge.

- In the previous section we saw that charges are surrounded by electric fields and we defined the electric field strength and direction as the force on a unit positive charge placed at the point, which we wrote as:

  $E = F/q$

  We can rearrange this equation as:

  $F = qE$

  This gives us an alternate expression for the force on a charge. If the magnitude and direction of the electric field are known, this expression can be used to find the force on a change instead of Coulomb's law.
- The electric field strength can also be defined in term of energy. We can express the electric field strength as a force per unit charge or as a measure of how much the potential energy per unit charge (V) changes per unit length in the field. That is:

$$E = \frac{V}{d}$$

where E is the electric field strength

V is the change in potential energy change per unit charge

d is the distance moved along the field line

- Electric potential differnce is defined for positive charge. If a positive charge moves in the direction of the electric field its potential energy will decrease. The SI unit of the volt is V and hence the SI units for the electric field can be written as NC^{-1} or Vm^{-1}. We will examine the change in energy of a charge moving in an electric field in more detail in the next section.

EXAMPLE 7

A small piece of paper of mass 0.1 g has a charge of + 6 μC. The piece of paper is observed to levitate in a vertical electric field. Find the magnitude and direction of the applied field.

Answer:

If an object is at rest all the forces on it must sum to zero

If the piece of paper does not fall down, the force of gravity must be balanced by an upwards force due to the electric field. As the charge is positive it will experience an upwards force if the electric field is also directed upwards. We can find the magnitude of the force by equating the two forces acting on the piece of paper:

$F_{gravity} = F_{electric}$

$mg = qE$

$E = mg/q = \frac{1\times10^{-4}\times9.8}{6\times10^{-6}} = 163.3\ NC^{-1}$ upwards

EXAMPLE 8

When a small charged ball of mass 5 g is suspended from a light string in a uniform, horizontal electric field of strength 400 NC^{-1}, the ball comes to rest when the string makes an angle of 30° to the vertical, as shown in Figure 11.17. Use this information to find the charge on the ball.

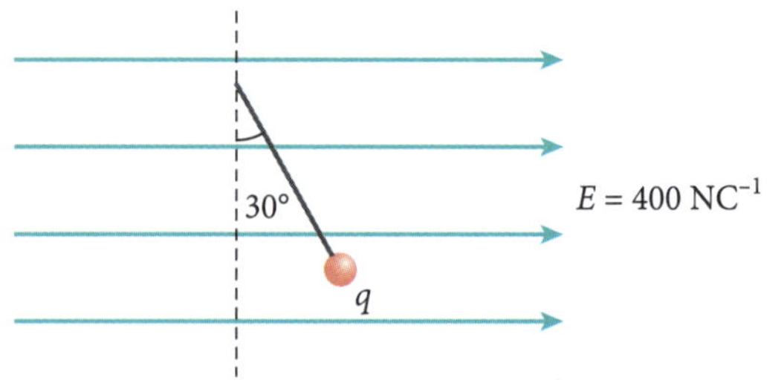

Figure 11.17 Charge suspended in a horizontal electric field

Answer:

As the object is at rest, the tension in the string must be equal to the sum of the other acting forces

As the ball is at rest, the net force must be zero on the ball. Hence the gravitational force ($F = mg$), electric force ($F = qE$) and the tension in the string must sum to zero. This enables us to draw the vector diagram shown in Figure 11.18.

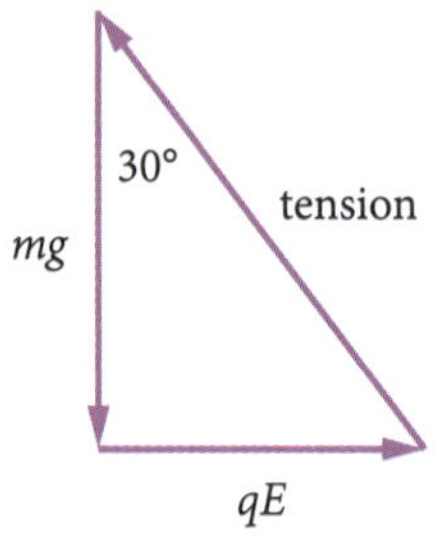

Figure 11.18 Vector sum of forces on the ball

From Figure 11.8 we see that $\tan 30° = qE/mg$ and hence:

$q = mg \tan 30°/E$

$= \frac{0.005\times9.8\times\tan 30°}{400}$

$= 7.07\times10^{-5}$ C

(We know the charge is positive as the force on the charge is in the same direction as the electric field.)

4 Potential difference and energy

» Students analyse the effects of a moving charge in an electric field, in order to relate potential energy, work and equipotential lines, by applying:

- $V = \frac{\Delta U}{q}$, where U is potential energy and q is the charge.

➔ When a charge moves in an electric field its potential energy changes. Consider the charge shown in Figure 11.19. There is a force on the charge to the right and, when it moves from point A to B, the field does work on the charge and hence the charge moves to a position where it has a lower potential energy. To move the charge from B to A would require work to be done because the charge must be moved against the electric force ($F = qE$).

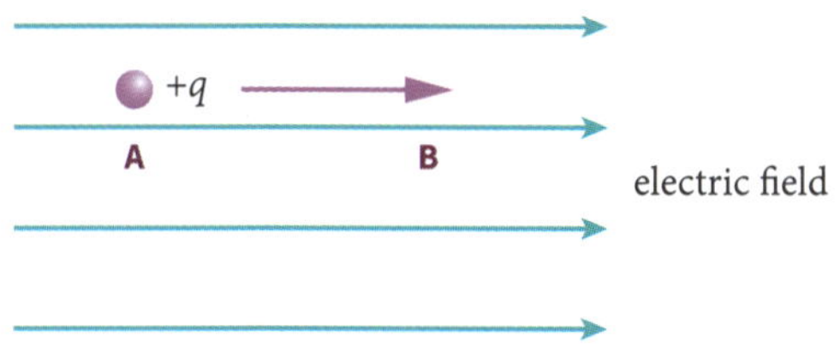

Figure 11.19 The charge is moved between two points A and B in an electric field

➔ To quantify the energy change between two points in an electric field we use **electrical potential difference** (V). The electrical potential difference between two points in an electric field is defined as the potential energy change a unit positive charge would experience if it moved between the two points. We can express this definition as:

$$V = \frac{\Delta U}{q} = \frac{\text{work}}{q}$$

where V is the electrical potential difference
ΔU is the change in potential or work done by the charge
q is the charge

electrical potential difference: the change in potential energy that a unit positive charge experiences when it is moved between two points in an electric field

➔ With reference to Figure 11.19, point A is at a higher electrical potential than point B because the electric force (qE) on the charge would do work on the charge as it moves the charge to point B. Hence the charge has lost potential energy moving from A to B. Conversely, B is at the lower electrical potential because a positive charge would require external energy to be moved from B to A against the electrical force. Note that electrical potential, like electric field, is defined for a positive charge. Even though it would require work to move a negative charge from A to B we still say A is at the higher potential because electrical potential energy is defined for a positive charge.

➔ Charges that move perpendicular to the electric field do not gain or lose potential energy and hence lines drawn perpendicular to the electric field are called **equipotential lines**. This is similar to moving a mass horizontally in a gravitational field. The mass would remain at the same height and hence would have the same gravitational potential energy. Figure 11.20 shows the equipotential lines between parallel charged plates and around a point charge and a dipole. Note that the equipotential lines are drawn closer together where the electric field strength is greater (i.e. where the electric field lines are closer together).

equipotential lines: lines drawn perpendicular to the electric field that link points of equal potential; if a charge moved along an equipotential line it would do no work and the field would do no work on the charge

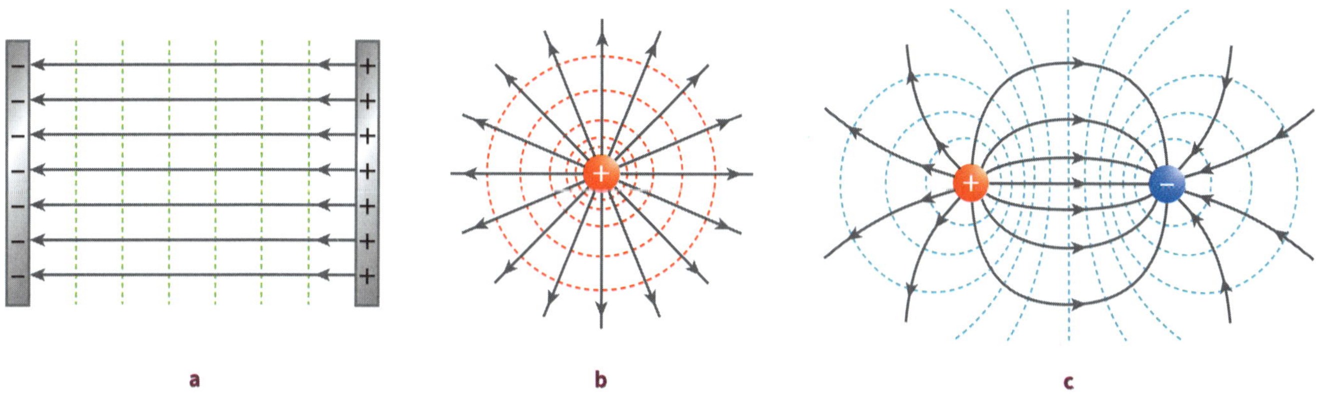

Figure 11.20 Equipotential lines (dotted) for **a** charged parallel plates, **b** a positive point charge and **c** a dipole

- The SI unit of potential difference is the volt. Note that one volt (V) equals one joule per coulomb. Hence if there is a potential difference of 200 volts between two points in an electric field, 200 joules of work would be done on one coulomb of charge if it was moved between the points. If 2 coulombs of charge was moved between the points, 400 joules of work would be done.
- We saw in the previous section that the electric field strength can be defined in terms of electrical potential difference ($E = V/d$). To derive this relationship, consider the charge shown in Figure 11.19. The work done by the field on the charge when the charge moves a distance (d) from point A to B in this uniform electric field is:

  Work = $Fd = qEd$

  But from our definition of electrical potential difference we have:

  $V = \Delta U/q$ or $\Delta U = qV$

  Note that in moving from A to B the charge loses potential energy and the change in potential energy is equal to the work done by the charge:

  Work done by the charge = $\Delta U = qV$

  Equating these two expressions for the work done on the charge when it moves between the two points in the field gives:

  $qEd = qV$

  Hence:

  $E = V/d$
- This relationship is useful for parallel charged plates. If a potential difference V was applied across a pair of parallel charged plates separated by a distance d, the electric field strength between the plates would be $E = V/d$. Remember the direction of the electric field is the direction in which the potential energy of a positive charge would decrease. That is, the potential decreases as we move in the direction of the electric field. With reference to Figure 11.19, this is why point A is at a higher electrical potential than B.

EXAMPLE 9

For parallel charged plates:

a **sketch a graph of the force on a positive charge against position from the positive plate to the negative plate**

b **sketch the potential energy of a positive charge against position from the positive plate to the negative plate.**

Answer:

Think about the force on a unit charge and the potential energy of a unit charge as it moves between the charged plates

a The electric field between parallel plates is a constant at all points and hence the force on the charge will also be constant ($F = qE$) between the plates, as shown in Figure 11.21.

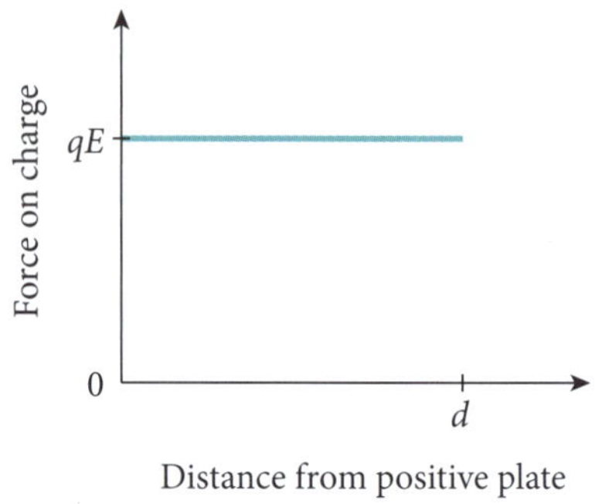

Figure 11.21 Electric field strength as a function of distance from one plate towards the other plate

b The potential energy of the charge will decrease because it is moving in the same direction as the electric field (Figure 11.22). As the force on the charge is constant, the potential energy of the charge will change at a constant rate between the plates. (Similar to the way that the gravitational potential energy of a mass changes as it is moved to a lower height h above the ground.)

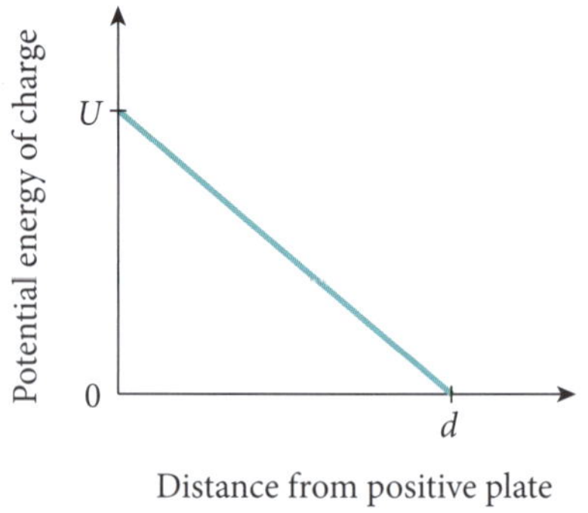

Figure 11.22 Potential energy of a positive charge as a function of distance from the positive plate towards the negative plate

EXAMPLE 10

A voltage of 100 V is placed across pair of parallel plates separated by 1 cm in a vacuum. If a proton (mass = 1.67×10^{-27} kg and charge $+1.6 \times 10^{-19}$ C) is released at the positive plate, find:

- **a** **the magnitude of the electric field between the plates**
- **b** **the force on the proton**
- **c** **the velocity of the proton just before it reaches the negative plate.**

When a charge is accelerated by an electric field, potential energy is converted to kinetic energy

Answer:

a The electric field strength will be given by

$$E = V/d = \frac{100}{0.01} = 1 \times 10^4 \text{ Vm}^{-1}.$$

b The force on the proton will be given by

$$F = qE = 1.6 \times 10^{-19} \times 10^4$$
$$= 1.6 \times 10^{-15} \text{ N towards the positive plate.}$$

c We can use the force on the charge, or the energy change, to solve this problem.

Using force, the acceleration of the proton is $a = F/m = qE/m$ and the final velocity can be calculated using $v^2 = u^2 + 2as$.

As $u = 0$, $a = qE/m$ and $s = d$,

the plate separation $v = \sqrt{2as} = \sqrt{\frac{2qEd}{m}}$.

But $V = Ed$ and hence $v = \sqrt{\frac{2qV}{m}}$.

Alternatively, as energy must be conserved, the loss of potential energy when the charge moves between the plates must be equal to the kinetic energy gained. From the definition, electrical potential difference $V = \Delta U/q$ and hence the change in potential energy when the charge moves between the plates must be $\Delta U = qV$. Equating the potential energy lost to the kinetic energy gained:

$$\Delta U = E_k$$

$$qV = \frac{1}{2}mv^2$$

Or $v = \sqrt{\frac{2qV}{m}}$, as we obtained above.

The final velocity is therefore:

$$v = \sqrt{\frac{2qV}{m}} = \sqrt{\frac{2 \times 1.6 \times 10^{-19} \times 100}{1.67 \times 10^{-27}}}$$
$$= 1.38 \times 10^5 \text{ ms}^{-1}$$

→ KEY QUESTIONS

8 **How is the work done by a charge when it moves between two points in an electric field related to the electrical potential difference between the points?**

9 **How is the force on a charge related to the electric field strength at the point?**

10 **How are potential difference and electric field strength related in a uniform electric field?**

Answers ➲ p. 151

CHAPTER SYLLABUS CHECKLIST

Are you able to answer these questions from the syllabus for this chapter? Tick each question as you go through the checklist if you are able to answer it. If you cannot answer a question, turn to the relevant page in the study guide to find the answer. For NESA key word meanings, go to www.educationstandards.nsw.edu.au and search 'key words'.

FOR A COMPLETE UNDERSTANDING OF THIS TOPIC:		PAGE NO.	✓
1	Can I explain how an object can become electrostatically charged?	137	
2	Can I describe the force between charged bodies?	137	
3	Can I explain why uncharged objects can be attracted to charged objects?	138	
4	Can I explain how electrostatic charge can be measured?	138–139	
5	Can I explain how the strength and direction of the electric field at a point in space is determined?	141	
6	Can I draw electric field lines around charged bodes?	141–142	
7	Can I relate the density of electric field lines to the electric field strength?	142	
8	Can I use Coulomb's law to make predictions and solve problems?	142–143	
9	Can I define electrical potential difference?	145	
10	Can I draw equipotential lines around charged objects?	145	
11	Can I relate electrical potential difference and electric field strength for a uniform field?	146	
12	Can I solve problems and make predictions about the electric field between charged parallel plates?	146	

YEAR 11 EXAM-TYPE QUESTIONS

Objective-response questions (1 mark each)

1 **Why are small neutral pieces of paper attracted to a charged Perspex rod?**

- A The opposite charge must be induced on the paper.
- B The same charge must be induced on the paper.
- C Polarisation of the charges in the paper results in an imbalanced force on the positive and negative charges in the paper.
- D Polarisation of the charges in the paper results in an equal and opposite force on the positive and negative charges in the paper.

2 **A positively charged ebonite rod is slowly moved close to and away from an electroscope that was initially uncharged. Which of the following best describes and explains the motion of the vane of the electroscope?**

- A The vane will become negatively charged and move away from the post as the rod is brought closer and remain in this position when the rod is removed.
- B The vane will become positively charged and move away from the post as the rod is brought closer and remain in this position when the rod is removed.
- C The vane will become positively charged and move away from the post as the rod is brought closer and return to the post when the rod is removed.
- D The vane will become negatively charged and move away from the post as the rod is brought closer and return to the post when the rod is removed.

3 **A student conducts an experiment in which the force (F) between two charges is measured as a function of the separation (r) between the charges. When the student plotted F against r, the graph was not a straight line. What could the student have plotted to produce a straight-line graph?**

- A F^2 against r
- B F against r^2
- C F against $1/r$
- D F against $1/r^2$

4 **A positive charge is moved from point X to point Y and then to point Z in the electric field produced by the two charged parallel plates shown in Figure 11.23. Which of the following best describes this situation?**

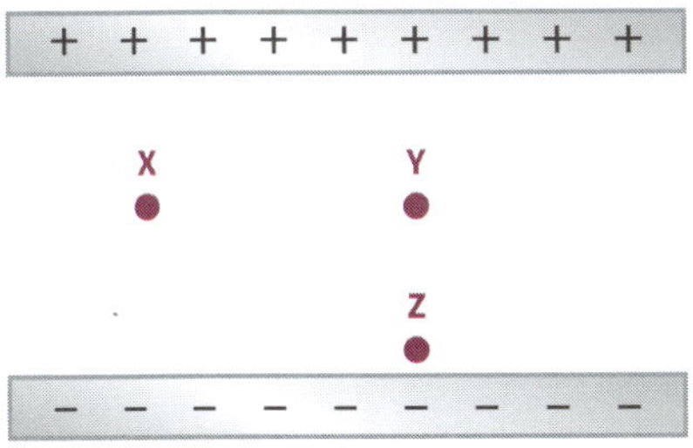

Figure 11.23 Three points between charged parallel plates

- A The potential energy of the charge remains constant when moving from X to Y and then decreases as the charge moves from Y to Z.
- B The potential energy of the charge remains constant when moving from X to Y and then increases as the charge moves from Y to Z.
- C The electrical potential difference is constant throughout the motion.
- D The potential energy decreases when moving from X to Y and then increases as the charge moves from Y to Z.

5 **Consider the electric field shown in Figure 11.24. Which of following statements best relates the electrical properties of the points X and Y in the figure?**

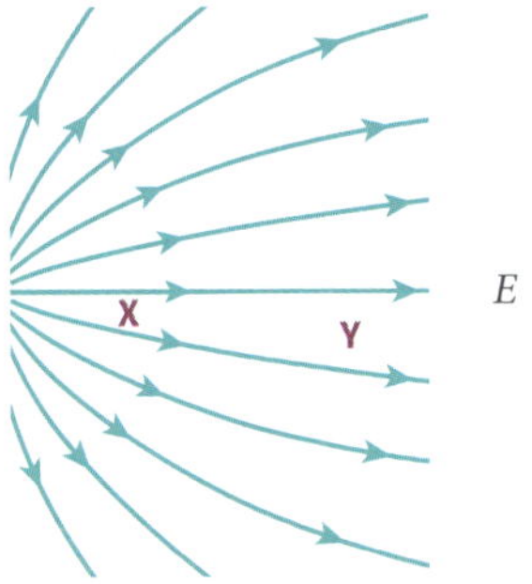

Figure 11.24 Two points X and Y in an electric field

- A X is at a higher electrical potential than Y and is in a region of greater electric field strength.
- B X is at a higher electrical potential than Y but is in a region of lower electric field strength.
- C X has the same electrical potential as Y but is in a region of greater electric field strength.
- D X and Y have the same electrical field strength but Y has a lower electrical potential than X.

Extended-response questions

6 **In an experiment the teacher gives you an electroscope, glass rod and some silk cloth.**

- a **Rubbing the glass rod with the silk cloth leaves the glass rod with a positive charge. With reference to conservation of charge, explain why rubbing the rod with the cloth leaves the rod charged.** (2 marks)
- b **When the charged glass rod is brought near a stream of water, the water is attracted to the rod. With the aid of a diagram, explain why this occurs.** (2 marks)
- c **An electroscope can be used to measure charge. Explain how the glass rod could be used to charge an electroscope.** (1 mark)

7 Two equal negative charges are placed near each other.

a Draw the charges 2 cm apart and sketch the electric field lines around the charges. (2 marks)

b On the same diagram, draw the equipotential lines around the charges. (1 mark)

8 Two equal charges, each with a mass of 2 g, are suspended from light-weight 0.2 m long strings, as shown in Figure 11.25.

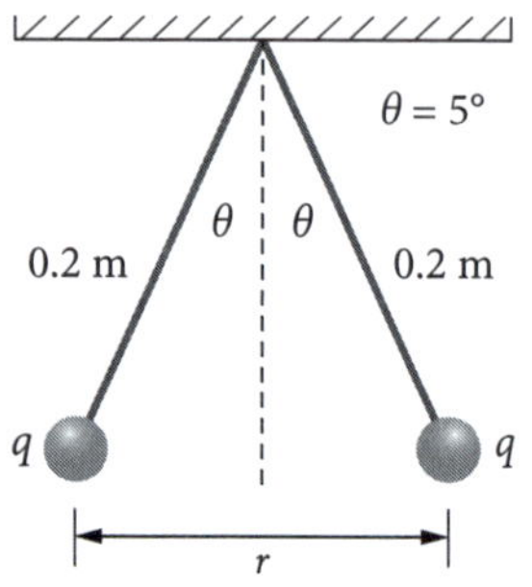

Figure 11.25 Two equal charges suspended from light strings

Given that both strings are displaced 5° from the vertical when the system is in equilibrium, find:

a the separation between the charges (r) (1 mark)

b the electrostatic force on the charges ($F_{electric}$) (2 marks)

c the size of the charge (q). (1 mark)

9 A charge of +4 μC is located at position X in a uniform electric field of strength 200 NC^{-1}, as shown in Figure 11.26.

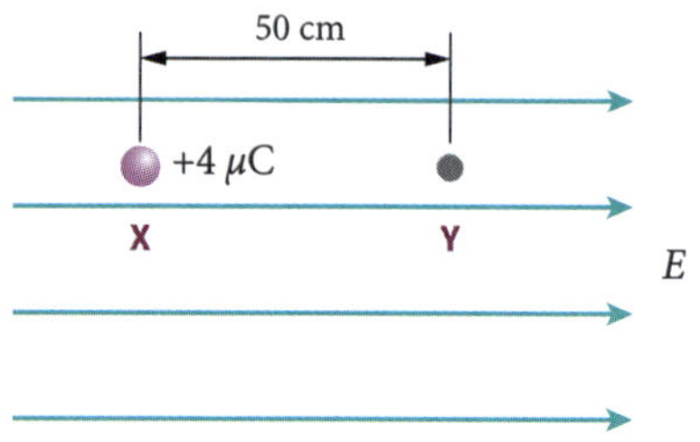

Figure 11.26 Charge in a uniform 200 NC^{-1} electric field

a Find the magnitude and direction of the force on the charge. (2 marks)

b Given that X and Y are separated by 50 cm, what is the potential difference between X and Y? (1 mark)

c Does X or Y have the higher potential? Explain your answer. (2 marks)

d If the charge was released at X, how much work would be done by the field when the charge moved from X to Y? (1 mark)

10 To provide thrust, an experimental rocket-propulsion system uses an accelerating voltage of 1500 V to eject xenon ions (charge = 1.6×10^{-19} C and mass 2.18×10^{-25} kg) at high speed from the back of the rocket.

a Find the work done by the accelerating voltage on each ion. (1 mark)

b Determine the final velocity of the ions. (1 mark)

c Calculate the change of momentum for each ion and, given that the acceleration occurs over a time of 1 μs, find the force exerted on each ion. (2 marks)

d What electric field strength would be required in the ion drive to provide this force? (1 mark)

ANSWERS

KEY QUESTIONS

Key questions ➲ p. 138

1 Charge can be transferred from one object to another by rubbing them together. If one object attracts surface electrons more strongly than the other, charges will move from one object to the other. This is called triboelectric charging.

2 Nearby charged objects can cause induced charge movement in conductors, and polarisation of charge in insulators. This charge separation leads to imbalanced forces on the opposite charges, which results in a net force of attraction between the charged objects and the nearby uncharged objects.

Key questions ➲ p. 141

3 Touching the top disk of an electroscope with a negatively charged rod will result in charges moving from the charged rod onto the electroscope. Touching with a postively charged rod will result in negative charges moving from the electroscope to the rod.

4 Bringing a charged rod near the electroscope, momentarily earthing the electroscope and then removing the charged rod, will induce the opposite charge onto the electroscope.

Key questions ➲ p. 142

5 We can visualise electric fields by drawing electric lines of force (electric field lines) to represent the force that would appear on a positive test charge placed at each point.

6 The electric field lines give us the direction of the field at each point, and the density of lines indicates the strength of the field.

Key question ➲ p. 143

7 Coulomb's law tells us that the electrostatic force between charges is proportional to the product of the charges and inversely proportional to the square of their distance apart.

Key questions ➲ p. 147

8 The electrical potential difference (V) between two points is equal to the potential energy change that a unit positive test charge would experience if it was moved between the points. Thus if a charge q was moved between the points, the work done on or by the charge would be qV. If a positive charge moved from the higher to the lower potential the field would do work on the charge, but if a negative charge moved from the higher to the lower potential, an external source of energy would have do work on the charge to move it.

9 As the electric field (E) is the force on a unit charge, the force on a charge q would be $F = qE$.

10 In a uniform electric field $E = V/d$ and hence the potential difference between two points separated by a distance d parallel to the field lines is given by $V = Ed$.

YEAR 11 EXAM-TYPE QUESTIONS

Objective-response questions

1 **C** describes polarisation of charge correctly. **A** and **B** are incorrect because a net charge cannot be induced on the paper as there is no way for the charge to reach or leave the paper. **D** is incorrect because equal and opposite forces would produce a net force on the paper of zero.

2 **C** correctly describes charge movement by induction. **A** and **D** are incorrect because the positively charged rod would cause positive charges to move towards the vane and post and negative charges to move towards the top of the electroscope. **B** is incorrect because it implies the electroscope is left with a net positive charge. This is not possible as there is no path for the charges to leave or join the electroscope.

3 **D** is correct because Coulomb's law states that the electrostatic force between charges is inversely proportional to the square of their distance apart ($F\ \alpha\ 1/r^2$). Answers **A**, **B** and **C** are all incorrect because the electrostatic force between charges is not proportional to r, r^2 or $1/r$.

4 **A** is correct as XY is an equipotential line and therefore the potential energy of the charge would not change moving from X to Y. The field would do work on a positive charge moving from Y to Z and hence the charge would lose potential energy. **B** is incorrect because the charge could not gain potential energy moving from Y to Z. **C** is incorrect because there is a potential energy difference between Y and Z changes. **D** is incorrect because XY is an equipotential line, hence the potential energy will not decrease moving from X to Y.

5 **A** is correct as a positive charge would have a greater potential energy at X and the field is stronger at X because the field lines are closer together. **B, C** and **D** are incorrect because the field is weaker at Y than at X.

Extended-response questions

6 EM This question tests students' understanding of the triboelectric effect, the electroscope and induced charge movement.

a The glass rod holds its surface electrons less tightly than the silk cloth, and when it is rubbed with the cloth, electrons move from the glass to the cloth, leaving the silk cloth negatively charged and the glass rod positively charged. ✓ The total amount of charge is conserved in this process and hence the silk cloth gains the same number of electrons as the glass rod loses. ✓

b A charged rod will produce an electric field that will cause the polar molecules to align with the negative side of the molecule closest to the positively charged rod, as shown in Figure A11.1. Because the negative charges are closer to the rod they will experience a greater force of attraction than the repulsive force experienced by the positive side of the molecules. This results in a net force of attraction on the molecules, which attracts the stream of water to the charged rod. ✓

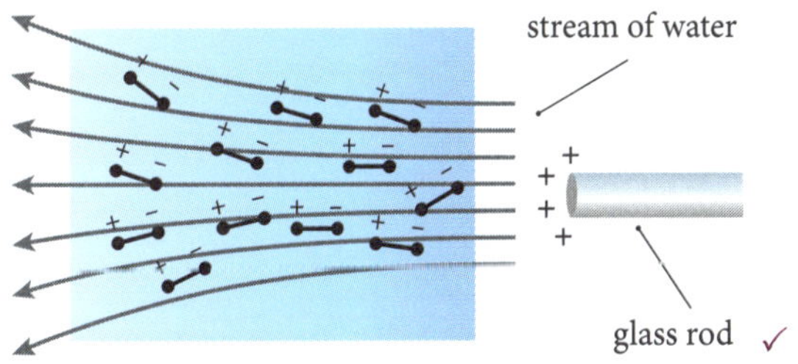

Figure A11.1 Polar water molecules aligning in the electric field produced by a positively charged rod

c Simply touching the top plate of the electroscope with the positive rod and then removing the rod will draw electrons onto the rod when it is in contact, which will leave the electroscope with a net positive charge when the rod is removed. ✓

7 **EM** This question tests students' understanding of electric field lines and equipotential lines.

a and **b**. See Figure A11.2, where one mark is given for the electric field lines drawn correctly, one mark for the correct direction on the electric field lines, and one mark for drawing equipotential lines at 90° to the electric field lines.

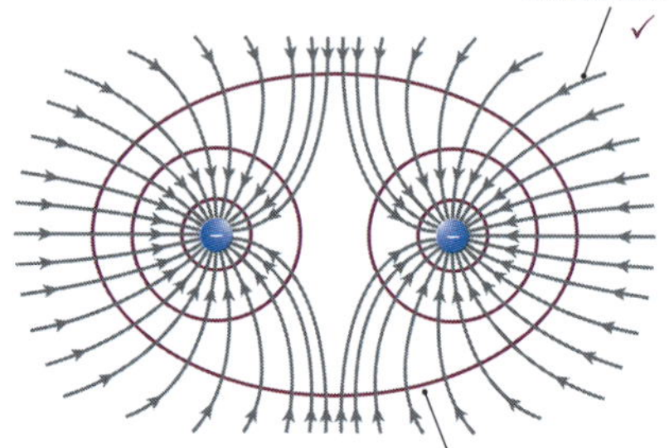

Figure A11.2 Electric field lines and equipotential lines around two negative charges

8 **EM** This question tests students' understanding of Coulomb's law and vectors.

a We can find r by solving the right-angle triangle formed by the string, which connects either of the charges and the vertical:

$\sin\theta = (r/2)/0.2$ and therefore $r = 0.4\sin 5° = 0.035$ m ✓

b As the charges are at rest, the gravitational force, electrostatic force and tension in the string must sum to zero. The vector addition of the three forces is shown in Figure A11.3.

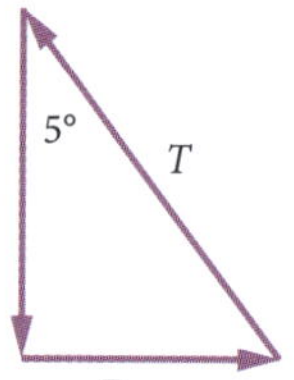

Figure A11.3 Vector sum of forces acting on the right-hand charge

With reference to Figure A11.3:

$\tan 5° = F_{electric}/mg$

Hence $F_{electric} = mg(\tan 5°) = 0.002 \times 9.8 \times \tan 5°$
$= 1.7 \times 10^{-3}$ N repulsion ✓

c Noting that $q_1 = q_2$ and applying Coulomb's law yields:
$F = (1/4\pi\varepsilon_0)(q^2/r^2)$

Or $q = \sqrt{4\pi\varepsilon_0 \times F \times r^2}$

$= \sqrt{\dfrac{1.7\times10^{-3}\times(0.035)^2}{9\times10^9}} = 1.5 \times 10^{-8}$ C ✓

9 **EM** This question tests students' understanding of the electric field and how electrical potential difference is related to the electric field.

a From the definition of the electric field:

$E = F/q$ and hence $F = qE = (4 \times 10^{-6}) \times 200 = 8 \times 10^{-4}$ N ✓ to the right ✓

b Using $V = -Ed = 200 \times 0.5 = -100$ V ✓

(The absolute potential between the points is 100 V and the negative sign just tells us that Y is at the lower potential.)

c X is at the higher potential ✓ because the field would do work on the charge if it moved from X to Y, and this would decrease the charge's potential energy. ✓ (Just like letting a rock fall down would decrease its gravitational potential energy.)

d From the definition of potential difference $V = \text{work}/q$ or work $= qV = (4 \times 10^{-6}) \times 100 = 4 \times 10^{-4}$ J ✓

10 **EM** This question tests students' understanding of charges in electric fields, electrical potential and students' ability to combine their understanding of forces, momentum and motion with electric fields.

a From the definition of electrical potential difference: $V = work/q$

Or work $= qV = (1.6 \times 10^{-19}) \times 1500 = 2.4 \times 10^{-16}$ J ✓

b The work done on the charge goes into increasing the charges kinetic energy:

Work $= w = \frac{1}{2}mv^2$

Hence $v = \sqrt{\dfrac{2w}{m}} = \sqrt{\dfrac{2\times2.4\times10^{-16}}{2.18\times10^{-25}}} = 4.69 \times 10^4$ ms^{-1} ✓

c Note that the initial velocity is zero and hence:

$\Delta p = mv - mu = (2.18 \times 10^{-25})(4.69 \times 10^4) = 1.02 \times 10^{-20}$ Ns ✓

Now impulse $= \Delta p = Ft$ and hence $F = \Delta p/t = \dfrac{1.02\times10^{-20}}{1\times10^{-6}}$

$= 1.02 \times 10^{-14}$ N ✓

d Using $E = F/q = \dfrac{1.02\times10^{-14}}{1.6\times10^{-19}} = 6.4 \times 10^4$ NC^{-1} or Vm^{-1} ✓

CHAPTER 12 ELECTRIC CIRCUITS

MODULE 4
ELECTRICITY AND MAGNETISM

INQUIRY QUESTION:

How do the processes of the transfer and the transformation of energy occur in electric circuits?

Electrical power sources such as batteries and generators use energy to separate opposite charges. Once separated, opposite charges produce electric fields that can be used to move charges in circuits and hence transfer electrical energy from the power source to other places in the circuit. Circuit components such as resistors, light globes and electric motors can then be used to transform electrical energy into a range of other forms of energy, including heat, light and motion.

1 Electric current

» Students investigate the flow of electric current in metals and apply models to represent current, including:

- $I = \frac{q}{t}$.

➔ In their solid states metals are made up of atoms that are neatly arranged in crystal lattices. Each metal atom has 1 or 2 electrons that are free to move through the lattice. We call these mobile electrons in metals, *valence electrons*. We can picture metals as a lattice of uniformly arranged atoms surrounded by a 'sea' of mobile valence electrons.

➔ Electrical power supplies like a battery or generators use energy from an external source to separate opposite charges. The positive terminal of the power supply has an excess of positive charge and the negative terminal has an excess of negative charge. We call the potential difference produced by a battery or generator the **electromotive force** (EMF). As EMF is a potential difference, it is measured in joules per coulomb or volts.

electromotive force: the potential difference (voltage) produced by a power supply such as a battery or electric generator

➔ When a metal wire is connected between the terminals of a power supply, a conducting path is created that allows electrons from the negative terminal of the battery to flow through the wire to the positive terminal. Connecting the power supply produces an electric field through the wire that will exert a force on all the charges in the wire. Because the valence electrons are free to move in the metal wire, electrons will flow from the negative terminal of the power supply, through the circuit to the positive terminal. We call this flow of charge an electrical **current**.

current: the rate at which charge moves past a point in a circuit, measured in coulombs per second (Cs^{-1}) or amperes (A)

➔ Because we used *positive charge* to define the electric field and electrical potential difference, we also define current in terms of positive charge flow. Conventional current flow is taken as the direction that positive charges would flow, if they could flow, in a circuit. Thus current would flow from the positive terminal of the power supply through the circuit to the negative side of the power supply, as shown in Figure 12.1.

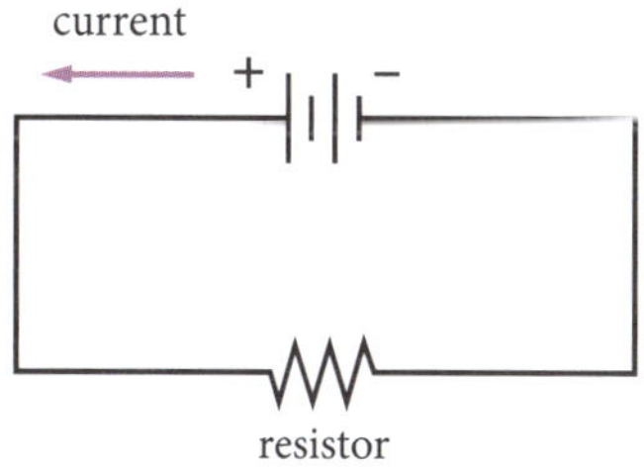

Figure 12.1 By convention, the direction of the current flow in a circuit is the direction that positive charges would flow if they could flow

➔ The current at any point in a circuit is defined as the rate at which charge moves past the point and hence:

$$I = \frac{q}{t}$$

where I is the current
q is the charge that has passed the point
t is the time over which the flow of charge was measured

The SI units of current are coulombs per second (Cs^{-1}) or amperes (A).

- Current flow in one direction is called *direct current* (DC), while oscillating current is called *alternating current* (AC). Batteries produce direct current, while mains electricity is an alternating current with an oscillation frequency of 50 Hz.

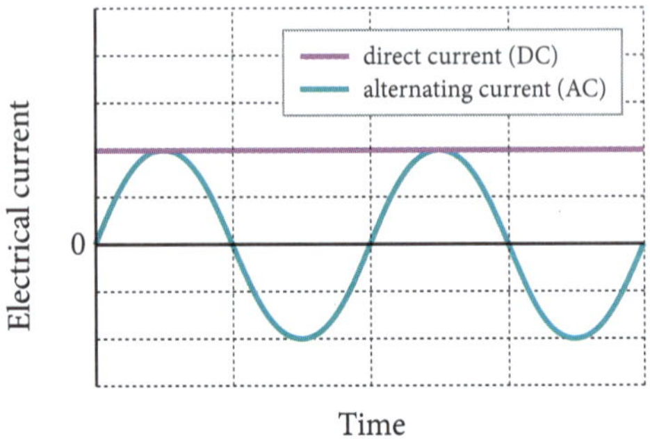

Figure 12.2 Direct current (DC) and alternating current (AC)

→ KEY QUESTIONS

1. **How is current related to charge?**
2. **What is the difference between alternating current and direct current?**

Answers ➲ p. 168

EXAMPLE 1

A current of 5 A flows in a copper wire of diameter 2 mm. Given that the charge on an electron is $e = -1.6 \times 10^{-19}$ C, and that each metre of the copper wire contains approximately 2.8×10^{23} valence electrons, find:

a **the charge that passes through the wire each second**

b **the number of electrons that pass through the wire each second**

c **the average speed of the electrons along the wire.**

Answer:

a The current is defined as $I = q/t$ and therefore $q = It = 5 \times 1 = 5\ Cs^{-1}$.

Recall that current is the amount of charge that flows per second

b The number of electrons per second = q/e

$= \frac{5}{1.6 \times 10^{-19}} = 3.125 \times 10^{19}$ electrons per second.

c As there are 2.8×10^{23} valence electrons per metre of wire, this many electrons would pass per second if they were moving at 1 ms^{-1}. If half this many electrons passed per second the electrons would be travelling at 0.5 ms^{-1}, and hence if 3.125×10^{19} electrons passed per second, the electron speed would be given by $\frac{3.125 \times 10^{19}}{2.8 \times 10^{23}} = 1.1 \times 10^{-4}\ ms^{-1}$.

Note that although the electrons are moving very fast in all directions in the wire due to their thermal energy, their drift velocity down the wire that produces the current is very slow. A current of 5 A means that the valence electrons would move 1 m along this wire every 2.5 hours.

2 Ohm's law

» Students investigate quantitatively the current–voltage relationships in ohmic and non-ohmic resistors to explore the usefulness and limitations of Ohm's law using:

- $W = qV$
- $V = IR$.

- You will recall from your previous studies that materials which contain mobile charge carriers are called *conductors*, while materials in which charges are not free to move are called *insulators*.
- Studies by Georg Ohm in the 1820s showed that for many conductors, provided the temperature is kept constant, the current that flows in the conductor (I) is proportional to the potential difference applied across the conductor (V). That is:

  $V \alpha I$

- Ohm called the constant of proportionality the **resistance** (R) as it is a measure of the material's resistance to the flow of current. Today we express **Ohm's law** as:

$$R = \frac{V}{I}$$

where V is the potential difference
I is the current
R is the resistance

Note resistance has the SI units of volts per amp (VA^{-1}) or ohms (Ω). The symbol used to represent the ohm, omega (Ω), is the last letter in the Greek alphabet. The uppercase version of omega is Ω, while the lowercase is ω.

resistance: the constant of proportionality in Ohm's law that is a measure of how much an object resists the flow of charge

Ohm's law: a law stating that for many conductors the current that flows through the conductor is proportional to the potential difference applied across the conductor

- ➔ A graph of the potential difference against the current for two materials with different resistances is shown in Figure 12.3. Note that the gradient (V/I) is equal to the resistance and hence the line with the steeper gradient represents the ohmic conductor with the higher resistance.

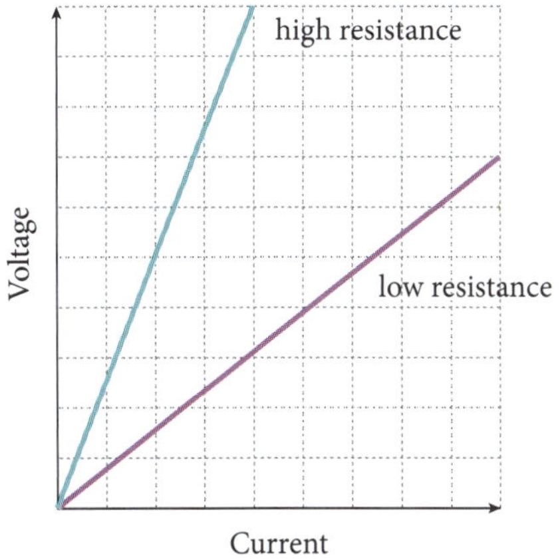

Figure 12.3 Graph of potential difference against current for two ohmic conductors with different resistances

- ➔ Often the current is plotted against the potential difference and in this case the gradient (I/V) is the inverse of the resistance and a steeper gradient would therefore represent a material with a lower resistance.
- ➔ Materials that obey Ohm's law are called **ohmic conductors**. Generally metals are ohmic conductors, while insulators and semiconductors are non-ohmic conductors. An example of a voltage against current graph for a non-ohmic resistor is shown in Figure 12.4. Note that the gradient (V/I) changes and therefore the current is not proportional to voltage making the resistor a non-ohmic resistor.

ohmic conductor: a conductor that obeys Ohm's law

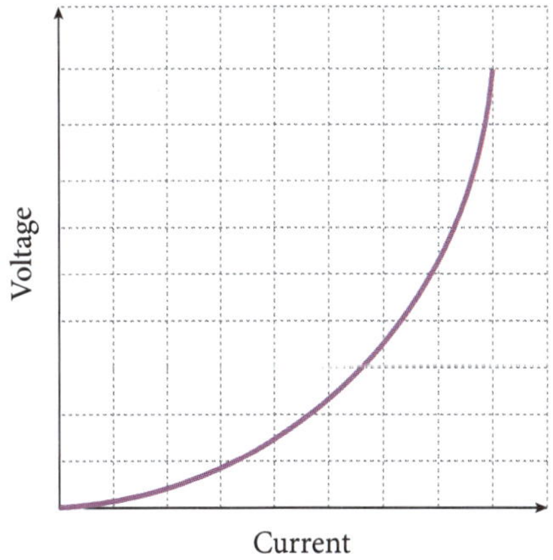

Figure 12.4 Graph of potential difference against current for a non-ohmic resistor

- ➔ The resistance of a conductor depends on the type of material used, the shape of the conductor and the temperature:
  - The resistance of a material is related to the density of free charge carried in the material. Insulators have high resistance and conductors have low resistance.
  - The resistance of a specific conductor is proportional to the length of the conductor and inversely proportional to the cross-sectional area of the conductor. That is:

    $R \alpha L/A$

    where R is the resistance
    L is the length
    A is the cross-sectional area
  - Heating metals increases their resistance because at higher temperatures the lattice atoms vibrate more energetically and the valence electrons have more collisions as they drift down the wire.

➔ KEY QUESTIONS

3 **How are current and potential difference related in ohmic conductors?**

4 **What factors determine the resistance of a specific ohmic conductor?**

Answers ➲ p. 168

EXAMPLE 2

Table 12.1 sets out information for a group of small cylindrical resistors.

Table 12.1 Record of resistor information

Resistor	Resistance (ohms)	Potential difference (volts)	Current (amps)
A	20		3
B		25	5
C		15	30
D	5	10	
E		40	1

a **Use Ohm's law to complete Table 12.1.**

b **Which of the above resistors (A to E) is the best conductor? Justify your answer.**

c **If the length and radius of resistor D was doubled, how would its resistance change?**

d **Is there enough information in the above table to determine if the resistors are ohmic? If not what other information would be needed?**

Answer:

Resistance depends on the shape, temperature and type of material used

a Rearrange Ohm's law as $R = V/I$, $I = V/R$ and $V = IR$, and substitute in the given variables to find each of the unknowns (see Table 12.2).

Table 12.2 Completed resistor information

Resistor	Resistance (ohms)	Potential difference (volts)	Current (amps)
A	20	60	3
B	5	25	5
C	0.5	15	30
D	5	10	2
E	40	40	1

b Resistor C has the lowest resistance and hence the highest conductivity.

c If the radius is doubled the cross-sectional area at the cylinder would increase by the square of the radius (as the area of a circle increases as the square of the radius). Now as $R \alpha L/A$ and, as the length is increased by two times but the cross-sectional area is increased by four times, the net change in the resistance will be $R \times \frac{2}{4} = \frac{R}{2}$. The resistance will halve.

d A resistor must obey Ohm's law to be an ohmic resistor. The information given in the table is insufficient to determine if the resistors are ohmic. To do this we would need to measure the resistance for a range of other potential differences to ensure the resistance remained constant as the potential difference changed.

EXAMPLE 3

A 240 V incandescent light globe draws 5 A of current the instant it is switched on, but the current rapidly decreases to a steady 0.5 A.

a Find the initial resistance of the globe filament.

b Find the final resistance of the globe filament.

c Account for the change in the resistance.

Answer:

The resistance of a metal increases with temperature

a Using Ohm's law $R = V/I = \frac{240}{5} = 48\ \Omega$.

b Using Ohm's law $R = V/I = \frac{240}{0.5} = 480\ \Omega$.

c As the filament heats up the resistance increases due to the increased frequency of collisions between the metal atoms and the conduction electrons.

FIRSTHAND INVESTIGATION 1

Ohm's law

We can investigate Ohm's law for a range of materials by measuring the current that flows through the material as a function of the potential difference applied across the material. A suitable circuit to make these measurements is shown in Figure 12.5.

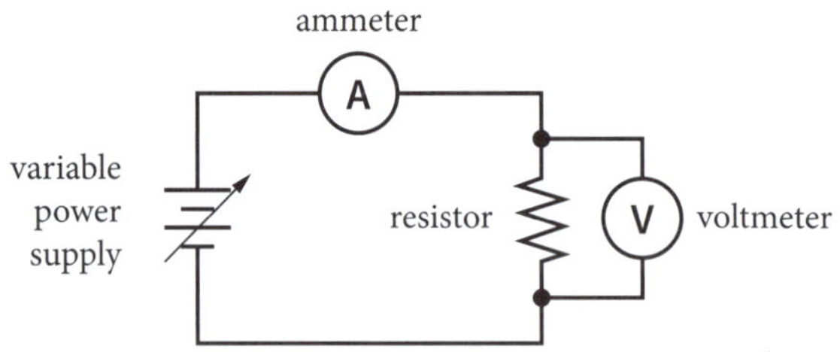

Figure 12.5 Circuit used to measure current as a function of applied potential difference

Sample results for a resistor kept at constant temperature (using a water bath) and for a lightbulb are shown in Figure 12.6.

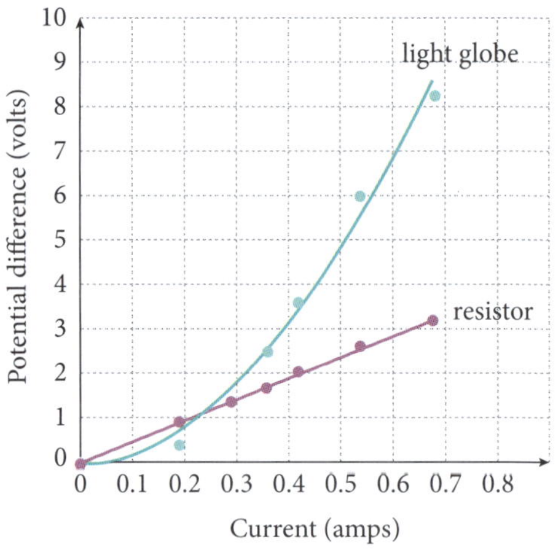

Figure 12.6 Plot of potential energy against current for an incandescent lightbulb and for a resistor at constant temperature

Note in Figure 12.6 that the resistor kept at constant temperature obeys Ohm's law and from the gradient we see that it has a resistance of $R = \text{rise/run} = \frac{3.2}{0.7} = 4.6\ \Omega$. In contrast, the gradient and hence the resistance of the light globe increases as more current flows through it and it gets hotter.

➔ We defined *potential difference* in the previous chapter as:

$$V = \frac{\Delta U}{q} = \frac{W}{q}$$

where V is the potential difference
ΔU is the change in potential energy
W is the work done
q is the charge

➔ When charges move through a resistor they collide with lattice atoms, transferring energy to the lattice and transforming electric energy to heat. By rearranging the definition of voltage we see that the work done by the charge q as it moves through a resistor will be related to the voltage across the resistor:

$W = qV$

➔ The work done by the charge represents the electrical energy transformed to heat in the resistor. This heat energy will increase the temperature of the resistor, which will increase the electromagnetic radiation emitted by the resistor. As we saw in Chapter 10, if the resistor gets hot enough it will radiate visible radiation (i.e. it will radiate heat and light).

EXAMPLE 4

Find the amount of charge that passes through a 100 W, 240 V lightbulb in 1 hour.

Answer:

Power = work/time and hence work = power × time.

Remember that power is related to current flow

Hence, $W = Pt = 100 \times (60 \times 60) = 3.6 \times 10^5$ J per hour.

Now rearranging $W = qV$ we can write:

$$q = W/V = \frac{3.6 \times 10^5}{240} = 1.5 \times 10^3 \text{ C}$$

3 Energy and power

» Students investigate quantitatively and analyse the rate of conversion of electrical energy in components of electric circuits, including the production of heat and light, by applying $P = VI$ and $E = Pt$ and variations that involve Ohm's law.

➔ You will recall from Chapter 2 that power is the rate at which work is done:

$P = W/t$ or $W = Pt$

Combining this equation for the work done with $W = qV$ yields:

$W = Pt = qV$

Hence $P = V(q/t)$

But we defined current in the first section of the chapter as $q/t = I$ and hence:

$$P = IV$$

where P is the power (rate at which energy is transformed)
I the current
V the potential difference

➔ We can apply Ohm's law to write the previous power equation in different ways. For example, substituting $I = V/R$ for I in $P = IV$ gives $P = V^2/R$. Similarly, substituting $V = IR$ into $P = IV$ gives $P = I^2R$.

➔ KEY QUESTIONS

5 **Explain why work is done when current flows through a resistor?**

6 **Explain how energy is conserved when charges move through a resistor?**

Answers ➲ p. 168

EXAMPLE 5

What resistance heater-element would be needed for a 240 V electric kettle to increase the temperature of 1 L of water from 20 °C to 100 °C in 2 minutes? You may assume all the heat goes into the water and none is lost to the environment.

Answer:

We first need to determine how much energy is required to boil the kettle:

Specific heat relates temperature change to energy added

Heat added = $Q = cm\Delta T = 4184 \times 1 \times 80 = 3.347 \times 10^5$ J

The power required to boil the water in 2 minutes would be:

$$P = E/t = \frac{3.347 \times 10^5}{2 \times 60} = 2.789 \times 10^3 \text{ W}$$

As we have the potential difference and power, the resistance required would be given:

$$P = V^2/R \text{ or } R = V^2/P = \frac{240^2}{2.79 \times 10^3} = 20.6\ \Omega$$

EXAMPLE 6

a A house supplied with 240 mains voltage is operating five 100 W light globes, a 2000 W heater, a 200 W refrigerator and a 50 W television. Find the total current drawn by the house.

b Find the resistance of one of the 100 W lightbulbs.

Answer:

a The total current will be the sum of the current drawn by each of the appliances.

The power used is related to the current

For each light globe, $I = P/V = \frac{100}{240} = 0.417$ A, and for five globes $I = 5 \times 0.417 = 2.08$ A.

For the heater: $I = P/V = \frac{2000}{240} = 8.33$ A

For the refrigerator: $I = P/V = \frac{200}{240} = 0.833$ A

For the television: $I = \frac{50}{240} = 0.208$ A

The total current drawn = 2.08 + 8.33 + 0.833 + 0.208
= 11.5 A

Alternatively, we can work out the total power used.

Total power used = 5 × 100 + 2000 + 200 + 50
= 2750 W

The total current draw is therefore:

$I = P/V = \frac{2750}{240} = 11.5$ A

b As $P = IV$ and $I = V/R$, $P = V^2/R$ or

$R = V^2/P = \frac{240^2}{100} = 576\ \Omega$

4 Series and parallel circuits

» Students investigate qualitatively and quantitatively series and parallel circuits to relate the flow of current through the individual components, the potential differences across those components and the rate of energy conversion by the components to the laws of conservation of charge and energy, by deriving the following relationships:

- $\Sigma I = 0$ (Kirchhoff's current law—conservation of charge)
- $\Sigma V = 0$ (Kirchhoff's voltage law—conservation of energy)
- $R_{series} = R_1 + R_2 + \ldots + R_n$
- $\frac{1}{R_{parallel}} = \frac{1}{R_1} + \frac{1}{R_2} + \ldots + \frac{1}{R_n}$.

Circuit components in series

→ When circuit components are placed in **series** (as shown in Figure 12.7) the current flows through each sequentially and hence there is only one path for the current to take.

series: a circuit that consists of only one path for the current and in which the same current passes through each component in the circuit

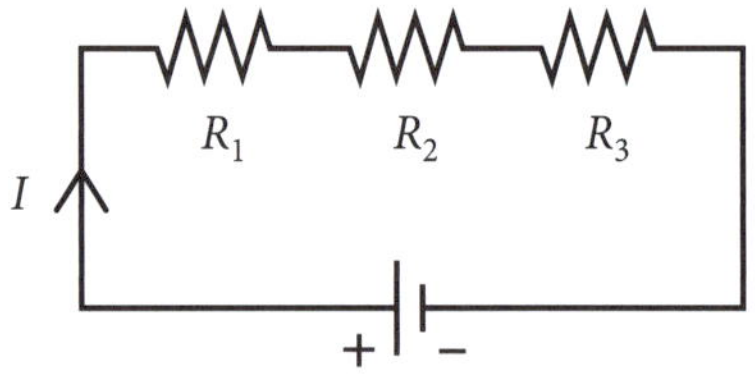

Figure 12.7 Three resistors in series with one another

→ When circuit components are in series the current that flows through each component must be the same. If this was not the case, charge would accumulate or disappear somewhere in the circuit. Thus for circuit components in series:

$$I_T = I_1 = I_2 = I_3$$

where I_T is the total current in the wire

I_1, I_2 and I_3 are the currents in each of the series components

→ To conserve energy, the sum of the potential difference across each component in a series circuit must be equal to the total potential difference across the components. We can understand this by thinking about the total work done as the sum of the work done in each resistor. We can express this relationship as:

$$V_T = V_1 + V_2 + V_3$$

where V_T is the total potential difference

V_1, V_2 and V_3 are the potential differences across each of the series components

→ For resistors in series we can use Ohm's law to replace the potential differences in the previous equation with $V = R/I$ and obtain:

$$R_T/I_T = R_1/I_1 + R_2/I_2 + R_3/I_3$$

As all the currents are equal we can cancel them out to obtain an expression for the total resistance of resistors in series:

$$R_T = R_1 + R_2 + R_3$$

where R_T is the single resistor that would be equivalent to the three series resistors

R_1, R_2 and R_3 are the resistances of each of the resistors

- Ammeters measure current and are placed in series in the branch of the circuit in which we wish to measure the current. Ammeters have a very low resistance to ensure that placing them in series does not reduce the current flow in the wire.
- When two or more sources of EMF (e.g. batteries) are wired in series, their EMFs are added.

Circuit components in parallel

- When circuit components are placed in **parallel** with one another, as shown in Figure 12.8, the current splits into separate paths, passes through the components and then recombines.

parallel circuit: a circuit in which the current splits into two or more branches and then recombines; components that are wired in parallel with one another have the same potential difference across them

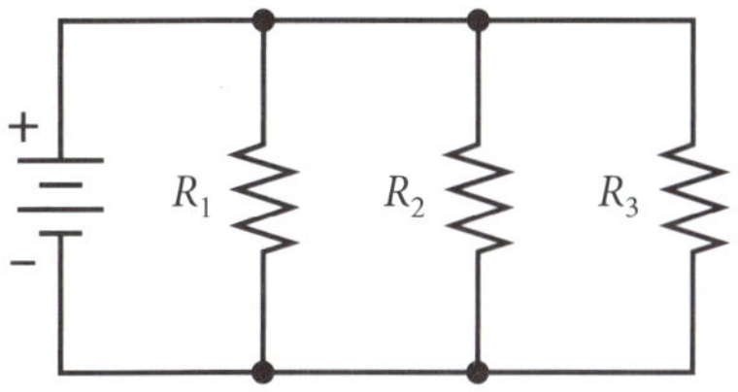

Figure 12.8 Resistors in parallel with one another

- When circuit components are in parallel the total current must be the sum of the current in each of the components. If this was not the case, charge would accumulate or disappear somewhere in the circuit, violating the law of charge conservation. Thus for circuit components in parallel:

  $I_T = I_1 + I_2 + I_3$

  where I_T is the total current in the wire

  I_1, I_2 and I_3 are the currents in each parallel branch
- The potential difference across each component in a parallel circuit must be the same as that across each branch, as each branch starts and ends at the same point. This is similar to the gravitational difference between two different heights, where the work done in moving between the points is independent of the path taken. We can express this relationship for parallel components as:

  $V_T = V_1 = V_2 = V_3$

  where V_T is the total potential difference

  V_1, V_2 and V_3 are the potential differences across each parallel component
- For resistors in parallel we can use Ohm's law to replace the current (I) in the parallel circuit equation above with $I = V/R$ to obtain:

  $V_T/R_T = V_1/R_1 + V_2/R_2 + V_3/R_3$

As all the voltages are equal for resistors on parallel, we can cancel them out to obtain an expression for the total resistance:

$$1/R_T = 1/R_1 + 1/R_2 + 1/R_3$$

where R_T is the single resistor that would be equivalent to the three parallel resistors

R_1, R_2 and R_3 are the resistances of each of the resistors

- Voltmeters measure potential difference and are wired in parallel across the two points where the potential difference needs to be measured. Voltmeters have a high resistance to prevent current flowing through the meter rather than through the circuit.

KEY QUESTIONS

7 **What happens to the total resistance when a resistor is added in series?**

8 **What happens to the total resistance of a circuit when a resistor is added in parallel?**

9 **Are voltmeters and ammeters connected in series or parallel?**

Answers ➲ p. 168

EXAMPLE 7

Figure 12.9 shows four resistors connected in series. Given that the current flowing from the battery is 0.5 A, find the value of the unknown resistor marked R.

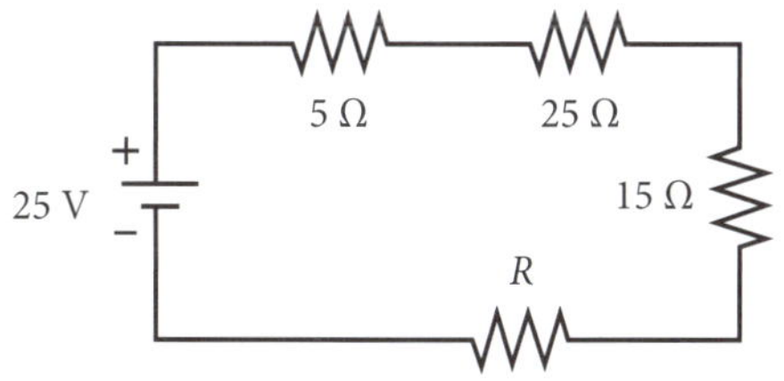

Figure 12.9 Resistors in series

Answer:

Resistors in series add to produce a greater total resistance

The series resistors could be replaced with a single resistor:

$R_T = R_1 + R_2 + R_3$

And hence $R_T = R + 5 + 25 + 15 = R + 45$

Now applying Ohm's law to the entire circuit:

$R_T = V_T/I_T$

And hence $(R + 45) = \frac{25}{0.5}$ and $R = \frac{25}{0.5} - 45 = 5\ \Omega$

EXAMPLE 8

Figure 12.10 shows three resistors in parallel. Calculate the current in the 5 Ω resistor and the total current in the circuit.

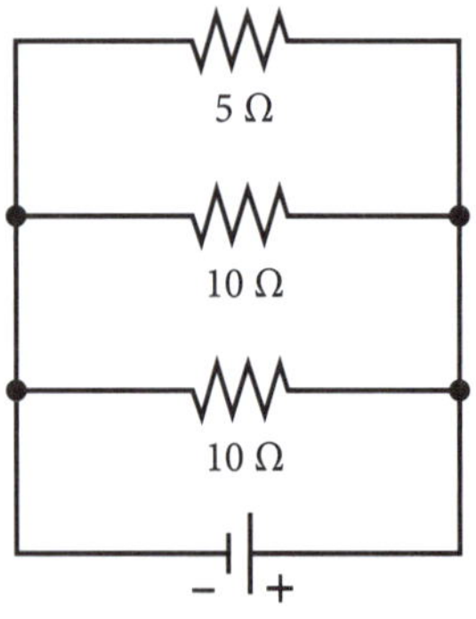

Figure 12.10 Resistors in parallel

Answer:

> Parallel resistors have the same potential difference across each resistor

Applying Ohm's laws to the 5 Ω resistor:

$I_5 = V/R = \frac{100}{5} = 20$ A

To find the total current we can replace all the resistors with a single resistor. For resistors in parallel:

$1/R_T = 1/R_1 + 1/R_2 + 1/R_3$

Hence $1/R_T = \frac{1}{5} + \frac{1}{10} + \frac{1}{10} = \frac{4}{10}$

$R_T = 2.5\ \Omega$

Applying Ohm's law to the entire circuit:

$R_T = V_T/I_T$

Hence $I_T = V_T/R_T = \frac{100}{2.5} = 40$ A

Alternatively, we could find the current in each of the 10 Ω resistors:

$I_{10} = V/R = \frac{100}{10} = 10$ A

The total current passing through the parallel resistors is given by:

$I_T = I_1 + I_2 + I_3 = 10 + 10 + 20 = 40$ A

EXAMPLE 9

Figure 12.11 shows a circuit with resistors in series and in parallel. Find the voltage across and the current in each of the resistors.

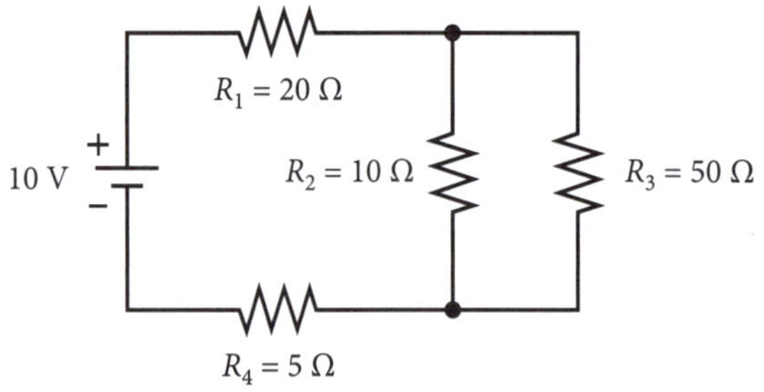

Figure 12.11 Resistors in series and parallel

Answer:

> The total potential difference is equal to the sum of the potential differences for resistors in series

The total current will pass through the 5 Ω and 20 Ω resistors, as they are in series with the power supply. We can find the total current by replacing the four resistors with a single resistor. To do this we first replace the two parallel resistors with a single resistor:

$1/R_T = \frac{1}{50} + \frac{1}{10} = \frac{6}{50}$

And hence the parallel resistors could be replaced with a resistor with a resistance of $R = \frac{50}{6} = 8.33\ \Omega$.

Now as the other resistors are in series, the total resistance for the circuit will be:

$R_T = 5 + 20 + 8.33 = 33.33\ \Omega$

Applying Ohm's law to find the total current gives:

$I_T = V_T/R_T = \frac{10}{33.33} = 0.3$ A

Now as this is the current through the 5 Ω and 20 Ω resistors we have:

$V_5 = IR = 0.3 \times 5 = 1.5$ volts, and $V_{10} = 0.3 \times 20 = 6$ volts

Replacing the parallel resistor with a single resistor ($R_P = 8.33\ \Omega$) and adding the voltages around the circuit will enable the voltage across the parallel resistors to be found:

$V_T = V_1 + V_2 + V_3$

$10 = 1.5 + 6 + V_P$

$V_P = 2.5$ volts

Finally the current in each of the parallel resistors can be calculated by applying Ohm's law to each resistor:

$I_{10} = V/R = \frac{2.5}{10} = 0.25$ A

$I_{50} = V/R = \frac{2.5}{50} = 0.05$ A

Note that the sum of the currents in the parallel resistors equals the total resistance, as expected.

FIRSTHAND INVESTIGATION 2

Series and parallel circuits

Students should use voltmeters and ammeters or multimeters to measure the potential difference (V) and current (I) in circuits with resistors in series, in circuits with resistors in parallel, and in circuits with resistors in series and parallel. The experimental results should be compared to the theoretical results calculated using the laws of series and parallel circuits, and any discrepancies between theory and experiment investigated.

Kirchhoff's circuit laws

- A useful technique for solving circuit problems is to apply Kirchhoff's circuit laws.
- **Kirchhoff's current law** states that for any node in the circuit the sum of the current moving into the node must be equal to the sum of the currents moving out of the node. If this was not the case, current would not be conserved at the node. Applying Kirchhoff's current law to the circuit shown in Figure 12.12, we see that:

  $I_1 + I_2 = I_3$

Kirchhoff's current law: a law stating that the sum of the currents entering and leaving any node in a circuit must be zero

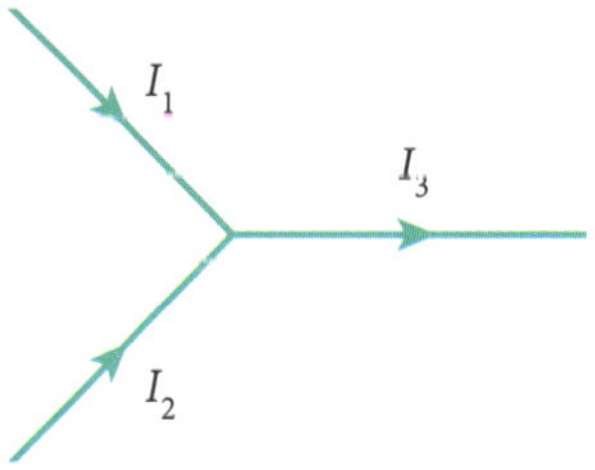

Figure 12.12 Example of currents moving into and out of a node in a circuit

- **Kirchhoff's voltage law** states that the voltage changes around any loop in a circuit must sum to zero. This law is a consequence of conservation of energy because when we return to the same point in the circuit we must be at the same potential. A simple example is shown in Figure 12.13. Appling Kirchhoff's voltage law to this circuit shows that the voltage increase (i.e. from the lower potential negative side of the battery to the higher potential positive side) moving across the battery must be equal to the voltage drop across the resistor and hence:

  EMF – $IR = 0$

  And hence EMF = IR

Kirchhoff's voltage law: a law stating that the sum of the voltage drops and rises around any loop in a circuit must be zero

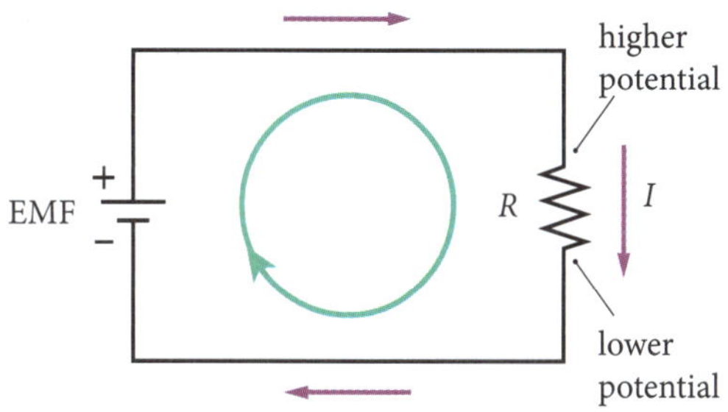

Figure 12.13 Using Kirchhoff's voltage law in a simple circuit

- To use Kirchhoff's laws in practice is a little more complex because we do not always know the current directions in the circuit. To get around this difficulty we make an educated guess about the current direction first. We then apply Kirchhoff's law by moving in one direction around the circuit. If we move across a resistor *in the direction* of the current the voltage drops ($V = -IR$), but if we move *in the opposite direction* to the current the voltage increases ($V = +IR$) across the resistor.
- If we move from negative to positive across the battery the EMF is positive, as we are moving from the lower to a higher potential. If we move from the positive to the negative across the battery we are moving to a lower potential and the EMF is negative. For example, if we assumed the current was in the opposite direction in Figure 12.13 and we applied Kirchhoff's law by moving around the loop in the direction of the inner arrow, the IR would be positive and the EMF would be positive and hence:

  EMF + $IR = 0$

  Hence $I = -(\text{EMF}/R)$

  The negative sign tells us that the current direction we initially guessed was incorrect.
- To solve circuits using Kirchhoff's laws we:
  - guess the current directions and label each of the currents I_1, I_2, I_3, etc.
  - write equations for the currents in the nodes in the circuit
  - follow the loops around in one direction and write equations for the voltages in each loop, remembering to follow the sign conventions
  - solve the equations simultaneously to find the unknown voltages and currents.

→ KEY QUESTIONS

10 How is Kirchhoff's circuit law related to conservation of charge?

11 How is Kirchhoff's voltage law related to conservation of energy?

Answers ➲ p.168

EXAMPLE 10

Use Kirchhoff's laws to solve the circuit problem previously solved using Ohm's law and series and parallel resistance laws in example 9.

Answer:

The sum of the potential gains and drops around any loop in a circuit must be zero

We first guess the direction of the current in each branch and label the currents, as shown in Figure 12.14.

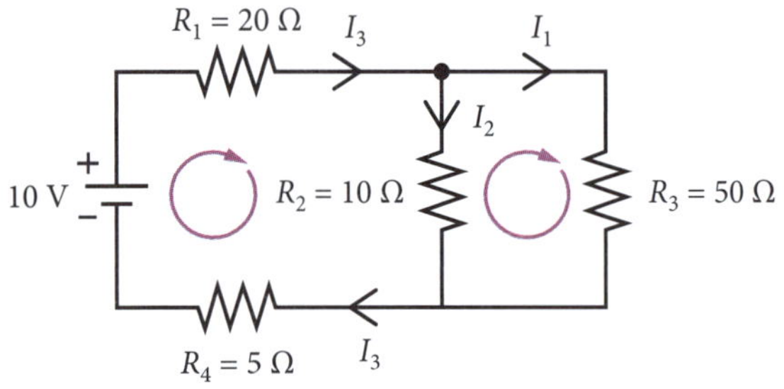

Figure 12.14 Solving a circuit problem using Kirchhoff's laws

Applying the current law to the note marked with the dot:

$I_3 - I_1 - I_2 = 0$ .... (1)

Now adding the voltage changes around each loop in the direction of the arrows in Figure 12.14 gives:

$10 - 20I_3 - 10I_2 - 5I_3 = 0$ .... (2)

$10I_2 - 50I_1 = 0$ .... (3)

Because there are three equations and unknowns we can solve these equations simultaneously. This can be done in many ways. For example, from equation (3) we have:

$I_2 = 5I_1$ .... (4)

Substituting this into equation (1) gives:

$I_3 = 6I_1$ .... (5)

Substituting these values for I_2 and I_3 into equation (2) gives:

$I_1 = 0.05$ A

Substituting this value for I_1 into equations (4) and (5) gives:

$I_2 = 0.25$ A and $I_3 = 0.3$ A

We can the find the voltage on each resistor, if required, by using Ohm's law ($V = IR$).

EXAMPLE 11

Find the currents in each branch of the circuit shown in Figure 12.15.

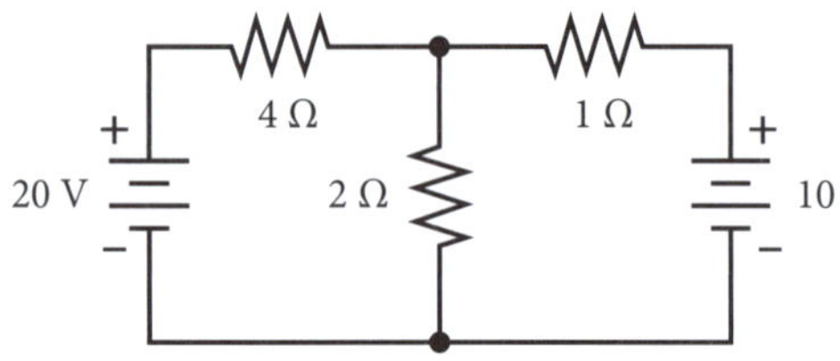

Figure 12.15 Circuit with two sources of *EMF*

Answer:

First guess the directions; if your answer shows a current is negative, your original guess was incorrect

We first guess current directions and label the currents as shown in Figure 12.16.

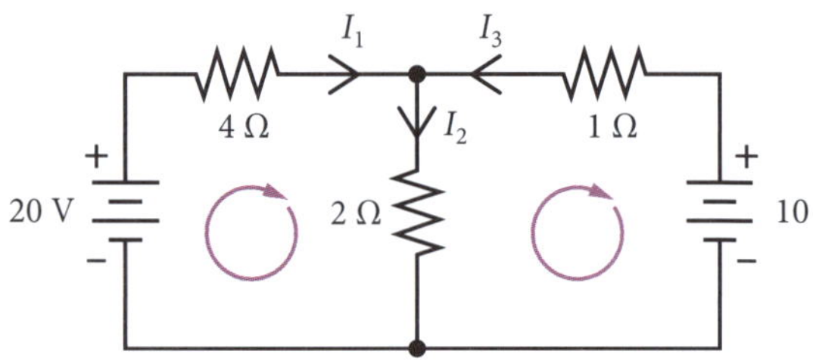

Figure 12.16 Assumed current directions

Appling the current law to the top node gives:

$I_1 + I_3 - I_2 = 0$

And hence $I_3 = I_2 - I_1$ .... (1)

Applying the voltage law to the left-hand loop, moving clockwise around the loop gives:

$20 - 4I_1 - 2I_2 = 0$

And hence $I_2 = 10 - 2I_1$ .... (2)

Applying the voltage law to the right-hand loop, moving clockwise around the loop gives:

$I_3 - 10 + 2I_2 = 0$

And hence $I_3 = 10 - 2I_2$ .... (3)

Note that the voltage change across the 1 Ω resistor and the 2 Ω resistor is positive because we are moving against the assumed direction of current flow, and the voltage change across the 10 V battery is negative because we are moving from the high to the low potential.

Equating relationships (1) and (3) to remove I_3 gives:

$I_2 - I_1 = 10 - 2I_2$

And hence $3I_2 = 10 + I_1$

Substituting for I_2 from equation (2) gives:

$3(10 - 2I_1) = (10 + I_1)$

And hence $I_1 = \frac{20}{7} = 2.86$ A

Substituting this value for I_1 into equation (2) gives:

$I_2 = 4.28$ A

Finally, substituting these values for I_1 and I_2 into equation (1) gives:

$I_3 = 1.42$ A

All the currents are positive, which tells us that the current does flow in the directions we initially assumed it would.

FIRSTHAND INVESTIGATION 3

Kirchhoff's laws

Students should use ammeters to investigate the currents entering and leaving nodes in a number of circuits to verify Kirchhoff's current law. Students should also use voltmeters to measure the changes in voltage across components in circuit loops to verify Kirchhoff's voltage law.

5 Electric currents and heat

» Students investigate quantitatively the application of the law of conservation of energy to the heating effects of electric currents, including the application of $P = VI$ and variations of this involving Ohm's law.

➔ We saw earlier in the chapter that the rate at which electrical energy is converted into heat in a resistor is related to the potential difference (V) across the resistor and the current (I) flowing in the resistor by:

$P = IV$

This is because the potential difference across the resistor is the energy transformed per coulomb of charge that passes through the resistor, and the current is the number of coulombs that move across the resistor per second. The product of the potential difference and the current therefore gives us the energy transformed per second or the power (P).

➔ By using Ohm's law in the form, $V = IR$ and substituting for V in the power equation we obtain:

$P = I^2R$

Similarly, by using Ohm's law in the form $I = V/R$ and substituting for I we obtain the expression:

$P = V^2/R$

➔ As power = energy/time, energy = power × time and, substituting the expression above for the electrical power we see that the heat produced by a resistor can be calculated from:

$$Q = I^2Rt$$

where Q is the heat
I the current
R the resistance
t is the time

➔ This relationship, called **Joule's law**, enables us to calculate the heat produced by a resistor.

Joule's law: a law stating that the heat produced by a resistor is related to the resistance, current and the time the current flowed by $Q = IR^2t$

FIRSTHAND INVESTIGATION 4

Resistance heating

Students investigate conservation of energy in an experiment in which they use electrical energy to heat water. Using the experimental arrangement shown in Figure 12.17, students measure the applied voltage and current, the time over which the electrical energy is applied and the change in the water temperature.

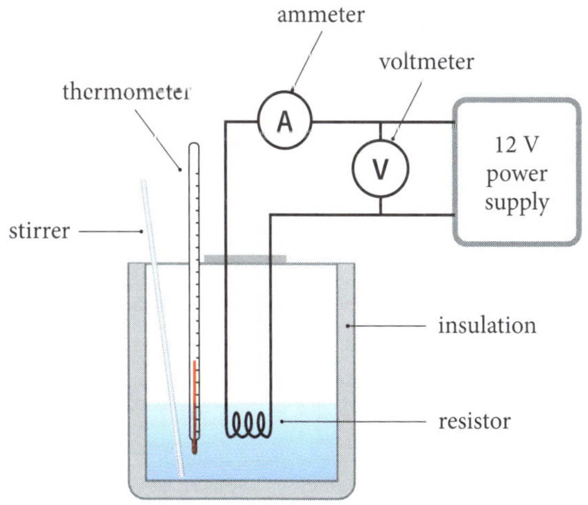

Figure 12.17 Experimental set-up to investigate the transformation of electrical energy to heat

Using the heat capacity of water (4184 $\text{Jkg}^{-1}\text{K}^{-1}$) students can calculate the heat transferred to the water using:

$Q = mc\Delta T$

This energy is then compared to the electrical energy:

Energy = power/time = IV/t

If all the electrical energy goes into heating the water these two quantities would be equal. Differences in the two quantities should be explained and investigated by the students.

EXAMPLE 12

A 240 V electric kettle uses a resistance heating-coil to boil water. If it takes 4 minutes for the kettle to boil 1.5 L of water that was initially at 25 °C, what is the resistance of the heating coil used in the kettle?

Answer:

We can use power and time to find the energy that goes into heating the water

The heat required to boil the kettle is given by:

$Q = mc\Delta T = 1.5 \times 4184 \times 75 = 4.707 \times 10^5$ J

The rate at which heat must be supplied to boil the water in 4 minutes is given by:

Power = energy/time = $\dfrac{4.707 \times 10^5}{4 \times 60} = 1.96 \times 10^3$ W

Electrical power is related to resistance by:

$P = V^2/R$ and hence $R = V^2/P = \dfrac{240^2}{1.96 \times 10^3} = 29.4\ \Omega$

CHAPTER SYLLABUS CHECKLIST

Are you able to answer these questions from the syllabus for this chapter? Tick each question as you go through the checklist if you are able to answer it. If you cannot answer a question, turn to the relevant page in the study guide to find the answer. For NESA key word meanings, go to www.educationstandards.nsw.edu.au and search 'key words'.

FOR A COMPLETE UNDERSTANDING OF THIS TOPIC:		PAGE NO.	✓
1	Can I define electric current?	153	
2	Can I define electromotive force?	153	
3	Can I outline Ohm's law and discuss its limitations?	153–154	
4	Can I define electrical resistance?	154	
5	Can I differentiate between ohmic and non-ohmic conductors?	155	
6	Can I outline the factors that determine the electrical resistance of an object?	155	
7	Can I interpret voltage against current graphs in terms of Ohm's law and resistance?	155–156	
8	Can I relate electric potential difference to work?	157	
9	Can I explain how the total voltage, current and resistance of resistors in series are determined?	158	
10	Can I explain how the total voltage, current and resistance of resistors in parallel are determined?	159	
11	Can I solve circuit problems by applying the laws for series and parallel resistors and Ohm's law?	159–160	
12	Can I outline Kirchhoff's circuit laws and explain how they are related to the conservation of charge and energy in a circuit?	161	
13	Can I solve circuit problems using Kirchhoff's laws?	162	
14	Can I relate the power transformed to heat in a resistor to the potential difference across and current in the resistor?	163	
15	Can I solve problems and make predictions about energy transfers and transformations in electrical circuits using conservation of energy?	163–164	

Objective-response questions (1 mark each)

1 If a current I passed through the heating filament of a 36 V incandescent lightbulb when 12 V as applied across the bulb, what current would flow if the voltage was increased to 36 V?

A 0
B $3I$
C more than $3I$
D less than $3I$

2 What can be implied about the resistors X and Y from the data shown in Figure 12.18?

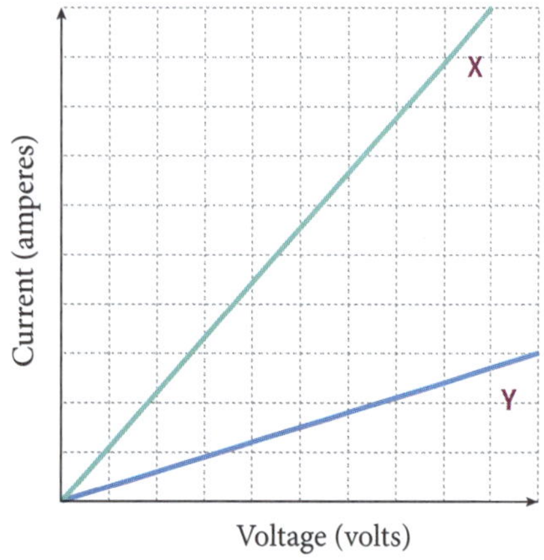

Figure 2.18 Current as function of applied voltage for two resistors, X and Y

A Both resistors are ohmic and X has more resistance than Y.
B Both resistors are non-ohmic and X has more resistance than Y.
C Both resistors are ohmic and Y has more resistance than X.
D Both resistors are non-ohmic and Y has more resistance than X.

3 A simple circuit containing three identical light globes is shown in Figure 12.19. If the switch S was opened, how would the brightness of the lightbulbs change?

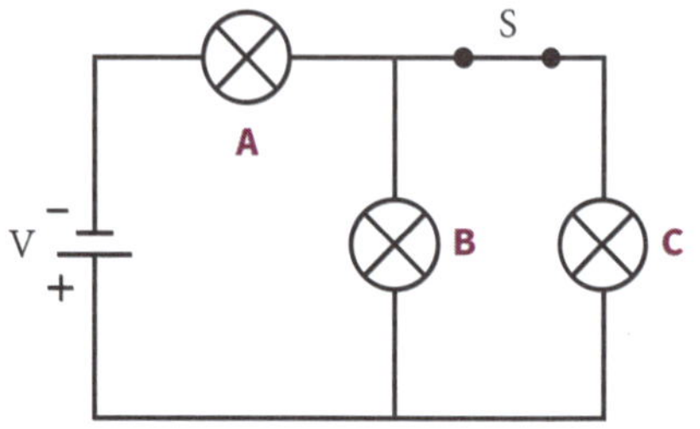

Figure 12.19 Circuit containing three lightbulbs

A C would go out and globes A and B would get brighter.
B C would go out and bulbs A and B would glow dimmer.
C C would go out, A would glow brighter and B would be dimmer.
D C would go out, B would glow brighter and A would be dimmer.

4 Which of the following best compares the potential difference across and current in each of the resistors shown in Figure 12.20?

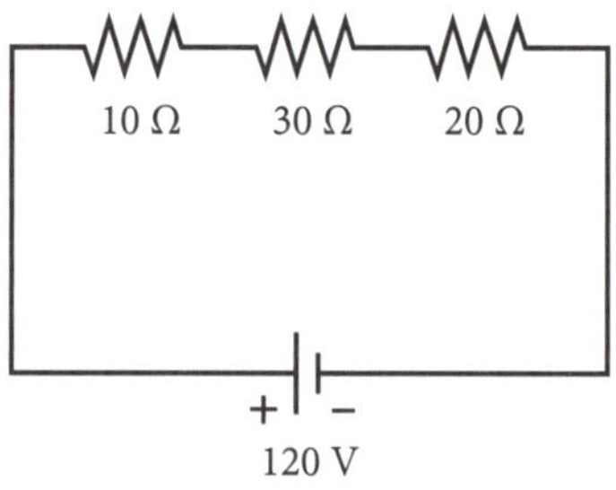

Figure 12.20 Circuit containing three resistors

A The current is least in the 10 Ω because it has the lowest resistance.
B The current is least in the 30 Ω resistor because it is the best conductor.
C The potential difference across each resistor is identical because they are in series.
D The potential difference across the 30 Ω is greatest because it has the highest resistance.

5 A 240 V electric kettle uses resistance wire to heat water and takes a time (T) to heat 1 L of water. If the operating voltage was doubled, how would it change the time the kettle would take to boil a specific volume of water? (You may assume the resistance remains constant.)

A T^2
B $T/2$
C $T/4$
D $\sqrt{T}$

Extended-response questions

6 This question refers to the circuit shown in Figure 12.21.

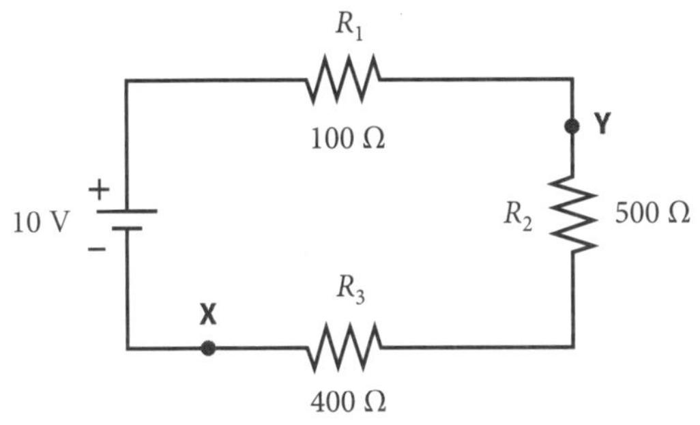

Figure 12.21 Resistors in a circuit

a Find the total resistance of the circuit. (1 mark)

b Determine the current flowing through R_1. (1 mark)

c Determine the potential difference across R_2. (1 mark)

d Find the potential difference between the point X and Y? (1 mark)

7 This question refers to the circuit shown in Figure 12.22.

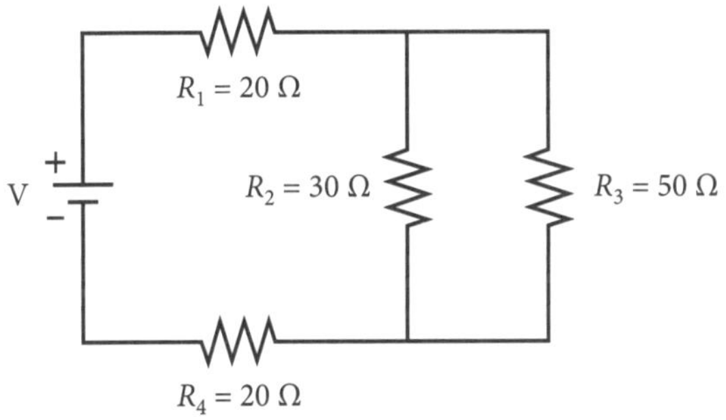

Figure 12.22 Resistors in a circuit

A current of 0.3 A flows through the 50 Ω resistor.

a Find the current in each of the other resistors. (2 marks)

b Find the voltage across each of the resistors. (2 marks)

c Find the voltage of the battery. (1 mark)

d Determine which resistor uses the least electrical power and find the power this resistor uses. (2 marks)

8 Use Kirchhoff's circuit laws to write three simultaneous equations that would enable the currents in the circuit shown in Figure 12.23 to be determined. Note that you do not have to solve the equations. (3 marks)

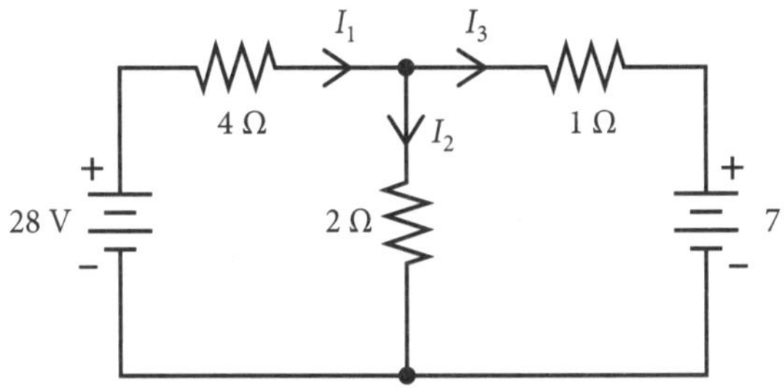

Figure 12.23 Circuit with two power supplies

9 A 240 V household water heater draws 20 A of current. When the heater is operated for 30 minutes it heats 200 L of water by 9.5 °C. Note the specific heat of water is 4184 $Jkg^{-1}K^{-1}$.

a Find the electrical energy consumed in the 30 minutes of operation. (1 mark)

b Use the temperature change of the water to find how much the thermal energy of the water is changed in the 30 minutes when the heater was operating. (1 mark)

c Determine the efficiency (useful energy out × 100/input energy%) of the water heater, and explain why the efficiency is less than 100%. (2 marks)

10 A farmer uses a 240 V electric pump to move water from a dam to a field 8 m above the dam. Assuming the pump is 100% efficient and draws a current 50 A when operating, and that 1 L of water as a mass of 1 kg, find:

a the energy required to lift a litre of water from the dam to the field. (1 mark)

b the power produced by the motor. (1 mark)

c the volume of water the pump would lift to the field each minute. (1 mark)

KEY QUESTIONS

Key questions ➲ p. 154

1 Current is a measure of the amount of charge that flows past a point in a circuit per unit time.

2 Direct current only flows in one direction in a wire, while alternating current oscillates backwards and forwards in a wire.

Key questions ➲ p. 155

3 In ohmic conductors at constant temperature, the current that flows is proportional to the potential difference placed across the conductor.

4 The resistance of an ohmic conductor depends on the shape of the conductor, the temperature of the conductor and the material from which the conductor is made. Resistance increases with length and decreases with the cross-sectional area. For metals the resistance increases with temperature.

Key questions ➲ p. 157

5 When current flows in a resistor the electrons collide with lattice atoms, transferring energy to the lattice and increasing the temperature of the resistor.

6 The charge loses potential energy as it moves across the resistor but this energy is transformed into thermal energy in the resistor.

Key questions ➲ p. 159

7 Adding a resistor in series increases the total resistance as it is the equivalent of increasing the length of a resistor.

8 Adding a resistor in parallel decreases the total resistance as it is equivalent to increasing the cross-sectional area of a resistor.

9 Ammeters are connected in series and voltmeters are connected in parallel.

Key questions ➲ p. 162

10 Kirchhoff's current law is a statement of conservation of charge for currents moving into and out of a circuit node. It simply states that the current entering a node must equal the current leaving the node, or charge would be created or destroyed at the node.

11 Kirchhoff's voltage law is a statement of conservation of energy because when a charge returns to the same point in a circuit it must return to the same potential energy. Work may be done on or by the charge as the charge moves around the circuit, but the net energy change must be zero or the charge would not be at the same potential when it returned to the starting point.

YEAR 11 EXAM-TYPE QUESTIONS

Objective-response questions

1 **D.** The resistance of the filament will be greater because the filament temperature will be higher when a greater voltage is applied across it. **B** is incorrect because this would only occur if the resistance remained constant. **C** is incorrect because it would require the resistance to decrease when more current flowed. **A** is incorrect because current would continue to flow through the resistor.

2 **C.** A straight-line graph of I against V proves the resistors are ohmic as the gradient is constant and hence the resistance is constant. Because the gradient of the graph is $I/V = 1/R$, the resistor with the steeper gradient (X) has the lower resistance. **B** and **D** are incorrect because the graphs are linear, proving that the resistors are ohmic. **A** is incorrect because the gradient is $1/R$ and hence X has a greater resistance than Y.

3 **D.** When the switch S is opened, no current would flow through light globe C and hence it would go out. The total resistance of the circuit would increase with one less parallel branch and hence the total current would decrease. This would decrease the voltage across bulb A, decreasing the brightness of bulb A, but would increase the voltage across bulb B, increasing its brightness. **A** is incorrect because the voltage across bulb A and hence its brightness would decrease. **B** is incorrect because the voltage and brightness of globe B would increase. **C** is incorrect because the voltage across and hence brightness of bulb A would decrease.

4 **D.** The same current would flow through all the resistors as they are in series and the voltage across the largest resistor is greatest as $V = IR$. **A** and **B** are incorrect because the current in each resistor must be identical as they are wired in series. **C** is incorrect as the voltage across each resistor will be different as it is given by $V = IR$.

5 **C.** Because the power transformed to heat the kettle is given by $P = V^2/R$, doubling the voltage would increase the power by a factor of $2^2 = 4$ times. This means the kettle will boil four times faster. **A**, **B** and **D** are incorrect as they are greater or smaller then $T/4$.

Extended-response questions

6 EM This question tests students' understanding of resistors in series, Ohm's law and potential drop around a circuit.

a As the resistors are in series $R_T = R_1 + R_2 + R_3$
$= 100 + 500 + 400 = 1000\ \Omega$ ✓

b The same current will flow through all the resistors as they are in series:
$I_1 = I_T = V_T/R_T = \dfrac{10}{1000} = 0.01$ A ✓

c Applying Ohm's law to resistor R_2, $V = IR = 0.01 \times 500 = 5$ V. ✓

d The potential difference between X and Y would be the sum of the potential differences across R_2 and R_3. Potential difference across $R_3 = IV = 0.01 \times 400 = 4$ V, and the potential difference across R_3 was 5 V. Therefore the potential difference between the points X and Y is $4 + 5 = 9$ V. ✓

7 EM This question tests students' understanding of parallel resistors and circuits involving series and parallel resistors.

a The potential difference across the 50 Ω resistor is $V = IR = 50 \times 0.3 = 15$ V. Now as R_2 is in parallel with the 50 Ω resistor, it has the same potential difference across it and $I_2 = V/R = \dfrac{15}{30} = 0.5$ A. ✓ The current in R_1 and R_4 will be the sum of the currents in R_2 and R_3 and hence $I_1 = I_4 = 0.3 + 0.5 = 0.8$ A. ✓

b The potential difference across R_2 and R_3 was 15 V. ✓
The potential difference across R_1 is $V_1 = IR = (0.8 \times 20) = 16$ V, and across R_4 is $V_4 = IR = (0.8 \times 20) = 16$ V. ✓

c The total potential difference will be the sum of the potential difference across the parallel combination and the potential difference across the two series resistors.
Hence $V_T = V_1 + V_2 + V_3 = 15 + 16 + 16 = 47$ V. ✓

d Applying $P = IV$ to each resistor we see that R_3 uses the least power, ✓ and $P = 0.3 \times 15 = 4.5$ W. ✓

8 **EM** This question tests students' ability to apply Kirchhoff's laws to a circuit containing two sources of EMF.

Applying Kirchhoff's circuit law to the top node:

$I_1 - I_2 - I_3 = 0$ or $I_1 = I_2 + I_3$ ✓

Applying Kirchhoff's voltage law by moving clockwise around the left-hand loop:

$28 - 4I_1 - 2I_2 = 0$ ✓

Applying Kirchhoff's voltage law by moving clockwise around the right-hand loop:

$2I_2 - I_3 - 7 = 0$ ✓ (or the equation below)

Alternatively students could apply Kirchhoff's voltage law to the outside loop to obtain:

$28 - 4I_1 - I_3 - 7 = 0$

9 **EM** This question tests students' understanding of electrical power, specific heat, conservation of energy and efficiency.

a Using power = energy/time, $E = Pt = (IV)t = 240 \times 2 \times (30 \times 60)$ $= 8.64 \times 10^6$ J ✓

b $Q = mC\Delta T = 200 \times 4184 \times 9.5 = 7.95 \times 10^6$ J ✓

c Efficiency = (useful output energy/useful input energy) × 100%

Efficiency = $\frac{7.95\times10^6}{8.64\times10^6}\times100$ = 92%. ✓ Some energy is lost to the environment or in heating the surrounds of the heater. ✓

10 a Energy = the increase in potential energy = $mgh = 1 \times 9.8 \times 8$ $= 78.4$ J ✓

b Power = $IV = 240 \times 50 = 1.2 \times 10^4$ W ✓

c Power = E/t, and hence, energy = $Pt = (1.2 \times 10^4 \times 60)$ $= 7.2 \times 10^5$ J ✓

Volume of water = $\frac{7.2 \times 10^5}{78.4}$

= 9183 litres min^{-1} ✓

CHAPTER 13 MAGNETISM

INQUIRY QUESTION:

How do magnetised and magnetic objects interact?

Magnets exert non-contact forces on other magnets and on objects that can be magnetised.

1 Magnets and magnetism

» Students investigate and describe qualitatively the force produced between magnetised and magnetic materials in the context of ferromagnetic materials.
» Students investigate and explain the process by which ferromagnetic materials become magnetised.

Magnetic force

➔ You will recall from previous studies that magnets have two poles; a north-seeking pole and a south-seeking pole. All investigations to date have confirmed that magnets always have two poles and magnetic monopoles do not exist. Magnets exert a non-contact magnetic force on one another. Like-poles repel one another and unlike-poles attract one another.

➔ In the same way that charged particles experience a force in electric fields, magnetic materials experience a force when they are placed in a magnetic field. We draw the magnetic field lines around a magnet from the **north-seeking pole** to the south-seeking pole, as shown in Figure 13.1. Like the electric field, the magnetic field is a vector field because it has a magnitude and direction in each point in space. Also like the electric field, the density of field lines reflects the strength of the field. We call a magnetic field a ***B* field.** The strength of the *B* field is called the **magnetic flux density**. It is a vector quantity with the SI units of tesla.

north-seeking pole: the pole of a magnet that points towards the Earth's North Pole when the magnet is freely suspended
***B* field:** a magnetic field
magnetic flux density: strength of the magnetic field, expressed using the tesla (T) SI units

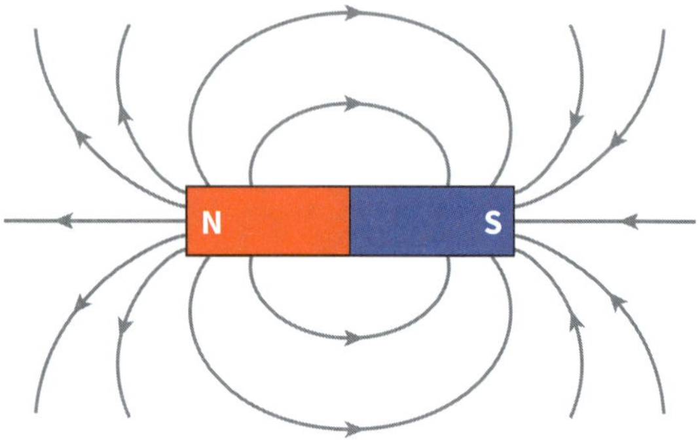

Figure 13.1 Magnetic field around a bar magnet

➔ The Earth has a magnetic field that is produced by currents in the liquid-iron outer core, deep in the Earth. A magnet suspended in the Earth's magnetic field will align with the field, with its north-seeking pole directed towards the Earth's North Pole. This phenomenon is utilised by magnetic compasses. The forces that cause a magnet to align with the field are shown in Figure 13.2.

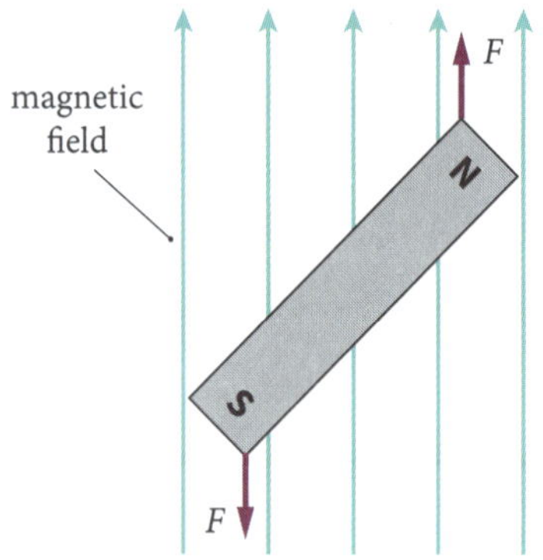

Figure 13.2 Illustration of forces that cause a magnet to align with an external magnetic field

➔ Note that because unlike-poles attract, the Earth's North Pole must be the opposite pole to the north-seeking pole of a magnet. That is, the Earth's North Pole is actually the south pole of a magnet. This idea is illustrated in Figure 13.3. The Earth's magnetic field has a magnitude of about 5×10^{-5} Tesla on the Earth's surface. Note also that the North Magnetic Pole is not coincident with the Earth's axis of rotation, and that the North Magnetic Pole actually drifts over time.

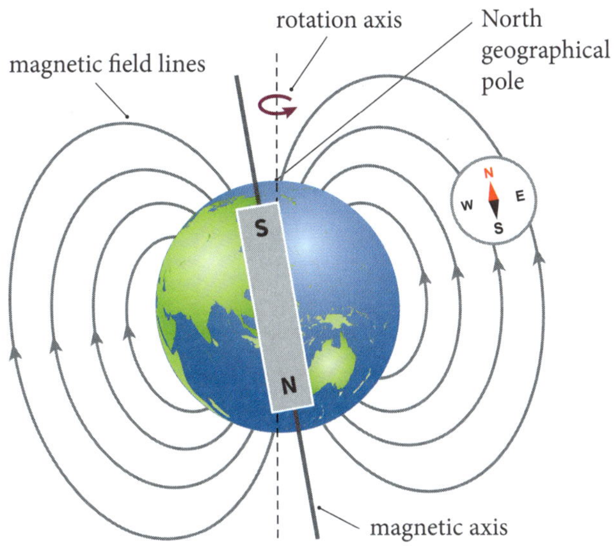

Figure 13.3 Earth's magnetic field

- If an object experiences a torque (turning force) when it is placed in a magnetic field, we say the object has a magnetic moment. Having a magnetic moment essentially means that the object is magnetic. Bar magnets, the Earth and loops of wire carry current; molecules and many subatomic particles have magnetic moments.
- Electrons possess a quantum mechanical property called **spin** that causes them to have a magnetic moment. Electrons therefore act like tiny bar magnets. Quantum mechanics shows that electrons can only have one of two spin states, called spin up and spin down. Electrons generally combine in opposite-spin pairs in atoms, which causes their magnetic moments to cancel out. When this occurs the atom does not have a net magnetic moment.

spin: quantum mechanical property that gives electrons a magnetic moment

KEY QUESTIONS

1. **How is the north pole of a magnet defined?**
2. **Can a magnet have a north-seeking pole alone?**
3. **Why does a magnetic field exert a turning force (torque) on a suspended magnet?**
4. **What is the direction of the magnetic field around a magnet?**

Answers ➲ p. 181

Ferromagnetic materials

- Magnets can exert forces on other magnets and materials that can be magnetised. We call a material that can be magnetised a **ferromagnetic material**. Ferromagnetic materials include iron, nickel, cobalt and their alloys, and some compounds of rare-earth metals. In these special materials some electrons are not paired and consequently they can produce regions within the material called **magnetic domains** that have a net magnetic moment.

ferromagnetic material: a material that contains domains that have net magnetic moments
magnetic domains: small magnetic regions found in ferromagnetic materials

- The magnetic domains in ferromagnetic materials are microscopic, ranging in size from 1 to 100 μm. Generally the magnetic moments of the domains are oriented randomly, as shown in Figure 13.4 (a), and the object has no net magnetic moment. But if a ferromagnetic material is placed in an external magnetic field, the magnetic moments of the domains align with the field and the object exhibits a net magnetic moment. This process is shown in Figure 13.4 (b).

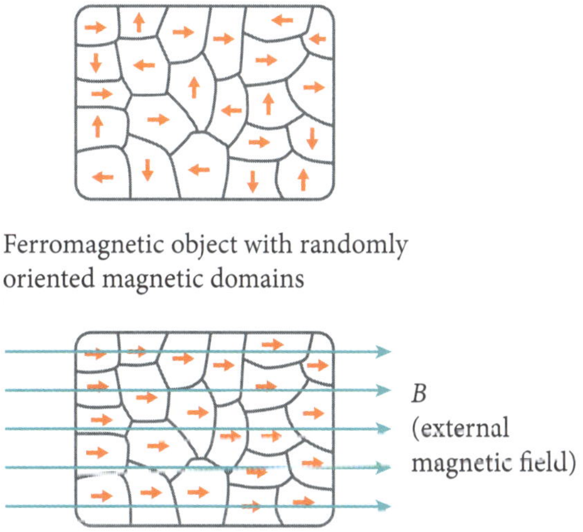

a Ferromagnetic object with randomly oriented magnetic domains

b Alignment of magnetic domains in a ferromagnetic object by an external magnetic field

Figure 13.4 Magnetic moments of domains in a ferromagnetic substance generally sum to zero but will align when they are placed in a strong external magnetic field

- The magnetic directions of the domains of some ferromagnetic materials, such as Neodymium–iron–boron and iron–neodymium maintain their orientation when the external magnetic field is removed, producing a permanent magnet. In other ferromagnetic materials, such as of iron and nickel–iron alloys, the orientation of the domains soon becomes random after the external magnetic field is removed. Ferromagnetic materials that maintain a net magnetic field are called **hard magnetic materials** and ferromagnetic materials that rapidly lose their net magnetic field are called **soft magnetic materials**.

→ The alignment of the magnetic moments of the magnetic domains in a permanent magnet can be disrupted by thermal agitation when a magnet is heated. The temperature at which a specific substance loses its magnetism is called the **Curie temperature**. The Curie temperature for iron is 1043 K and for nickel is 627 K. Mechanical shock waves moving though a permanent magnet—for example, by dropping it onto a hard surface—will also disrupt some of the alignment of the domains and reduce the strength of the magnet.

hard magnetic materials: ferromagnetic materials that maintain their magnetism after an external magnetic field is removed
soft magnetic materials: ferromagnetic materials that do not maintain their magnetism after an external magnetic field is removed
Curie temperature: temperature at which a permanent magnet ceases to be magnetic

→ KEY QUESTIONS

5 **What is a magnetic domain?**

6 **Why are ferromagnetic materials attracted to magnets?**

7 **How can an object be magnetised?**

Answers ➲ p. 181

FIRSTHAND INVESTIGATION 1

Mapping magnetic fields produced by permanent magnets

Students should examine magnetic fields around permanent magnets by using iron filings on a sheet of paper above a magnet (shown in Figure 13.5), and by using a compass (shown in Figure 13.6).

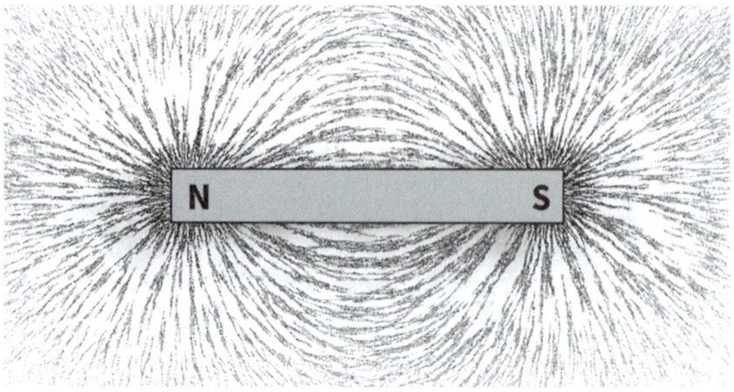

Figure 13.5 Using iron filings on a piece of cardboard or clear plastic placed above magnets to visualise magnetic fields

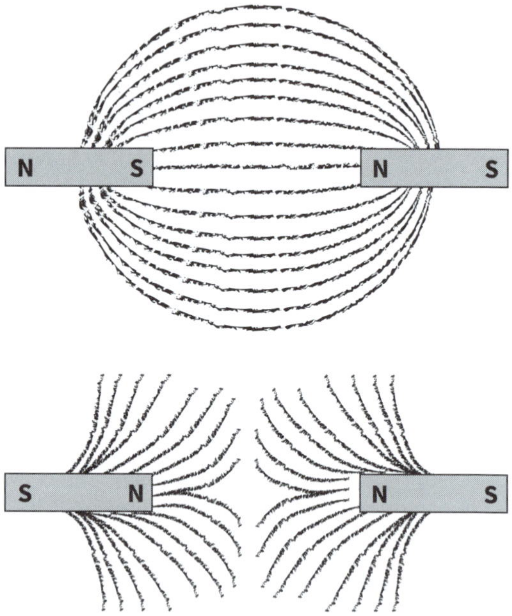

Figure 13.5 (Continued)

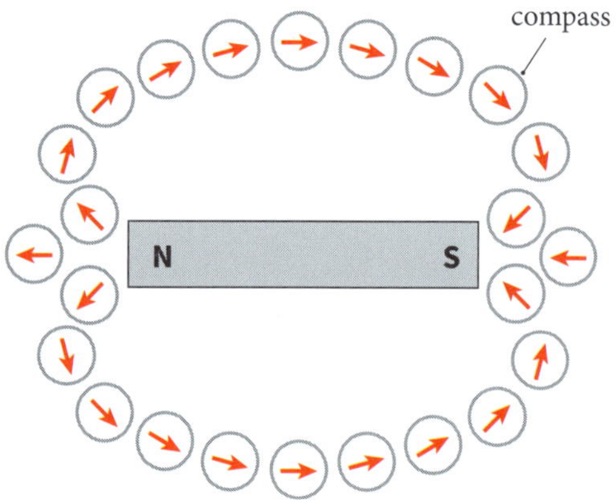

Figure 13.6 Mapping magnetic fields with a magnetic compass

EXAMPLE 1

If a magnet was cut in half, would one-half be a north pole and the other half be a south pole? (Justify your answer.)

Answer:

Recall that permanent magnets are made of a large number of tiny magnetic domains

No, the result would be two smaller magnets, both with a north and a south pole. We can understand this by recalling that the magnetic field is produced by the alignment of the magnetic field of a large number of tiny magnetic domains in the magnet. Cutting the magnet in half will not affect the domains, which will remain aligned, producing a north pole at one end and a south pole at the other of both halves of the magnet, as shown in Figure 13.7.

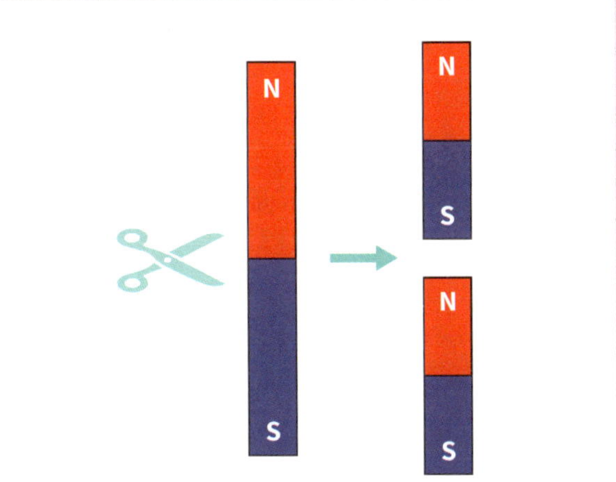

Figure 13.7 Cutting a magnet in half produces two smaller magnets

EXAMPLE 2

Explain how a permanent magnet could be made.

Answer:

Placing a hard magnetic material in a strong external magnetic field will align the magnetic moments of the domains. Because a hard magnetic material was used, the domains will remain aligned when the field is removed, leaving the objects with a net magnetic moment.

The magnetic moment of a magnetic domain will align with an applied magnetic field

EXAMPLE 3

Explain why permanent magnets should not be dropped or exposed to high temperatures.

Answer:

Permanent magnets rely on the alignment of electrons in magnetic domains.

The alignment of magnetic domains can be disrupted by collisions at the atomic level

Dropping magnets passes mechanical shock waves through the magnet, which may reorient some of the spinning changes in the domains, wakening the net magnetic moment of the magnet. Similarly, heating a permanent magnet causes motion in the atomic lattice, which will reorient some of the magnetic moments and reduce the net magnetic moment of the magnet.

2 Magnetism and electric current

» Students use magnetic field lines to model qualitatively the direction and strength of magnetic fields produced by magnets, current-carrying wires and solenoids, and relate these fields to their effect on magnetic materials that are placed within them.

→ When a ferromagnetic material is placed in a magnetic field the magnetic moments of the magnetic domains in the material align with the field, which gives the object a net magnetic moment. This is why magnets attract ferromagnetic objects. The ferromagnetic objects become magnetised by the field of the magnet and then get attracted to the magnet, as shown in Figure 13.8.

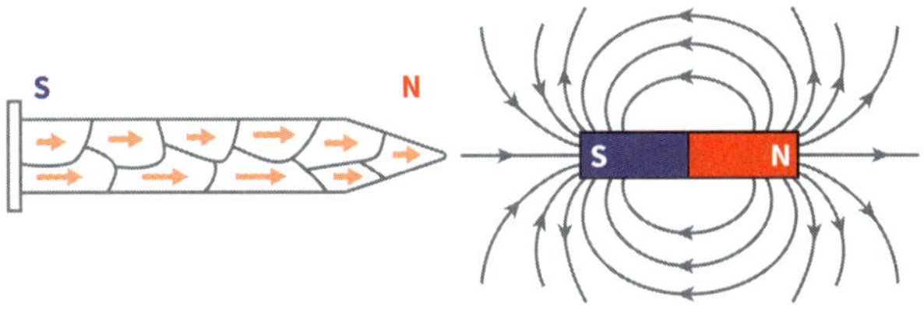

Figure 13.8 The magnetic domains in a nail are aligned by a nearby magnet, making the nail magnetic and causing it to be attracted to the magnet

→ In 1820 Hans Christian Oersted discovered that current in a wire produces a magnetic field around the wire that is directed in concentric circles around the wire. The direction of magnetic field is given by the **right-hand grip rule**. To apply the rule we grip the wire with the right hand, placing the thumb in the direction of the current. The fingers then point in the direction of the magnetic field as shown in Figure 13.9.

right-hand grip rule: a rule to determine the direction of the magnetic field around a current-carrying wire; we grip the wire with the thumb in the direction of the current and the fingers then point in the direction of the magnetic field

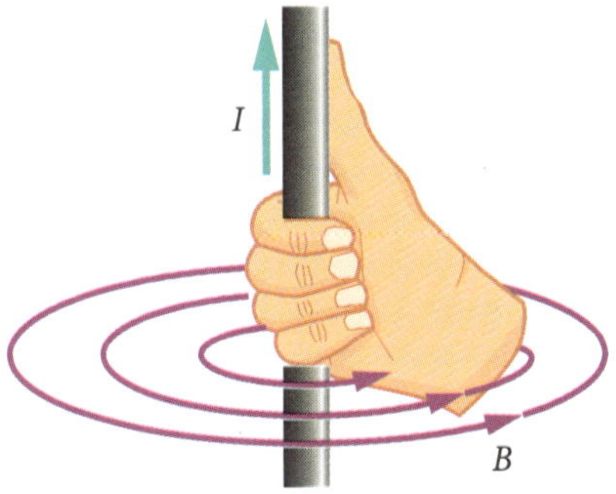

Figure 13.9 The right-hand grip rule gives the direction of the magnetic field around a wire carrying a current

→ When displaying fields and current-carrying wires in three dimensions, we use a cross to represent a field or current into the page, and a dot to represent a current or field out of the page. To remember this convention, think about the point of an arrow coming towards you (i.e. out of the page), and the tail feathers of the arrow going away from you (i.e. into the page). Figure 13.10 shows the use of this convention for wires carrying current out of the page and into the page. Note that the concentric circles are drawn closer together near the wire to represent the increased magnetic field strength close to the wire.

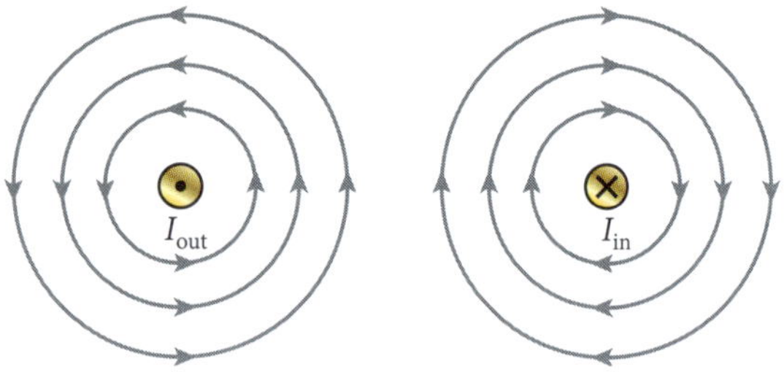

Figure 13.10 Magnetic fields around current-carrying wires; note that a dot is used to represent current into the page and a cross to represent current out of the page

→ If a current-carrying wire is wound into a helical coil, the magnetic field around each wire adds in the middle of the coil to produce a strong, uniform magnetic field. We call this device a **solenoid** or sometimes an air solenoid. Outside the solenoid the field is similar to the field around a bar magnet but inside the solenoid the symmetry of the windings produces a very uniform field. The magnetic field produced by a solenoid is shown in Figure 13.11. Note that the equally spaced magnetic field lines within the solenoid reflect the consistency of the field in this region.

solenoid: a helical coil of wire that produces a uniform magnetic field inside the solenoid and a field similar to a bar magnet outside the solenoid

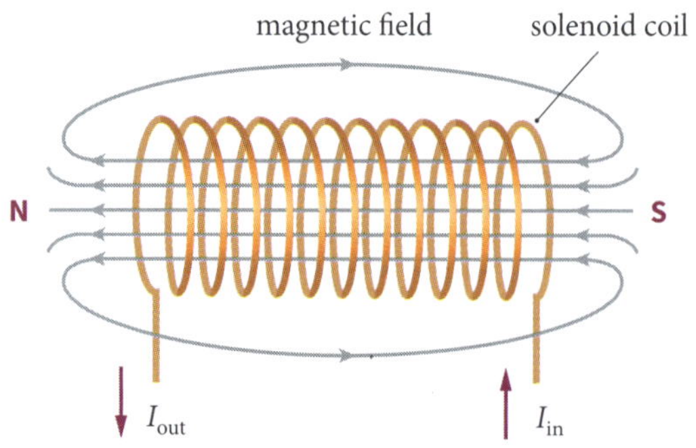

Figure 13.11 Magnetic field produced by a solenoid

→ The direction of the magnetic field in a solenoid can be determined by applying the right-hand grip rule to one of the wires in the solenoid coil or by putting the fingers around the solenoid in the direction of the current, as shown in Figure 13.12.

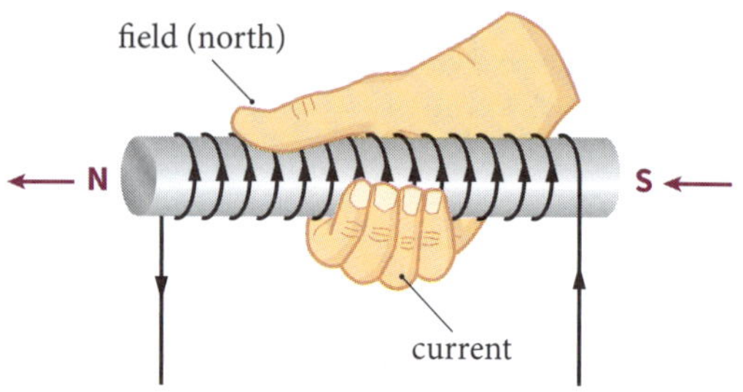

Figure 13.12 Finding the direction of the magnetic field produced by a solenoid

→ If a soft magnetic material (like soft iron) is placed inside a solenoid, the magnetic field of the solenoid will cause the magnetic field produced by the magnetic domains in the iron core to align with the field. The magnetic field produced by the aligned magnetic domains in the iron core will then add to the solenoid's magnetic field, greatly increasing the size of the magnetic field. This device is called an **electromagnet** and its magnetic field strength can be as much as 200 times greater than the magnetic field of the air solenoid alone. Because soft iron is used in the core, the magnetic moments of the domains in the core rapidly return to random orientations when the external magnetic field is switched off. This enables the magnetic field from the electromagnet to be switched off simply by turning off the current in the solenoid. An electromagnet is shown in Figure 13.13.

electromagnet: a solenoid with a soft iron core

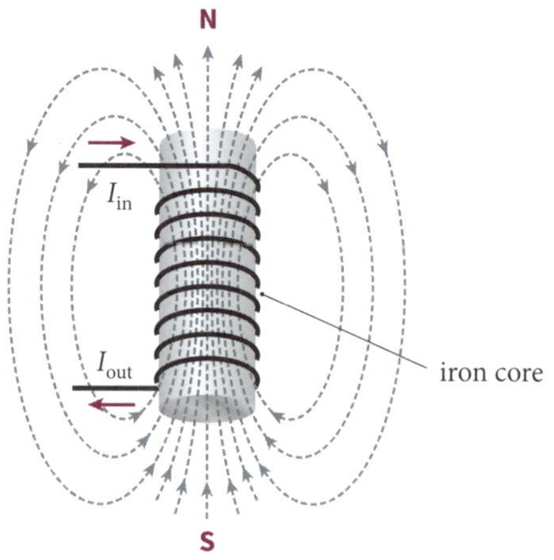

Figure 13.13 Magnetic field produced by an electromagnet

→ KEY QUESTIONS

8 **What does the density of magnetic field lines tell us about the field?**

9 **How does placing a soft iron core in a solenoid increase the magnetic field of the solenoid?**

10 **How can we determine the direction of the magnetic field around a current-carrying wire?**

Answers ➲ p. 181

FIRSTHAND INVESTIGATION 2

Solenoids and electromagnets

Students should construct solenoids and electromagnets and investiate the factors that determine the strength of fields produced by the devices. Students should also verify the right-hand grip rule and use a compass to map the magnetic field lines around current-carrying wires, soleniods and electromagnets.

Invetsigating ferromegnetic materials

Students should investigate the magnetic hardness of a variety of ferromagnetic materials, such as soft iron and steel, by measuring how long they retain their magnetism after an external magnetic field is removed.

EXAMPLE 4

Explain why magnets do not exert a force on all substances.

Answer:

Many substances cannot be magnetised

Magnets will only exert a significant force on other magnets and materials that can be magnetised (i.e. ferromagnetic materials). Many materials do not exhibit magnetic properties because they have paired electons and hence no magnetic domains.

EXAMPLE 5

Explain what will happen to the suspened magnet in Figure 13.14 when the solenoid is switced on.

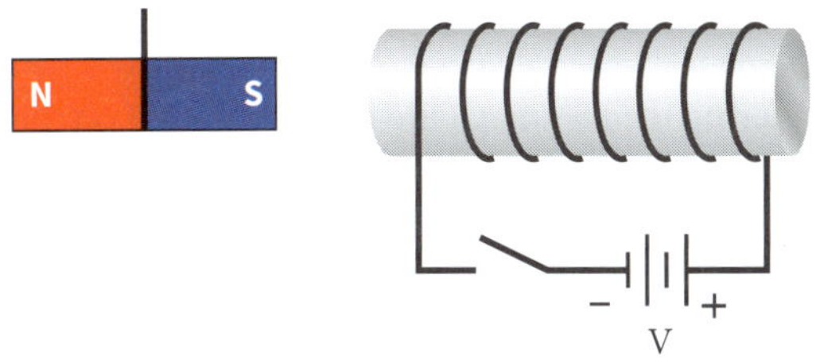

Figure 13.14 A bar magnet that is free to turn adjacent to a solenoid wrapped around a cardboard tube

Answer:

A magnetic field exerts a force on a north pole of a magnet in the direction of the field

The solenoid will produce a magnetic field directed from left to right through the solenoid. The magnet will therefore turn 180° to align with the solenoid's magnetic field, leaving the magnet with its north pole on the right-hand side.

3 Magnetic fields

» Students apply models to represent qualitatively and describe quantitatively the features of magnetic fields.

» Students conduct investigations into and describe quantitatively the magnetic fields produced by wires and solenoids, including:

- $B = \dfrac{\mu_0 I}{2\pi r}$
- $B = \dfrac{\mu_0 NI}{L}$.

➔ Oersted showed that the magnetic field a distance r from a long straight wire, as shown in Figure 13.15, is given by **Oersted's law**:

$$B = \mu_0 I / 2\pi r$$

where I is the current

r is the distance from the wire

$\mu_0 = 4\pi \times 10^{-7}\ \text{TmA}^{-1}$ is a constant called the permeability of free space

Oersted's law: a law that gives the magnetic field at a distance r from a long straight wire ($B = \mu_0 I / 2\pi r$)

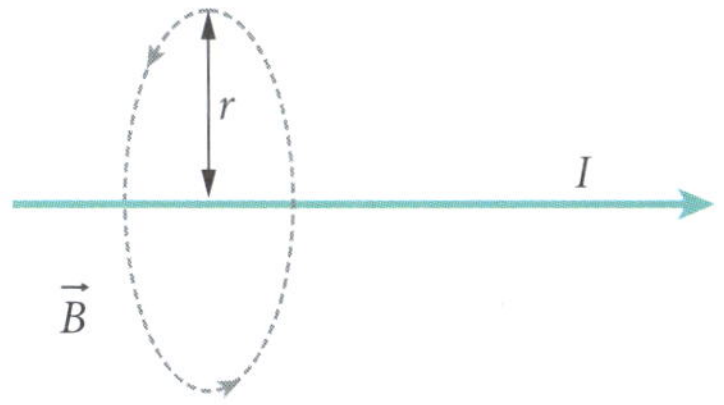

Figure 13.15 Magnetic field a distance r from a long straight wire

➔ The **magnetic permeability** (μ) reflects the degree to which the material/space is magnetised by the field. The field strength decreases as the inverse of the distance from the wire because the energy of the field is spread across the surface of a cylinder and, as the surface area of a cylinder is proportional to the radius of the cylinder, the field intensity decreases as the inverse of the radius.

This is quite different to the field from a point charge, which we saw in Chapter 11 decreased as the inverse square of the distance, because the energy was spread over the surface of a sphere surrounding the point.

magnetic permeability: a constant for a specific material that is related to how readily the material supports and enhances a magnetic field; its symbol is the italic lowercase Greek letter, *mu* (μ)

→ The magnetic field inside a long solenoid is uniform throughout the solenoid and is given by:

$$B = \mu_0 NI/L$$

where μ_0 is the magnetic permeability
I is the current
N/L is the number of turns per unit length on the solenoid

→ KEY QUESTION

11 **How can the magnetic field in an air solenoid be increased?**

Answers ➲ p. 181

EXAMPLE 6

Two long parallel wires on a horizontal plane separated by 10 cm carry current in the same direction, as shown in Figure 13.16. Find the magnitude and direction of the magnetic field at the points marked X, Y and Z.

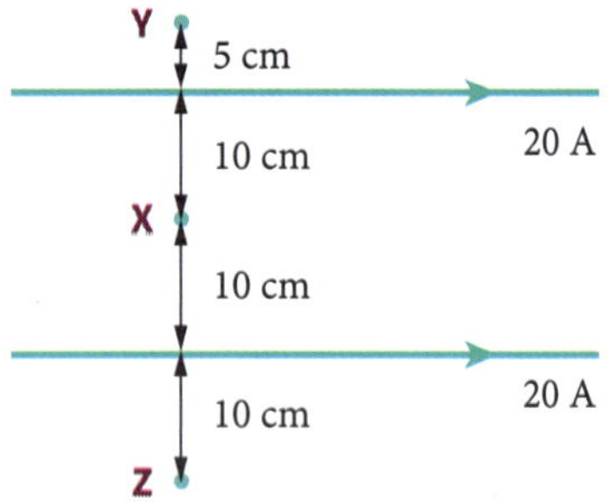

Figure 13.16 Two long, parallel current-carrying wires

Answer:

Magnetic fields are vector fields and their direction must be taken into account when adding them

- For position X

  Because the wires carry equal current and X is equidistant from each wire, the magnitude of the magnetic field due to each wire will be equal at X. But the right-hand grip rule shows the magnetic fields from each wire are in opposite directions at X and hence the magnetic field will cancel out at X. Thus the magnetic field at X is zero.

- For position Y

  The magnetic field due to the wire closest to Y is upwards and given by:

  $B = (\mu_0/2\pi)I/r = \dfrac{2\times10^{-7}\times 20}{0.05} = 8 \times 10^{-5}$ T upwards

  The magnetic field due to the wire furthest from Y is also upwards and given by:

  $B = (\mu_0/2\pi)I/r = \dfrac{2\times10^{-7}\times 20}{0.25} = 1.6 \times 10^{-5}$ T upwards

  Hence the net magnetic field at Y is 9.6×10^{-5} T upwards.

- For position Z

  The magnetic field due to the wire closest to Z is downwards and given by:

  $B = (\mu_0/2\pi)I/r = \dfrac{2\times10^{-7}\times 20}{0.10}$

  $= 4 \times 10^{-5}$ T downwards

  The magnetic field due to the wire furthest from Z is also downwards and given by:

  $B = (\mu_0/2\pi)I/r = \dfrac{2\times10^{-7}\times 20}{0.30}$

  $= 1.33 \times 10^{-5}$ T downwards

  Hence the net magnetic field at Z is 5.33×10^{-5} T downwards.

EXAMPLE 7

A solenoid wound with 10 turns per centimetre is held horizontal and oriented with its central axis directed in the east–west direction. When the solenoid is connected to a 12 V power pack the needle of a magnetic compass placed in the centre of the solenoid rotates from north to N70°. Given the Earth's magnetic field at the point has a strength of 5×10^{-5} T, find the current flowing in the solenoid.

Answer:

If two or more magnetic fields operate at the same point, the net field at the point is the vector sum of the fields acting at the point

We must first find the magnetic field strength of the solenoid by considering the two magnetic fields operating on the compass, as shown in Figure 13.17.

From Figure 13.17 we see that:

$B_{solenoid} = (5 \times 10^{-5}) \tan 70° = 1.374 \times 10^{-4}$ T east

As $B = \mu_0 NI/L$, the current in the solenoid will be given by:

$I = BL/\mu_0 N = 1.374 \times 10^{-4} \times 0.01/10 \times 4\pi \times 10^{-7} = 0.11$A

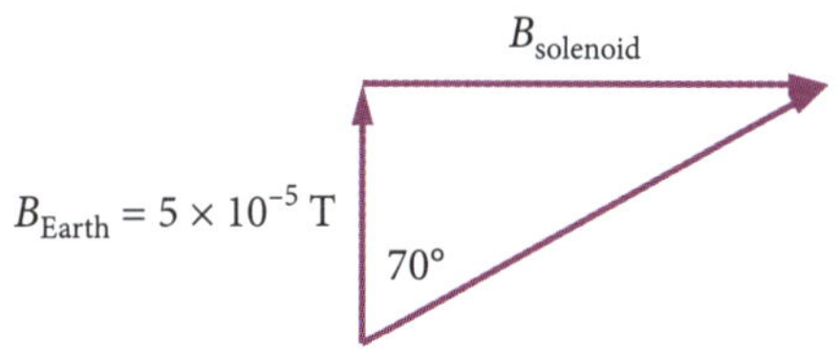

Figure 13.17 Magnetic field applied to magnet in solenoid

FIRSTHAND INVESTIGATION 3

Measuring the Earth's magnetic field

Students can measure the Earth's magnetic field strength by using a compass to compare the Earth's magnetic field with a known field. This can be done by placing a compass in a solenoid or by placing a compass above a north–south-oriented current-carrying wire, as shown in Figure 13.18. When the switch is turned on, the compass needle will be deflected towards the east. By measuring the distance to the compass and the current in the wire, the magnetic field at the compass due to the wire can be calculated. The angle of deflection can then be used to construct a vector diagram, as shown in Figure 13.19, to find the Earth's magnetic field.

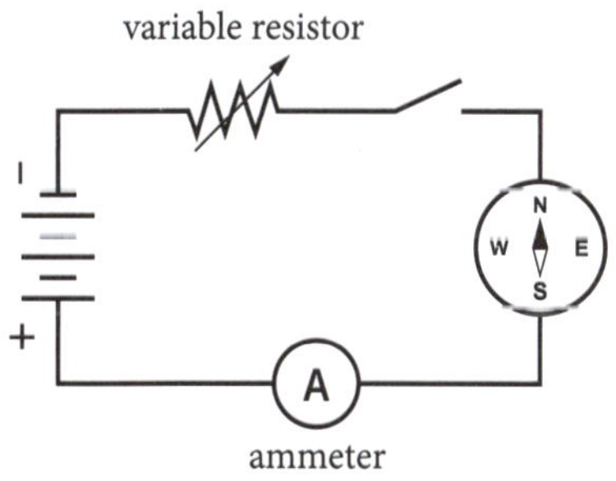

Figure 13.18 Experimental arrangement used to measure the Earth's magnetic field

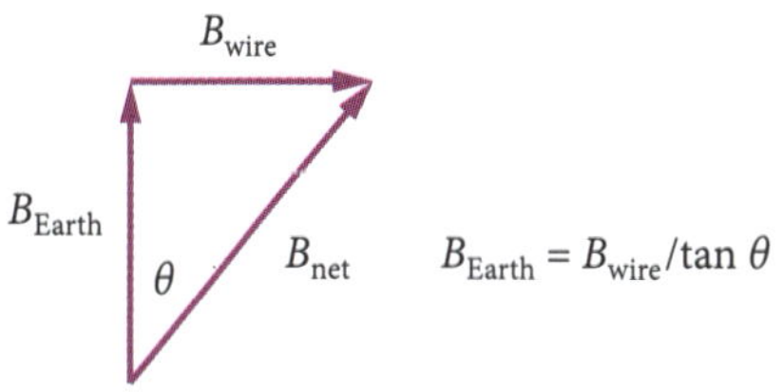

Figure 13.19 Vector diagram used to calculate the Earth's magnetic field

CHAPTER SYLLABUS CHECKLIST

Are you able to answer these questions from the syllabus for this chapter? Tick each question as you go through the checklist if you are able to answer it. If you cannot answer a question, turn to the relevant page in the study guide to find the answer. For NESA key word meanings, go to www.educationstandards.nsw.edu.au and search 'key words'.

FOR A COMPLETE UNDERSTANDING OF THIS TOPIC:		PAGE NO.	✓
1	Can I explain how the Earth's magnetic field is used to define the north-seeking pole of a magnet?	170	
2	Can I recall that magnetic monopoles do not exist?	170	
3	Can I describe the forces that act between magnets and why these forces cause a suspended magnet to align with an external magnetic field?	170	
4	Can I define the term ferromagnetic?	171	
5	Can I explain how ferromagnetic materials can be magnetised in terms of magnetic domains?	171	
6	Can I explain why some ferromagnetic materials cannot be used to make permanent magnets?	171	
7	Can I explain how magnets attract ferromagnetic materials and why they do not attract other materials?	171	
8	Can I draw magnetic field lines around permanent magnets, the Earth, current-carrying wires, solenoids and electromagnets?	172	
9	Can I use the right-hand grip rule to find the direction of the magnetic field around a long straight wire, an electromagnet and an air solenoid?	173	
10	Can I explain why an iron core greatly increases the magnitude of the magnetic field produced by a solenoid?	174	
11	Can I outline the key factors that determine the magnitude of the magnetic field produced by an electromagnet?	174	
12	Can I determine the magnitude and direction of the magnetic field at a distance r from a long straight wire?	175	

YEAR 11 EXAM-TYPE QUESTIONS

Objective-response questions (1 mark each)

1 **Which of the following statements best explains why a steel nail is attracted to a magnet?**

A The electrons in iron have a magnetic moment.

B Iron is a ferromagnetic material because it has magnetic domains due to paired, opposite-spin electrons.

C The magnetic domains in iron align to oppose an applied magnetic field.

D The magnetic domains in iron align with an applied magnetic field, making the iron magnetic.

2 **Two parallel wires P and Q are separated by a distance 2*r* and carry equal current in opposite directions, as shown in Figure 13.20. The net magnetic field at a distance *r* to the right of wire Q at position Y is *B* Tesla. Which of the following statements best describes the magnetic field midway between the wires at the point marked X?**

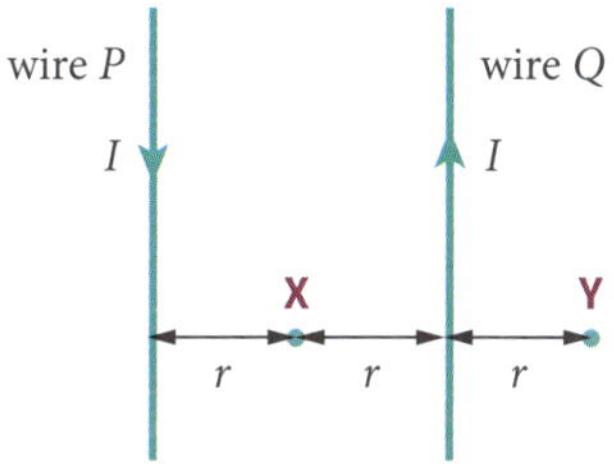

Figure 13.20 Two parallel current-carrying wires separated by a distance 2*r*

A The magnetic fields cancel out and hence the net field is zero.

B The fields add and hence the field between the wires is 2*B*.

C The field is more than 2*B*.

D The field is greater than zero but less than 2*B*.

3 **In three separate experiments a compass was placed above a north–south-directed wire and different currents were passed through the wire. The direction the compass pointed when current was passed through the wire in each experiment is shown in Figure 13.21.**

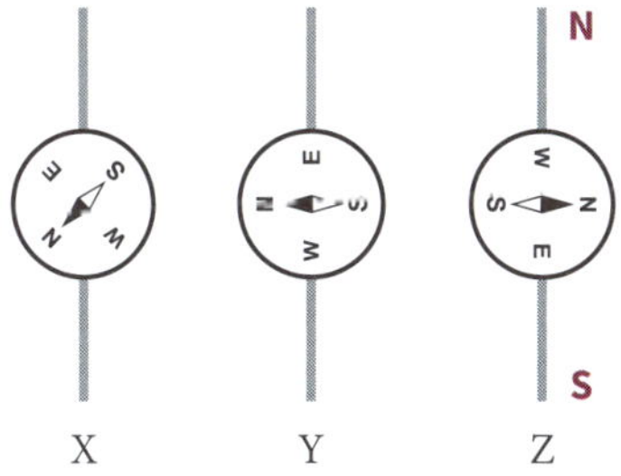

Figure 13.21 Three experiments conducted with a compass above a wire when different currents flowed in the wire

What can we say about the current in the wire in each experiment?

A The current is towards the north in Y and X and towards the south in Z.

B The current is towards the south in X and towards the north in Y and Z.

C The current in Y is greater than the current in X.

D The current in X is greater than the current in Y.

4 **A magnet is placed near a solenoid, as shown in Figure 13.22.**

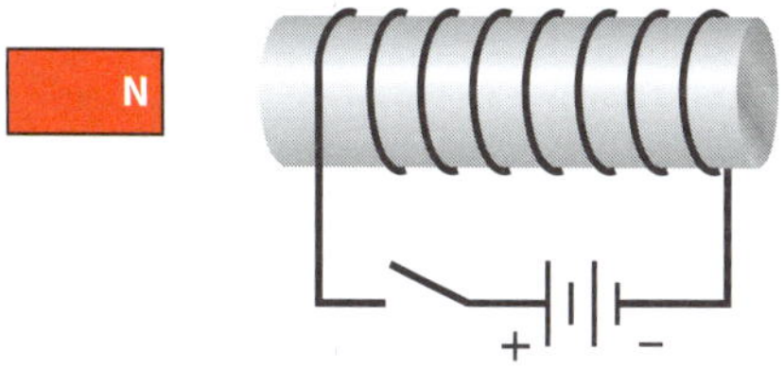

Figure 13.22 Magnet placed near a solenoid

Which of the following options best describes what will happen when the switch is closed?

A The solenoid will produce a magnetic field that will repel the magnet.

B The solenoid will produce a magnetic field within the solenoid but the magnet will not be affected by the field.

C The solenoid will produce a magnetic field that will attract the magnet until it reaches the solenoid.

D The magnet will be drawn into the solenoid and oscillate briefly before coming to rest.

5 **Which of the following graphs in Figure 13.23 best shows the magnetic field strength (*B*) as a function of distance (*r*) from a long straight wire?**

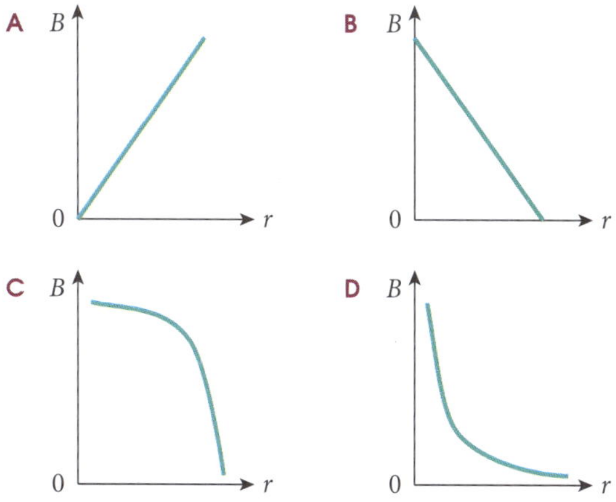

Figure 12.23 Magnetic field strengths

Extended-response questions

6 a Sketch the magnetic field produced by a bar magnet and by an air solenoid. Label the magnetic poles and show the current direction in the solenoid. (3 marks)

b Compare how the magnetic field in a bar magnet is produced with how the magnetic field is produced in an air solenoid. (2 marks)

7 An air solenoid produces a magnetic field of 6 mT when it carries a current of 15 A.

a Find the number of turns of wire per centimetre on the solenoid. (1 mark)

b If a soft iron core is placed inside the magnet, would the magnetic field outside the solenoid increase or decrease? Justify your answer. (2 marks)

8 A student investigating permanent magnets notices that the magnet does not attract an aluminium nail but does attract a steel nail. The student also notices that after working with the steel nail and magnet for some time the steel nail is left with some residual magnetism.

a Outline the process that occurs in the steel nail that causes it to be initially attracted to the magnet. (1 mark)

b Explain, in terms of magnetic domains, why the magnet exerts a force on the steel nail but not on the aluminium nail. (1 mark)

c Using the concept of magnetic domains, explain why the steel nail remains magnetised. (1 mark)

9 A magnetic compass is found to point N40° rather than north when placed below a DC power cable that runs north–south. Given that the Earth's magnetic field is 4.5×10^{-5} Tesla north at the point, determine:

a the direction of the current in the wire (1 mark)

b the magnitude and direction of the magnetic field produced by the current-carrying wire at the point. (2 marks)

10 Two long, straight parallel wires carry current in the same direction and are separated by 9 cm (see Figure 13.24). Wire X carries 20 A and is located vertically above wire Y, which carries a current of 10 A.

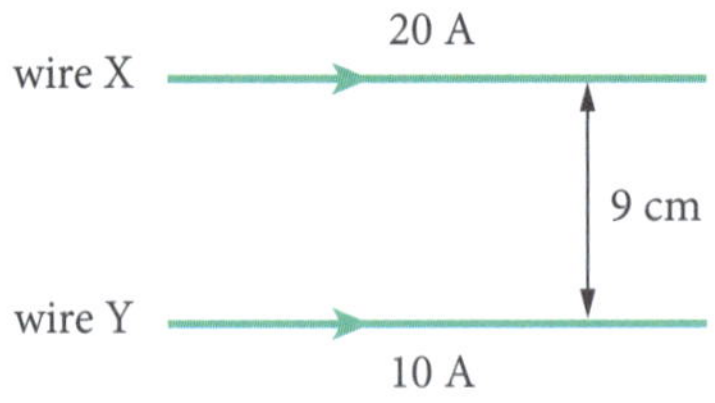

Figure 13.24 Two parallel current-carrying wires

a Find the position between the wires where the magnetic field is zero. (2 marks)

b Find the magnitude and direction of the magnetic field 1 cm above wire X. (2 marks)

ANSWERS

KEY QUESTIONS

Key questions ➲ p. 171

1 The north-seeking pole is the magnetic pole that would point towards the Earth's North Pole if the magnet was freely suspended.

2 No, as far as we know magnetic monopoles do not exist and magnets must have two poles; a north pole and a south pole.

3 The north-seeking pole experiences a force in the direction of the magnetic field, while the south-seeking pole experiences a force in the opposite direction of the field. These two forces turn the magnet until it is aligned with the field, with the north-seeking pole pointing in the direction of the applied field.

4 The magnetic field is directed from the north-seeking pole to the south-seeking pole outside the magnet.

Key questions ➲ p. 172

5 Magnetic domains are small regions in a ferromagnetic material that have a net magnetic moment due to unpaired electrons in the atoms within the domain.

6 Ferromagnetic materials have magnetic domains that are aligned by externally applied magnetic fields. When the magnetic moments of the domains align they give the material a net magnetic moment.

7 A ferromagnetic material can be magnetised by using an external magnetic field to align the magnetic moments of the domains in the material. If it is a hard magnetic material it will maintain its magnetism (i.e. the magnetic moments of the domains will stay aligned) after the field is removed.

Key questions ➲ p. 174–175

8 The density of magnetic field lines is related to the strength of the magnetic field at the point. A higher density of field lines indicates a stronger magnetic field.

9 Applying an external magnetic field causes the magnetic moments of the domains in the core to align and these produce a net magnetic field in the same direction as the applied field. The two fields add together, producing a stronger field than the field that was initially produced by the current in the windings.

10 We use the right-hand grip rule. To do this you use your right hand to grip the wire with the thumb in the direction of the current in the wire; the fingers then give the direction of the magnetic field around the wire.

Key questions ➲ p. 176

11 The magnetic field can be increased by using a higher number of turns per metre along the solenoid, by using a higher current, or by placing a soft iron core in the solenoid.

YEAR 11 EXAM-TYPE QUESTIONS

Objective-response questions

1 **D.** Iron has unpaired electrons, which result in the formation of magnetic domains with magnetic moments that can be aligned by placing the material in a magnetic field. These aligned domains give the nail a net magnetic moment. **A** is incorrect because all electrons have a magnetic moment. **B** is incorrect because the magnetic moment of the electrons cancels out when they are paired with opposite spin. **C** is incorrect because the domains align with the external magnetic field, not against it.

2 **C.** The right-hand grip shows that the field from both wires midway between the wires is upwards, but at the right of the wire, the field from each wire is in an opposite direction and will be lower than at a distance r from one wire alone. Thus the field from each wire at the midway point will be greater than B, giving a net field greater than $2B$ at the midway point. **A** is incorrect because the fields are in the same direction and hence add (rather than subtract) in the region between the wires. **B** is incorrect because the field from each wire a distance r from the wire must be greater than B. **D** is incorrect because the field from each wire at the midway point must be greater than B and their sum must therefore be greater than $2B$.

3 **C.** The magnet above Y is deflected by a greater angle than the magnet above X. **A** and **B** are incorrect as the right-hand grip rule gives the current as south in wires X and Y and to the north in Z. **D** is incorrect because if this was the case, the compass above wire X would be deflected more than the compass above Y.

4 **A.** The right-hand rule gives a B field in the solenoid directed towards the left. This would repel the magnet as the left-hand side of the solenoid would be like the north pole of a bar magnet. **B** is incorrect because the magnetic field produced by the solenoid will extend beyond the solenoid and exert a force on the magnet. **C** is incorrect by the right-hand grip rule. **D** is incorrect because the solenoid will produce a constant rather than an oscillating force.

5 **D.** $B = (\mu_0/2\pi)I/r$ and hence, $B \propto 1/r$ and the graph will be a hyperbola. **A**, **B** and **C** are incorrect as none of the graphs show a hyperbolic relationship between B and r.

Extended-response questions

6 EM This question tests students' understanding of how to draw magnetic fields, the relationship between the poles of a magnet and the direction of the magnetic field and the right-hand grip rule.

a The answer is illustrated in Figure A13.1.

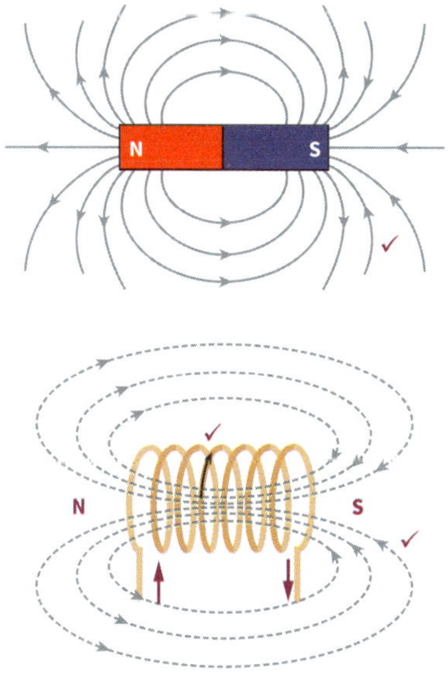

Figure A13.1 Magnetic field around a bar magnet and a solenoid b A magnetic field in a bar magnet is produced by the alignment of the magnetic moments of small magnetised regions called magnetic domains in the magnet. ✓ In an air solenoid the magnetic field is produced by the current passing through the wire-turns of the solenoid. ✓

7 EM This question tests students' knowledge of SI prefixes and their understanding of electromagnets and the solenoid equation.

a Applying the solenoid equation, $B = \mu_0 NI/L$,

$N/L = B/\mu_0 I = 6 \times 10^{-3}/\ 15 \times 4\pi \times 10^{-7} = 318$ turns per metre or 31.8 turns per centimetre. ✓

b The magnetic field would increase ✓ because the magnetic domains in the iron core would align with the applied field and add their magnetic moments to the magnetic field of the solenoid. ✓

8 EM This question tests students' understanding of the difference between ferromagnetic material and other materials, and their understanding of 'hard' and 'soft' magnetic materials.

a The magnet's magnetic field causes the magnetic domains in the steel nail to align with the field, making the nail a small magnet. This will cause the magnet and the nail to attract one another. ✓

b Aluminium does not contain unpaired electrons that can produce magnetic domains and hence there are no magnetic domains to become aligned in aluminium. ✓ Aluminium is therefore not a ferromagnetic material and will not be attracted to the magnet.

c In some 'hard' magnetic materials, such as steel, the magnetic moment of the domains remains aligned after the externally applied magnetic field is removed. ✓

9 EM This question tests students' ability to add magnetic fields in two dimensions.

a Because the power cable is above the magnet, the right-hand grip rule applied to a north–south cable shows that the resulting magnetic field is directed either east or west. As the compass needle is deflected to the east, the current must be running towards the south. ✓

b Figure A12.2 shows how the two magnetic fields add to produce the net field at the compass.

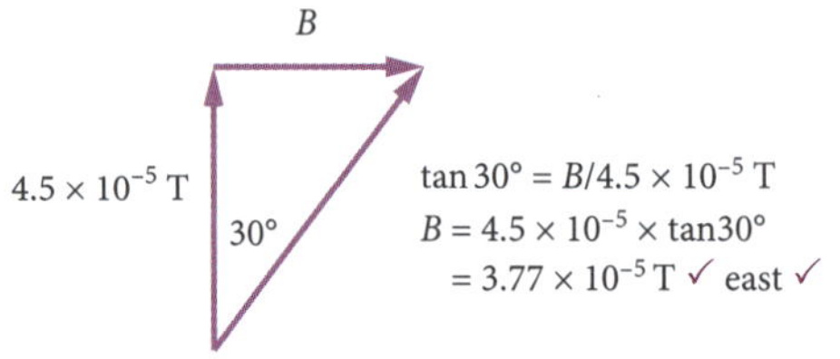

Figure A13.2 Vector diagram showing addition of two magnetic fields

10 EM This question tests students' understanding of the right-hand grip rule, Oersted's law and the addition of two vector fields.

a The right-hand rule shows that the magnetic field from each wire will be in opposite directions between the wires. Calling the distance to this point r from wire X, the distance to the point from wire Y becomes $(0.09 - r)$. The net field will be zero between the wires at the point where the magnitude of the field from each wire is equal. ✓ Equating the magnetic fields from each wire gives:

$B_X = B_Y$

$(\mu_0/2\pi)I/r = (\mu_0/2\pi)I/(0.09 - r)$

$20/r = 10/(0.09 - r)$

$r = 0.06$ m or 6 cm from wire X ✓

b To find the field strength above X we note the field from both wires will be out of the page by the right-hand rule and so we add the field due to each wire.

Hence the net B field will be $= (\mu_0/2\pi)20/0.01 + (\mu_0/2\pi)10/0.1 = 4.2 \times 10^{-4}$ T ✓ out of the page. ✓

SAMPLE YEAR 11 EXAMINATION 1

Try to complete these papers as if they are the real thing. These are the instructions you need to follow in the Year 11 exam:

General instructions

- Reading time: 5 minutes
- Working time: 3 hours
- Write using black pen.
- Draw diagrams using pencil.
- NESA-approved calculators may be used.
- Use the data sheet and formulae sheet in this book.

Total marks: 100

Section I: 20 marks

Section II: 80 marks

- Attempt all questions.

Section I: 20 Marks

Allow about 40 minutes for this section.

Objective-response questions (1 mark each)

Choose the letter A, B, C or D that corresponds to the best answer.

1 **What force must be applied to a 2000 kg car travelling at 20 ms^{-1} to bring it to rest in a distance of 40 m?**

A 10^4 N
B 5 ms^{-2}
C -10^4 N
D -5 ms^{-2}

2 **Which of the following statements best describes how we jump upwards from the ground?**

A We lift our legs upwards.
B We push the ground upwards.
C We push the ground downwards, producing an upwards reaction force.
D We reduce our acceleration due to gravity by pushing the ground downwards.

3 **Consider the velocity against time graph for a skateboarder in Figure E1.1. The skateboarder was initially travelling north at 20 ms^{-1}. Which of the following statements best describes the journey made by the skateboarder?**

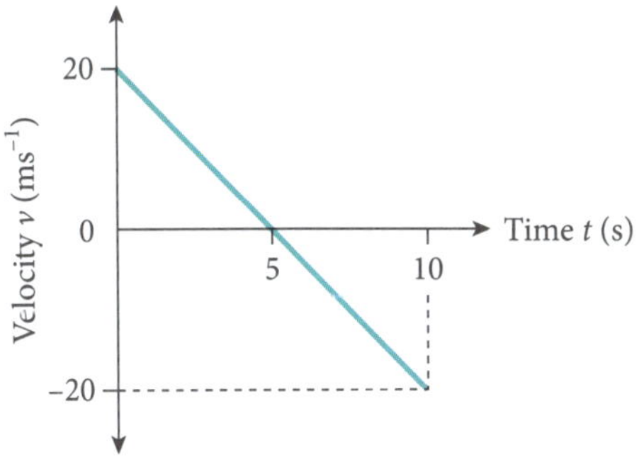

Figure E1.1 Velocity against time graph

A The skateboarder travelled a total distance of 50 m south.
B The skateboarder travelled a total distance of 40 m.
C The skateboarder travelled a total distance of 100 m north.
D The skateboarder travelled a total distance of 100 m.

4 **A 10 kg box placed on a frictionless plane inclined at 20° to the horizontal on a planet is observed to accelerate down the plane at 2 ms^{-2} (Figure E1.2). What is the acceleration due to gravity and the weight of the box on the planet?**

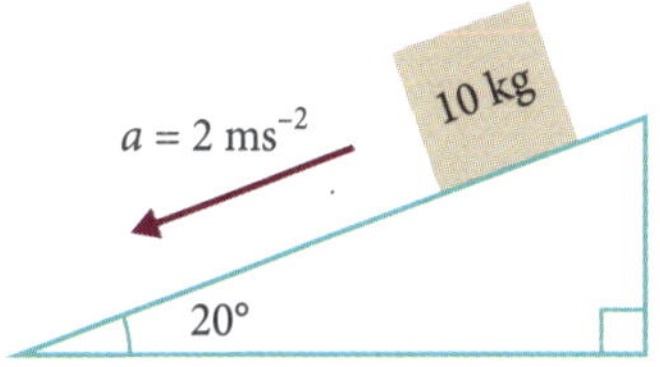

Figure E1.2 Box on a frictionless plane

A 5.8 ms^{-2} and 58 N
B 9.8 ms^{-2} and 98 N
C 5.8 ms^{-2} and 58 kg
D 9.8 ms^{-2} and 10 kg

5 **A car travelling east at 30 ms^{-1} approaches a motorcyclist travelling south at 40 ms^{-1}. What is the speed of the car relative to the motorcyclist?**

A 70 ms^{-1} N37°
B 50 ms^{-1} N37°
C 50 ms^{-1} N53°
D 50 ms^{-1} NE

6 A force (F) of 60 N is applied to a 50 kg sled at 30° above the horizontal, as shown in Figure E1.3. If the coefficient of friction between the sled and the snow is μ, which of the following expressions could be used to calculate the friction on the sled when 60 N force is applied?

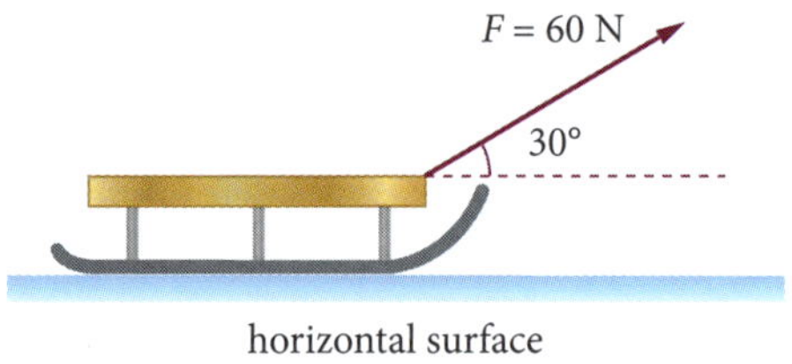

Figure E1.3 Force applied to a sled

A $\mu(50\text{ g} - 60\cos 30°)$
B $\mu(50\text{ g} + 60\sin 30°)$
C $\mu(50\text{ g})$
D $\mu(50\text{ g} - 60\sin 30°)$

7 If 12 crests of a water wave pass a point each minute and the waves have a wavelength of 15 m, what is the wave velocity?

A 3 ms^{-1}
B 1.25 ms^{-1}
C 75 ms^{-1}
D 0.8 ms^{-1}

8 Two football players running at 5 m^{-1} in opposite directions collide head on. If one player had twice the mass of the other player, which of the following statements best describes the change in momentum of the players during the collision?

A The change in momentum of the heavier player would be half that of the lighter player.
B The change in momentum of the lighter player would be half that of the heavier player.
C The change of momentum of both players would be equal but in opposite directions.
D The change of momentum of both players would be equal and in the same direction.

9 If the thermal conductivity (k) of the wall material used in a refrigerator was doubled and the thickness of the wall was halved, how would the rate of heat loss from the refrigerator change?

A The rate of heat loss would remain constant.
B The rate of heat loss would double.
C The rate of heat loss would halve.
D The rate of heat loss would be four times greater.

10 Which of the following conditions are necessary for a light ray to undergo total internal reflection when it encounters a boundary? Note that θ_C is the critical angle of incidence.

	Angle of incidence (i)	Refractive index in mediums (n)
A	$\theta_1 > \theta_C$	$n_1 < n_2$
B	$\theta_1 > \theta_C$	$n_1 > n_2$
C	$\theta_1 < \theta_C$	$n_1 < n_2$
D	$\theta_1 < \theta_C$	$n_1 > n_2$

11 The following table shows the intensities of two stars viewed from Earth.

Star	Light intensity (Wm^{-2})
Sirius	20
Polaris	80

Given that Sirius is twice as far from the Earth as Polaris, what can we deduce about the absolute energy radiated (the luminosity) from each star?

A The luminosity of Sirius is equal to the luminosity of Polaris.
B Polaris has twice the luminosity of Sirius.
C Polaris has four times the luminosity of Sirius.
D Polaris has half the luminosity of Sirius.

12 Consider the three blocks in Figure E1.4. The blocks are on a smooth horizontal plane and are being pushed with a force F.

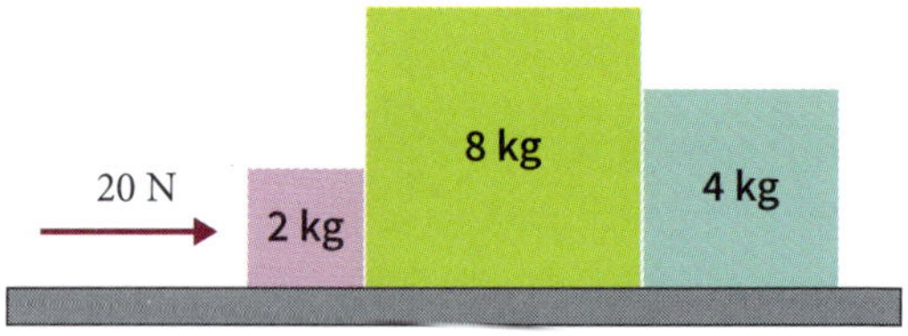

Figure E1.4 Pushed blocks

Which of the following correctly describes the forces between the blocks?

A The net force on each block is $\frac{20}{3}$ N.
B The net force on each block is 20 N.
C The net force on the 2 kg block is greater than the net force on the other blocks.
D The net force on the 8 kg block is greater than the net force on the other blocks.

13 **What can we conclude about the refracted light ray in Figure E1.5?**

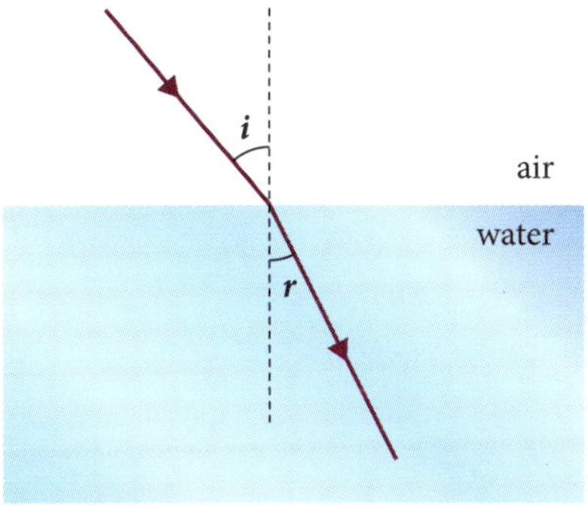

Figure E1.5 Refracted light ray

- **A** The frequency of the refracted ray is less than the frequency of the incidence ray.
- **B** The wavelength of the refracted ray is less than the wavelength of the incident ray.
- **C** The wavelength of the refracted ray is greater than the wavelength of the incident ray.
- **D** The refractive index of air is greater than the refractive index of water.

Questions 14 and 15 refer to the image of a sound wave shown in Figure E1.6.

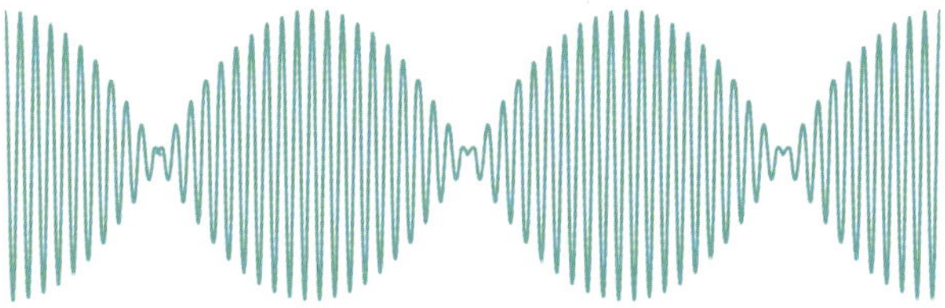

Figure E1.6 Image of a sound wave

14 **What sort of sound would this image represent?**

- **A** a sound that changes its loudness periodically
- **B** a sound wave that increases and decreases its period
- **C** a sound that gradually increases its loudness
- **D** a sound with a pitch that increases and decreases periodically

15 **What could produce a sound wave like the one shown in Figure E1.6?**

- **A** a tuning fork and a piano sounding the same frequency together
- **B** the superposition of a loud sound with a soft sound
- **C** two identical tuning forks sounding together
- **D** two tuning forks with slightly different frequencies sounding together

16 **The parallel plates shown in Figure E1.7 are separated by 2 cm and have a potential difference of 20 V across them. What would happen to a dust particle of mass 1 mg and charge $q = -1\ \mu$C if it was placed midway between the plates? (You may neglect gravity.)**

Figure E1.7 Parallel plates

- **A** The particle will accelerate upwards at 10^3 ms^{-2}.
- **B** The particle will accelerate upwards at 1 ms^{-2}.
- **C** The particle will accelerate downwards at 10^3 ms^{-2}.
- **D** The particle will accelerate downwards at 1 ms^{-2}.

17 **In the circuit shown in Figure E1.8, the potential difference across the 2 Ω resistor is 4 V. What is the potential difference across the 6 Ω resistor?**

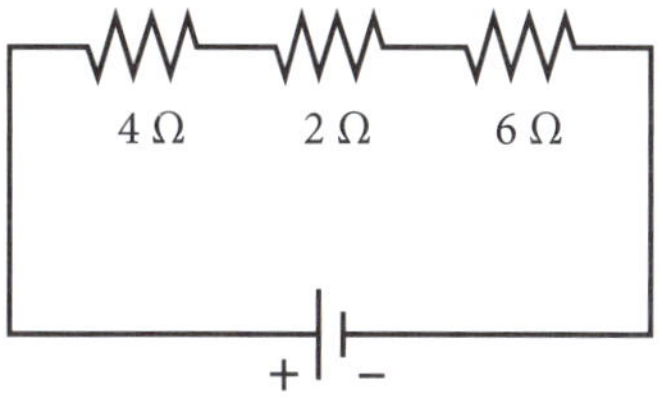

Figure E1.8 Circuit

A 12 V **B** 6 V **C** 8 V **D** 4 V

18 **Which of the following are all ferromagnetic materials?**

- **A** hard iron, soft iron and copper
- **B** nickel, iron and cobalt
- **C** nickel, aluminium and iron
- **D** soft iron, nickel and carbon

19 **Which of the following diagrams in Figure E1.9 best represents the magnetic field lines around a current-carrying wire?**

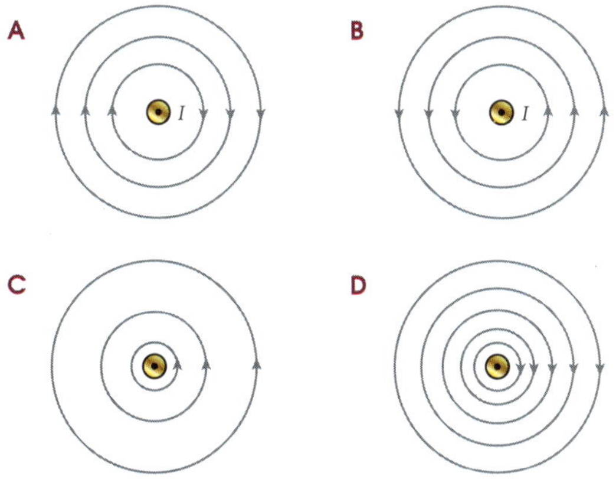

Figure E1.9 Current-carrying wires

20 If the temperature of a hot piece of metal was doubled, how would the electromagnetic radiation emitted from the object change?

- **A** The intensity of the radiation would increase and the wavelength at which most of the radiation was emitted would decrease.
- **B** The intensity of the radiation would decrease because there would be more heat lost by conduction and convection.
- **C** The intensity of the radiation would increase and the frequency at which most radiation was emitted would decrease.
- **D** The intensity of the radiation would not change but the radiation would be emitted at a shorter wavelength.

Section II: 80 Marks

Allow about 2 hours and 20 minutes for this section.

21 The motion of a student who ran up a narrow hallway was measured. The position was graphed as a function of time, shown in Figure E1.10. Note that the runner initially ran in a northerly direction.

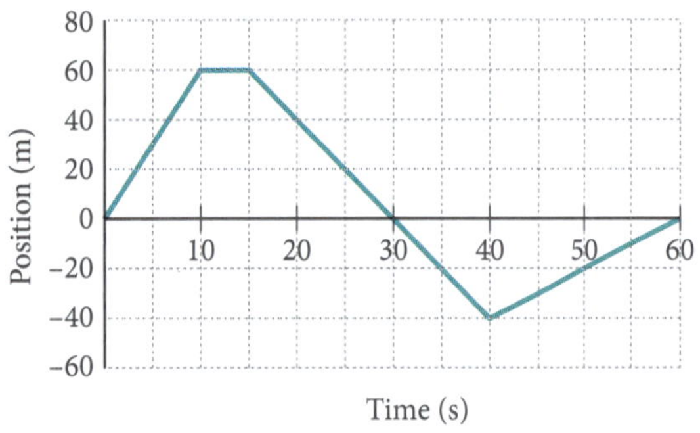

Figure E1.10 Position against time graph

- **a** What is the total distance covered by the runner? (1 mark)
- **b** What is the average speed and average velocity for the journey? (2 marks)
- **c** What was the maximum velocity achieved by the runner and when did it occur? (2 marks)
- **d** Sketch a graph of velocity against time for this motion and include numerical values. (2 marks)
- **e** A physics student who examined this data had doubts about the graph. Explain why this graph is unlikely to represent the actual motion of the runner. (2 marks)

22 A car of mass 2000 kg at accelerates at 5 ms^{-2} for 8 s and then travels with constant velocity. The car was initially at rest and travelled in a straight line in an easterly direction throughout the journey and the frictional force on the car was 2500 N.

- **a** Calculate the net force on the car when it is accelerating at 5 ms^{-2}. (1 mark)
- **b** Calculate the force applied by the engine to the car when it is accelerating. (1 mark)
- **c** Determine the final velocity of the car after 8 s. (1 mark)
- **d** Find the work done against friction in 8 s. (1 mark)
- **e** Determine the total work done by the engine in 8 s and the power produced by the engine. (2 marks)
- **f** Find the power used by the engine when the car travels at constant velocity. (1 mark)

23 A student on a distant planet uses a sonic ranger to measure the vertical velocity (v_y) as a function of time for a bouncing ball of mass 0.2 kg and has plotted the velocity as a function of time (t), shown in Figure E1.11.

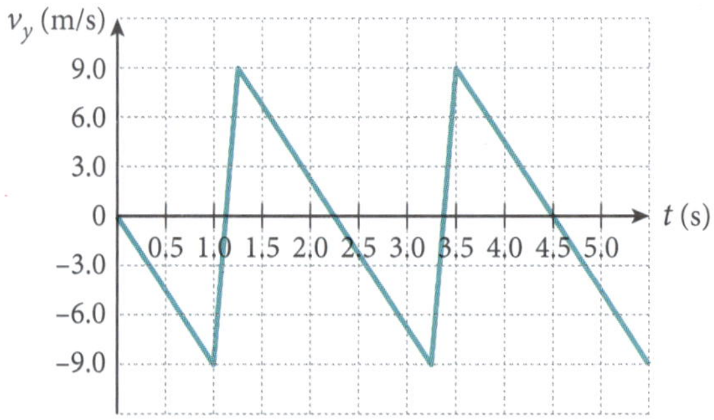

Figure E1.11 Velocity against time graph

- **a** Use the graph to determine how high the ball will bounce above the ground. (1 mark)
- **b** Use the graph to determine the acceleration due to gravity on the planet. (1 mark)
- **c** Is the ball undergoing an elastic or inelastic collision with the ground? Justify your answer. (2 marks)
- **d** Describe qualitatively how the kinetic energy of the ball changes between $t = 0$ s and $t = 2.25$ s. (2 marks)
- **e** Describe qualitatively how the gravitational energy of the ball changes between $t = 0$ s and $t = 2.25$ s. (2 marks)
- **f** How does the total energy of the ball change between $t = 0$ and $t = 2.25$ seconds? (1 mark)

24 Car A with a mass of 2000 kg and Car B with a mass of 3000 kg have a head-on collision and become entangled. The momentum of the cars, measured before and after the impact, is shown in Figure E1.12. Car A was travelling north and its direction is shown as positive in the diagram while Car B was travelling south and its initial momentum is shown as negative.

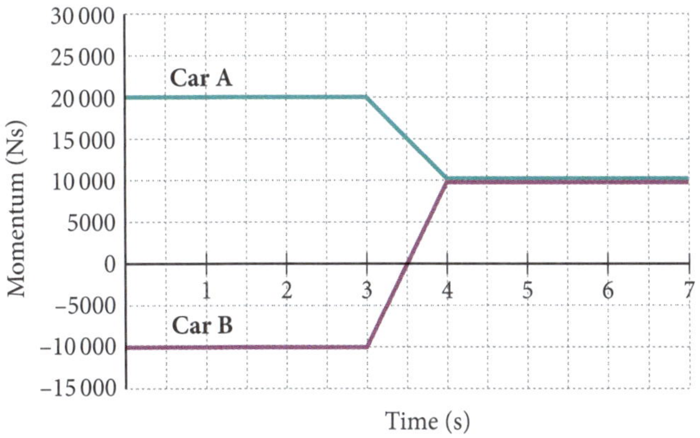

Figure E1.12 Momentum of the cars

a Find the velocity of the wreck after the collision. (1 mark)

b Was momentum conserved in this collision? (Justify your answer.) (2 marks)

c Was kinetic energy conserved in this collision? (Justify your answer.) (2 marks)

d Find the magnitude and direction of the change of momentum of Car A during the collision. (2 marks)

e Given the change of momentum occurred over a time of 1 s, find the force Car B exerted on Car A during the collision. (1 mark)

25 Figure E1.13 shows the wavefronts of a soundwave moving from medium A, where the speed of sound is 340 ms^{-1}, to medium B, where the speed of sound is 200 ms^{-1}.

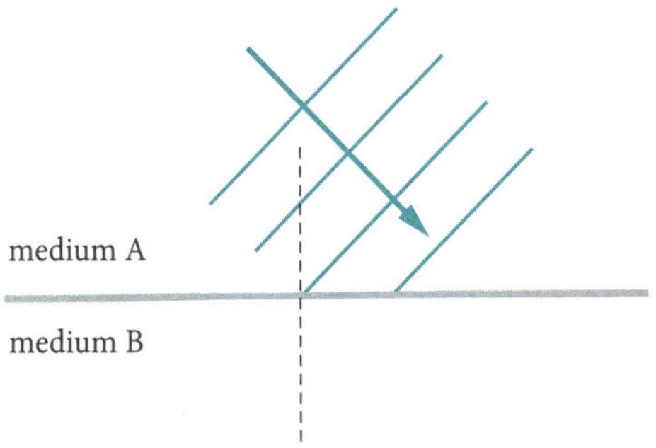

Figure E1.13 Waves through two media

a Sketch the reflected ray and draw three wavefronts on the reflected ray. (2 marks)

b Sketch the refracted ray on the diagram and three wavefronts in medium B. (2 marks)

c If the angle of incidence was 40°, find the angle of refraction in medium B. (1 mark)

26 An ambulance uses a siren that emits a loud sound at 550 Hz. Shortly after the ambulance overtakes a car that was travelling at 50 kmh^{-1}, the car driver measures the frequency of the ambulance siren to be 500 Hz. Both the car and the ambulance were travelling in an easterly direction.

a Find the speed of the ambulance. (2 marks)

b Find the velocity of the car relative to the ambulance. (2 marks)

27 A 25 kg box slides down a plane inclined at 25° above the horizontal (Figure E1.14). The plane is 40 m long and the coefficient of friction is $\mu = 0.2$.

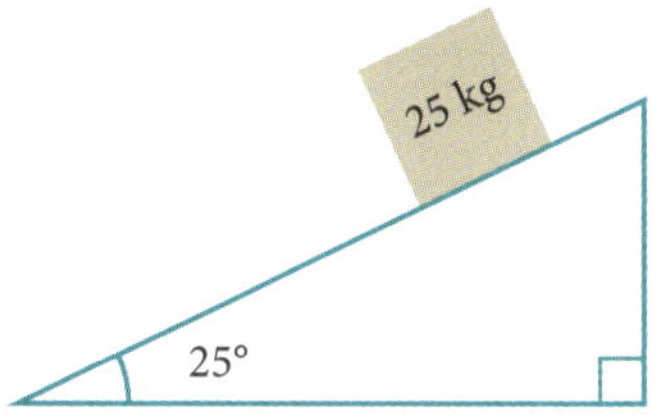

Figure E1.14 Box on an inclined plane

a Find the components of the weight force into the plane and down the plane. (2 marks)

b Determine the force of friction on the block. (1 mark)

c Find the acceleration of the block down the plane. (2 marks)

28 Consider the circuit shown in Figure E1.15.

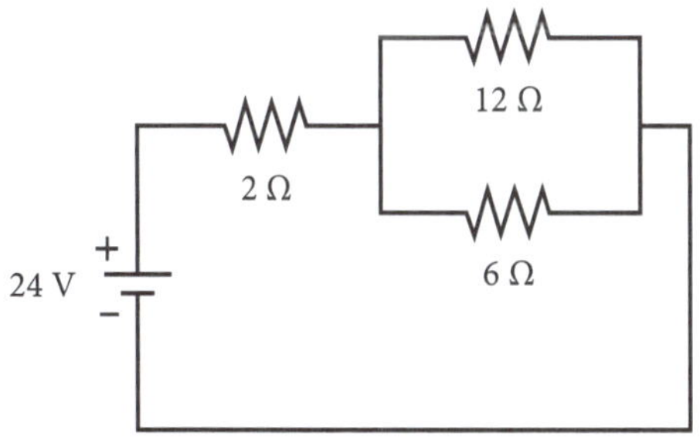

Figure E1.15 Circuit

a Find the total resistance of the circuit. (1 mark)

b Find the potential difference across and current in the 6 Ω resistor. (2 marks)

c Find the power dissipated in the 12 Ω resistor. (1 mark)

d Sketch a diagram to show how a voltmeter and ammeter could be connected into the circuit above to measure he potential difference across and current in the 12 Ω resistor. (2 marks)

29 A voltage of 240 V is applied to a resistance element in a coffee cup to heat 200 mL of water in the cup. The cup has a surface area of 250 cm^2 and is made of 2 mm thick polystyrene. Note that polystyrene has a thermal conductivity of 0.03 $Wm^{-1}K^{-1}$. (The heat capacity of water is $C = 4184$ $Jkg^{-1}K^{-1}$.)

a Assuming that no heat is lost from the water when the water is being heated, how much energy would be required to heat the water from 20 °C to 90 °C? (1 mark)

b If a 25 Ω resistor was used, how long would it take to heat the water from 20 °C to 90 °C? (2 marks)

c Find the rate at which heat would be lost from the cup via conduction when the water was at 90 °C and the outside temperature was 20 °C. (2 marks)

30 A potential difference of 200 V is placed across a pair of horizontal, parallel charged plates separated by 5 cm. A small charged plastic microsphere of mass 50 mg is observed to levitate between the plates, as shown in Figure E1.16.

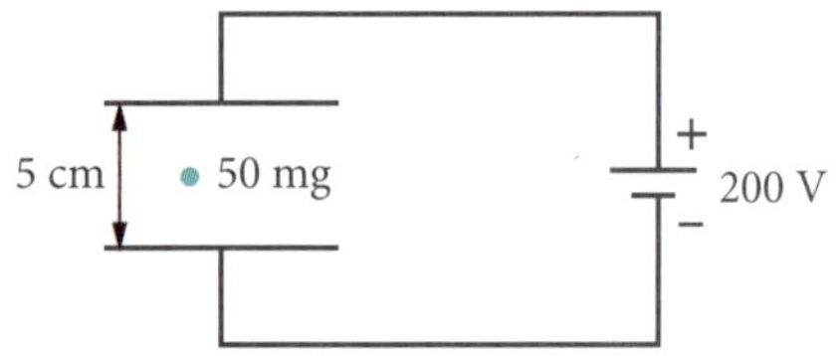

Figure E1.16 Parallel charged plates

a Find the magnitude and direction of the electric field between the plates. (2 marks)

b Find the gravitational force on the microsphere. (1 mark)

c Find the charge that would have to be placed on the microsphere for it to remain stationary between the plates. (2 marks)

31 Consider the electromagnet illustrated in Figure E1.17.

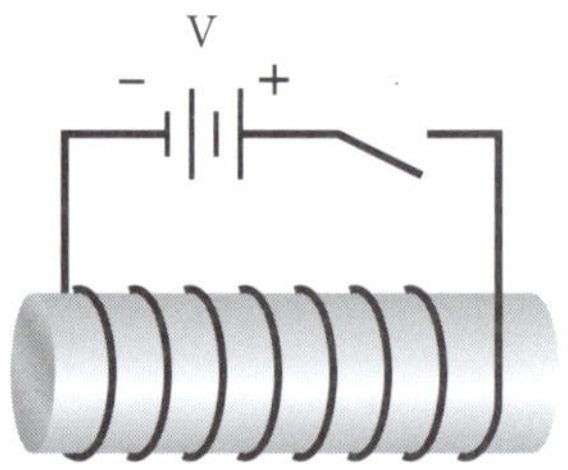

Figure E1.17 Electromagnet

a Sketch the magnetic field that would appear around the magnet when the switch was closed, and mark the north pole and south pole on the magnet. (3 marks)

b Explain how the magnetic field outside the electromagnet would change and why it would change, if the iron core was removed. (3 marks)

32 A compass points N45° when placed 4 cm above a north–south running current-carrying wire, as shown in Figure E1.18. Note the Earth's magnetic field at the compass is 5×10^{-5} T north.

Figure E1.18 Compass over a current-carrying wire

a Find the direction of current in the wire. (1 mark)

b Determine the magnitude and direction of the magnetic field at the compass produced by the wire. (1 mark)

c Find the magnitude of the current in the wire. (2 marks)

33 A truck of mass 12 000 kg travelling horizontally at 60 kmh^{-1} east collides and becomes entangled with a parked car with mass 2000 kg that was initially at rest. The collision occurred at the bottom of a hill that had an incline of 15°, as shown in Figure E1.19.

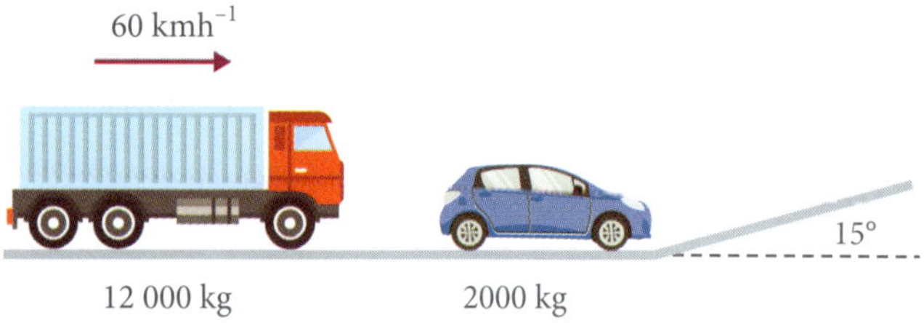

Figure E1.19 Velocity and mass of two vehicles

a Find the velocity of the wreckage immediately after the collision. (2 marks)

b Determine the change in kinetic energy of truck–car system during the collision. (2 marks)

c Assuming no energy is lost to friction; calculate the distance up the incline the wreckage would reach before coming to rest. (3 marks)

SAMPLE YEAR 11 EXAMINATION 2

Section I: 20 Marks

Allow about 40 minutes for this section.

Objective-response questions

(1 mark each)

Choose the letter A, B, C or D that corresponds to the best answer.

1 Consider the motion of the car illustrated in Figure E2.1.

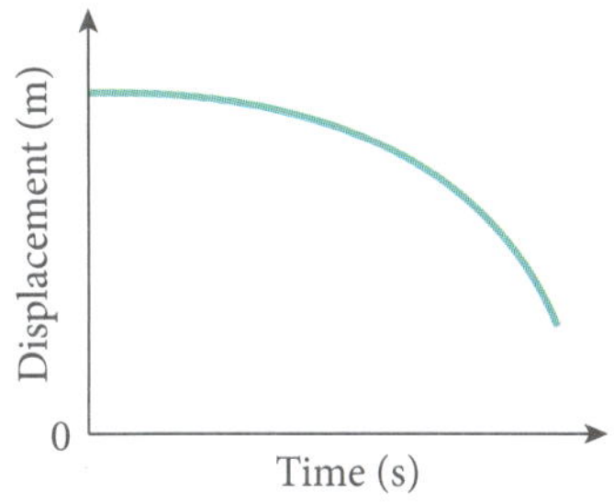

Figure E2.1 Motion of a car

Which of the following statements best describes the motion of the car?

A The car started with a positive velocity and then accelerated in the positive direction.

B The car started at rest and then drove at constant velocity in the negative direction.

C The car started with a positive velocity and then decelerated.

D The car started at rest and then accelerated in the negative direction.

2 A lift, initially moving upwards, decelerates at 2 ms^{-2}. During the time the lift was slowing down, what would the weight of a 50 kg passenger be?

A 390 N **B** 490 N

C 50 kg **D** 39.8 kg

3 Consider the suspended masses shown in Figure E2.2.

Given that the mass of $M_2 = 3M_1$, what would the net force on M_1 be?

A M_1 g

B $4M_1$ g

C $3M_1$ g

D 0

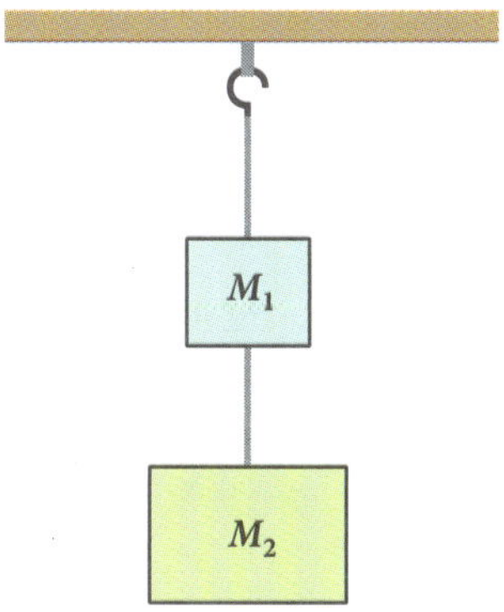

Figure E2.2 Suspended masses

4 In a simple experiment a student applies two different forces for 5 s to a toy car that was initially at rest. In the first experiment (A) the student applied a changing force, and in the second experiment (B) the student applied a constant force. The forces used by the student in each experiment are graphed as functions of time in Figure E2.3.

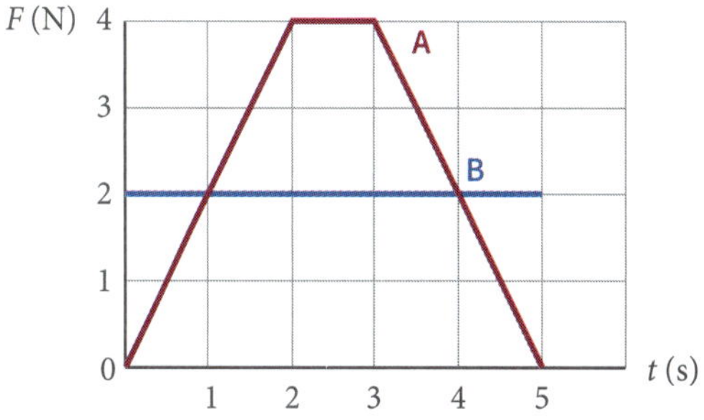

Figure E2.3 Force as a function of time graph

Assuming there was no friction, what can we say about the final velocity of the toy car after the force is applied in each experiment?

A The toy car will have the same final velocity in both experiments.

B The final velocity will be 20% greater in experiment A than in experiment B.

C The final velocity will be greater in experiment B.

D The final velocity will be twice as great in experiment A as it would be in experiment B.

5 Two masses are connected together with a string that runs over a very light, frictionless pulley.

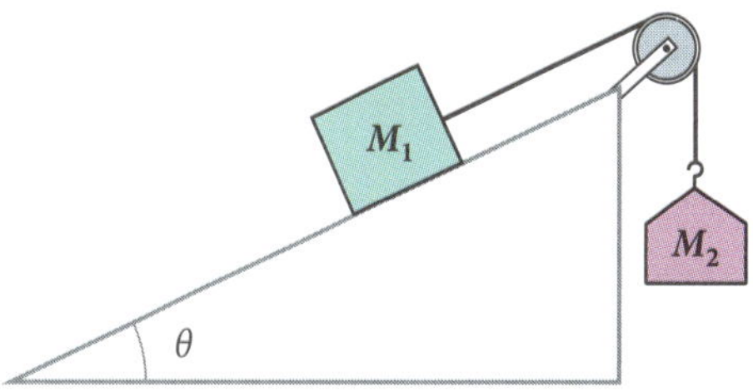

Figure E2.4 Two masses on a pulley

Assuming the plane is frictionless, which of the following expressions would enable us to calculate the acceleration of M_2?

A $M_2g/(M_1 + M_2)$

B $(M_2g - M_1g\sin\theta)/M_2$

C $(M_2g - M_1g\cos\theta)/(M_1 + M_2)$

D $(M_2g - M_1g\sin\theta)/(M_1 + M_2)$

6 A helicopter was initially seen to be flying at 30 ms^{-1} north by an observer on the ground. When a wind started to blow, the observer noticed that the helicopter began to fly at 42.4 ms^{-1} in a north-westerly direction. What was the wind velocity?

A 10 ms^{-1} NW
B 30 ms^{-1} W
C 30 ms^{-1} E
D 10 ms^{-1} W

Questions 7 and 8 refer to the pressure versus time graphs for two sound waves shown on the oscilloscope screen in Figure E2.5. The horizontal scale represents 2.5 ms per division and the vertical scale represents 0.2 V per division.

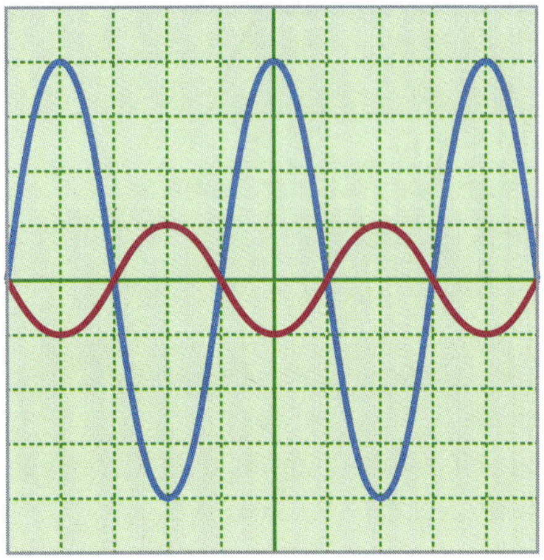

Figure E2.5 Two sound waves

7 What is the frequency of the larger wave?

A 0.1 Hz
B 400 Hz
C 100 Hz
D 0.01 s

8 If these two waves were superimposed, what would be the amplitude of the resulting wave?

A 1.0 V
B 0.6 V
C 1.2 V
D 2.0 V

9 When loud music is playing in a car with an open window, why do people on the side of the road tend to hear the low-frequency (bass) tones rather than the high-frequency (treble) sounds?

A Long wavelengths are diffracted more than short wavelengths.
B Short wavelengths are diffracted more than long wavelengths.
C Long wavelengths are refracted more than short wavelengths.
D Lower frequencies reflect more readily than higher frequencies.

10 Consider the two points A and B in a uniform electric field in Figure E2.6.

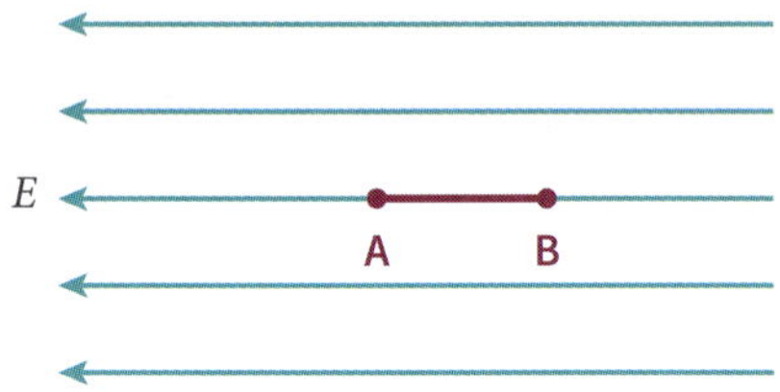

Figure E2.6 Two points in a uniform electric field

Which of the following statements best describes a negative charge moving from point A to point B?

A The field does work on the negative charge when it moves from A to B, but point A has a higher electric potential than point B.
B The field does work on the negative charge when it moves from A to B, but point B has a higher electric potential than point A.
C The work must be done by an external source to move the charge from A to B, but it has a higher electric potential at A.
D The work must be done by an external source to move the charge from A to B, but it has a higher electric potential at B.

11 In the investigation illustrated in Figure E2.7, a student sounds a tuning fork above a pipe that is initially filled with water. The student then gradually pours water out of the tube until the intensity of sound suddenly increases. What is the most likely length of the air column in the pipe when the sound gets louder?

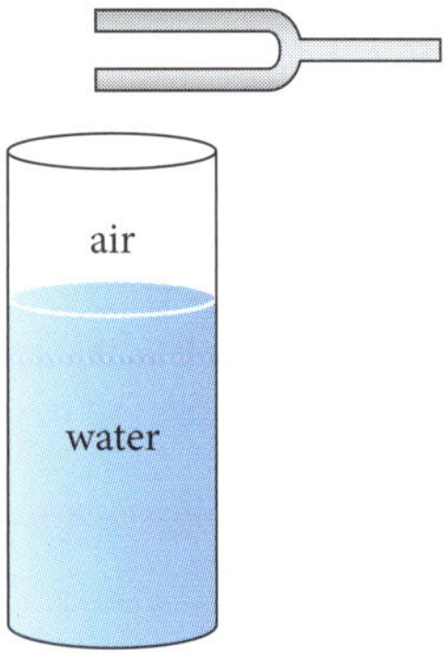

Figure E2.7 Investigating standing waves in an air column

A One-half of the wavelength of the sound produced by the tuning fork.
B The same as the length of the wavelength of the sound produced by the tuning fork.
C One-quarter of the wavelength of the sound produced by the tuning fork.
D It is impossible to tell from the information provided.

12 **What is the difference between a virtual image and a real image?**

- **A** A real image can be created only by a convex mirror or convex lens.
- **B** A virtual image can only be produced by a convex mirror or convex lens.
- **C** A virtual image can be projected onto a screen but a real image cannot.
- **D** A real image can be projected onto a screen but a virtual image cannot.

13 **The graph in Figure E2.8 shows the relationship between the potential difference across and the current within two circuit elements *A* and *B*. What can we deduce from the graph?**

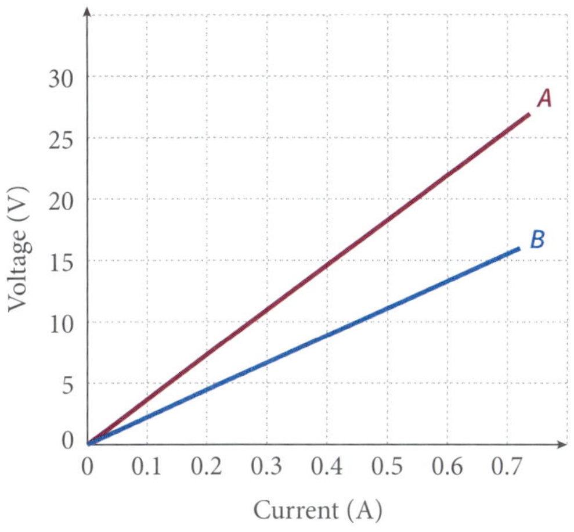

Figure E2.8 Voltage–current relationship for two circuit elements

- **A** Both circuit elements are ohmic resistors and *A* has a greater resistance than *B*.
- **B** Both circuit elements are ohmic resistors and *B* has a greater resistance than *A*.
- **C** Neither circuit element is an ohmic resistor as the current in each is variable.
- **D** Both circuit elements are ohmic resistors with the same resistance.

14 **The intensity of the Sun's light reaching Saturn is 13 Wm^{-2}. Given that Saturn is ten times further from the Sun than the Earth, what is the intensity of light reaching the Earth from the Sun?**

- **A** 1300 W
- **B** 130 Wm^{-2}
- **C** 1300 Wm^{-2}
- **D** 0.13 Wm^{-2}

15 **Two electromagnets are placed adjacent to each other, as shown in Figure E2.9.**

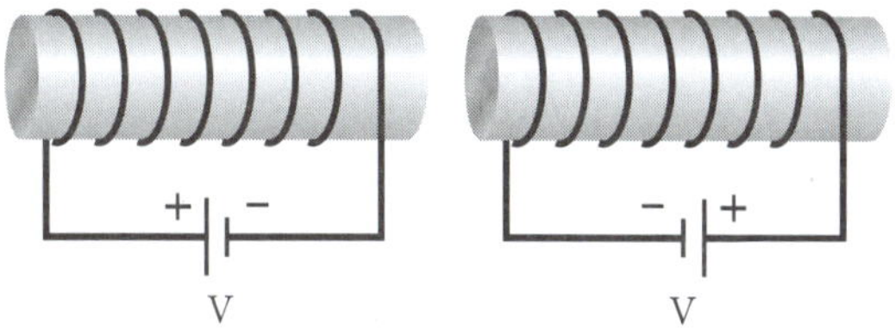

Figure E2.9 Two electromagnets side by side

Which of the following statements best describes the poles and forces on the magnets?

- **A** The magnets will have two north-pole faces adjacent to each other and hence repel each other.
- **B** The magnets will have two south-pole faces adjacent to each other and hence repel each other.
- **C** The magnets will have opposite-pole faces adjacent to each other and hence repel each other.
- **D** The magnets will have two opposite-pole faces adjacent to each other and hence attract each other.

16 **Consider sunlight flowing through the raindrop, as illustrated in Figure E2.10. Which of the following wave properties enable water droplets to produce rainbows?**

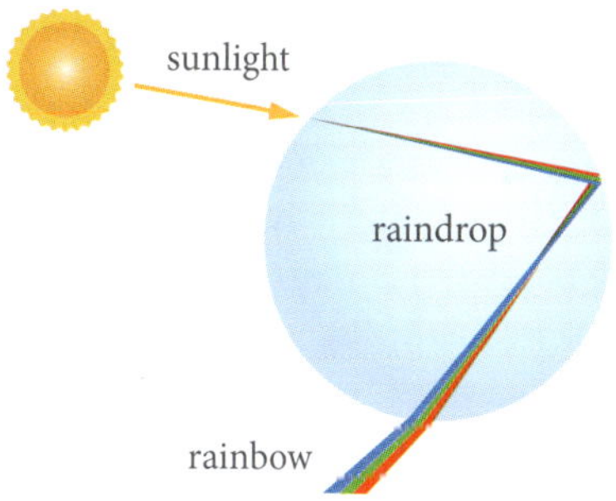

Figure E2.10 Light through a raindrop

- **A** reflection, refraction and scattering
- **B** total internal reflection and interference
- **C** total internal reflection, refraction and dispersion
- **D** total internal reflection and interference

17 **A student found that when the voltage across a light globe was doubled from 3 V to 6 V, the current passing through the globe changed from 100 mA to 150 mA. What is the most probable cause of this result?**

- **A** The globe is an ohmic resistor.
- **B** The resistance of the globe is lower at 6 V than at 3 V because the globe is hotter.
- **C** The resistance of the globe is higher at 6 V than at 3 V because the globe is hotter.
- **D** The student probably made an error in taking the measurements.

18 Consider the three circuit resistors in Figure E2.11.

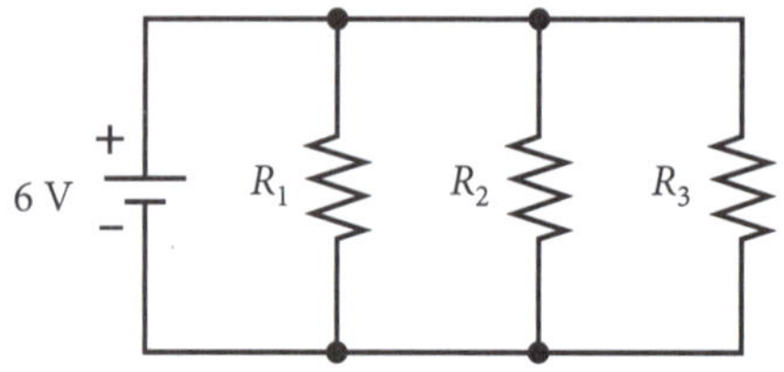

Figure E2.11 Three resistors in a circuit

Given that $R_1 = R_2$ and $R_3 = 2R_1$, what would the total current in the circuit be given by?

A $15/R_1$ **B** $12/R_1$ **C** $24R_1$ **D** $4R_1$

19 Figure E2.12 shows a tugboat towing a large barge. Which of the following statements best describes the forces involved when the tugboat accelerates the barge at 2 ms^{-1}?

Figure E2.12 Tugboat towing a barge

A The barge exerts more force on the tugboat than the tugboat exerts on the barge.

B The tugboat exerts more force on the barge than the barge exerts on the tugboat.

C The friction on the barge is equal to the force applied to the barge by the tugboat.

D The tugboat exerts the same force on the barge as the barge exerts on the tugboat.

20 What is the magnitude and direction of the current in the circuit shown in Figure E2.13?

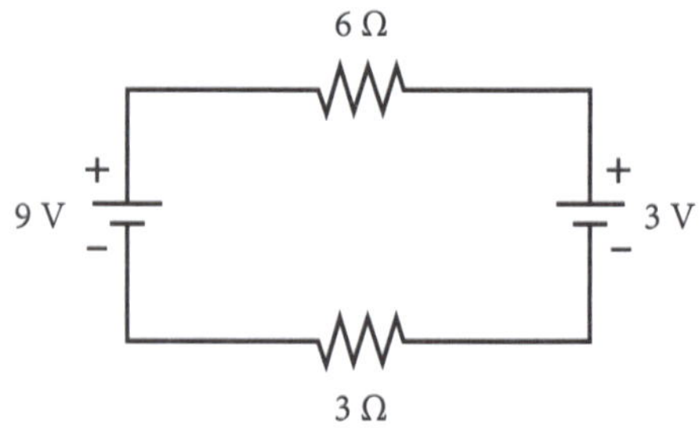

Figure E2.13 Circuit diagram

A 0.67 A, clockwise **B** 0.67 A, anticlockwise

C 1.33 A, clockwise **D** 1.33 A, anticlockwise

Section II: 80 Marks

Allow about 2 hours and 20 minutes for this section.

Extended-response questions

21 A student measures the velocity of a drone as a function of time. The drone was initially flying in a northerly direction. The results obtained are plotted on the graph in Figure E2.14. You may assume the motion is in a straight line.

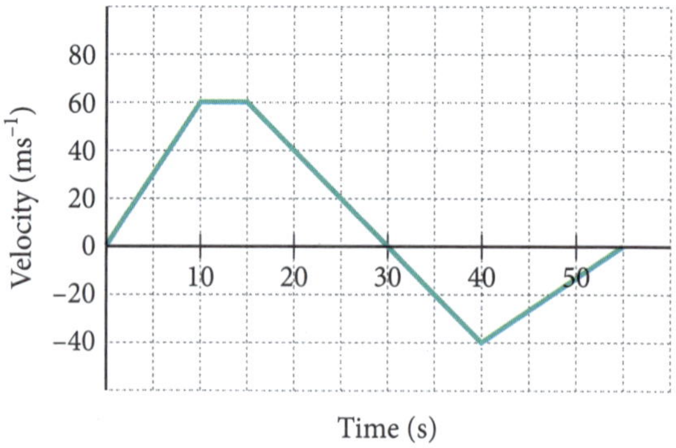

Figure E2.14 Velocity as a function of time graph

a Find the magnitude and direction of the drone's maximum acceleration. (2 marks)

b Find the distance travelled by the drone in the first 30 s. (1 mark)

c Find the total displacement of the drone after 50 s. (2 marks)

d Determine the average velocity after 50 s. (1 mark)

22 A helicopter flies directly east at 40 ms^{-1} near a plane travelling at 50 ms^{-1} north.

a What is the velocity of the helicopter relative to the plane? (2 marks)

b If a wind started blowing at 20 ms^{-1} from the south to the north, what would the new velocity of the helicopter be as measured by an observer on the ground? (2 marks)

23 A racing car of mass 1500 kg accelerates in an easterly direction from rest to 300 kmh^{-1}. The car travels 400 m while accelerating and then moves at constant velocity.

a Find the acceleration of the car. (2 marks)

b Find the net force on the car during the period of acceleration. (1 mark)

c Draw a diagram to show the direction and magnitude of the forces acting on the car when it is accelerating. You may consider the friction acting on the car to be a constant force of 1000 N, and numerical values for the forces acting should be included in the diagram. (4 marks)

d What power would the car engine need to supply to keep the car moving at a constant 300 kmh^{-1}? (You may assume that the frictional force is a constant 1000 N.) (1 mark)

24 Radio waves are electromagnetic waves and hence travel at 3×10^8 ms^{-1}. Figure E2.15 shows the electric field as a function of distance for a radio wave.

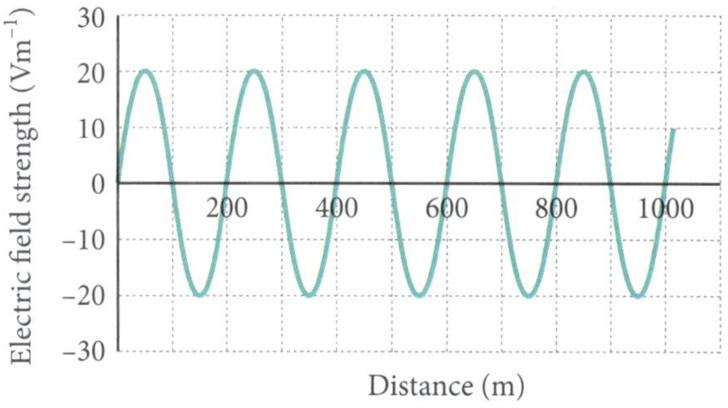

Figure E2.15 Radio wave

a Determine the amplitude of the wave. (1 mark)

b Determine the wavelength of the wave. (1 mark)

c Calculate the frequency of the wave. (2 marks)

25 When the neck of a 400 Hz tuning fork is touched on the end of a guitar string, the string starts to oscillate in its 2nd harmonic.

a Sketch the guitar string oscillating in its 2nd harmonic, marking the nodes and antinodes. (3 marks)

b Given the guitar string is 40 cm long, find the speed of the waves in the string. (2 marks)

26 Consider the lens illustrated in Figure E2.16.

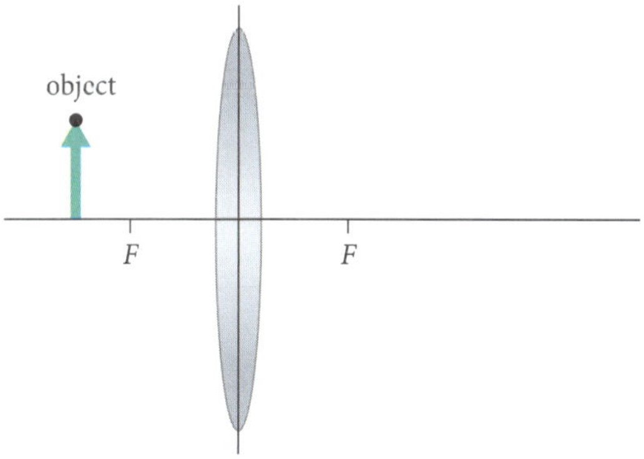

Figure E2.16 Lens

a Redraw the lens and use ray-tracing to show where the image of the object would appear. (3 marks)

b Use your drawing to determine if the image is:

i real or virtual

ii upright or inverted

iii magnified or diminished. (3 marks)

27 A 200 g block of ice at −20 °C is thrown into a large bath of hot water at 90 °C. You may assume that the temperature of the bath water remains constant. Use the data below to calculate how much heat would be required to raise the temperature of the 200 g block of ice from −20 °C to 90 °C. (3 marks)

Data:

- Specific heat capacity of ice = 2108 $J K^{-1} kg^{-1}$
- Latent heat of fusion of water = 3.34×10^5 $J kg^{-1}$
- Specific heat capacity of water = 4186 $J K^{-1} kg^{-1}$

28 Two small charged balls of mass 5 g are suspended from 20 cm long light strings, as shown in Figure E2.17. Both balls have a positive charge q.

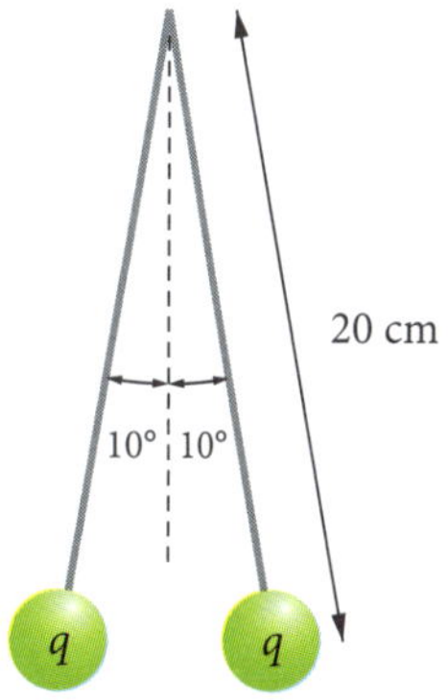

Figure E2.17 Charged balls on string

a Find the distance between the balls. (1 mark)

b Determine the horizontal force required to displace a 5 g ball in this way. (2 marks)

c Calculate the charge q on the balls. (1 mark)

d If the charge on each ball was doubled, what angle would the strings make with the vertical? (2 marks)

29 A student measures the force required to draw a long bow as a function of how far the string of the bow is pulled back. The results are graphed in Figure E2.18. (Note that the area under such a graph is equal to the work done.)

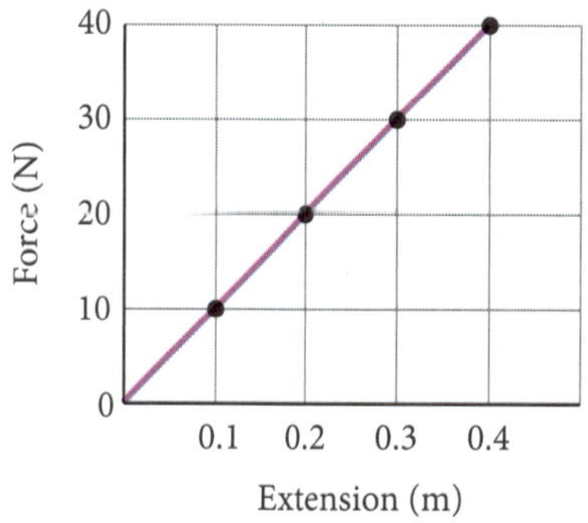

Figure E2.18 Force applied to pull back a bowstring

a Find the gradient of the graph and its units. (1 mark)

b What property of the bow is the gradient measuring? (1 mark)

c Find the area under the graph and its units. (1 mark)

d If the entire bow's stored potential energy was transferred to an arrow of mass 20 g, at what speed would the arrow leave the bow? (2 marks)

e If the arrow was fired vertically upwards, outline the energy transformations that would occur from the time when the arrow was at rest in the drawn bow until it reached the ground again. (You may ignore air friction and assume the bow transferred all its energy to the arrow.) (3 marks)

30 A student conducts a refraction experiment in which he measures the angle of refraction (θ_2) as a function of the angle of incidence (θ_1) for a ray of red light moving from a glass block to air. The results obtained by the student are plotted in Figure E2.19. Note the speed of light in air is 3×10^8 ms^{-1}.

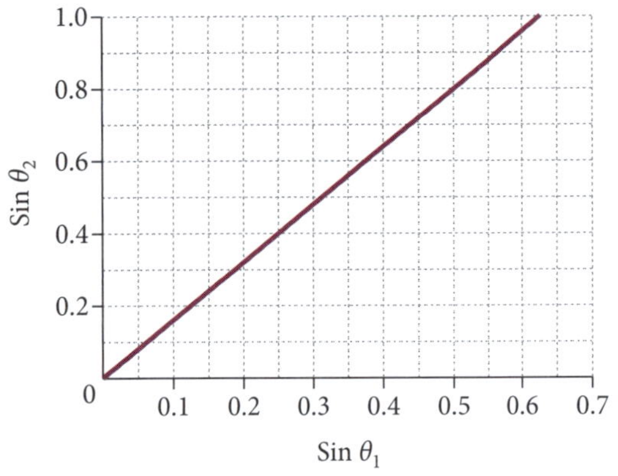

Figure E2.19 Experiment results

a Find the gradient of this graph and explain how the gradient is related to Snell's law. (2 marks)

b Use the graph to find the speed of light in glass and the refractive index of the glass used in the experiment. (2 marks)

c Explain how the graph would change if the student had used blue light rather than red light for the experiment. (1 mark)

31 The potential difference across an 80 Ω resistor is 10 V in the circuit in Figure E2.20.

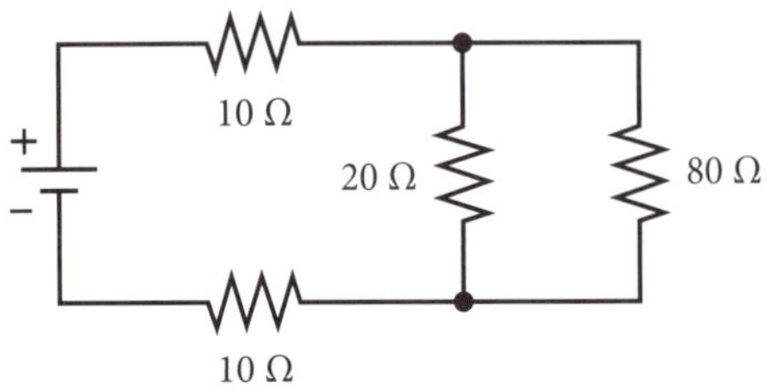

Figure E2.20 Circuit diagram

a Find the current in the 80 Ω resistor. (1 mark)

b Find the current in the 10 Ω resistor. (1 mark)

c Find the potential difference across the battery. (2 marks)

d If both of the 10 Ω resistors were replaced with 5 Ω resistors, what would be the total resistance of the new circuit? (2 marks)

32 Two wires carrying equal currents of 5 A in opposite directions in and out of the page are separated by 10 cm, as shown in Figure E2.21.

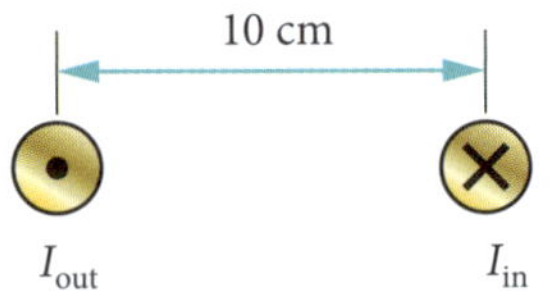

Figure E2.21 Two current-carrying wires

a Sketch the magnetic field lines around the wires. (2 marks)

b Find the magnitude and direction of the magnetic field midway between the wires. (2 marks)

c If the current in the wire carrying current into the page was reversed and doubled, what would be the magnetic field strength midway between the wires? (2 marks)

33 A pair of parallel plates separated by 4 cm has a potential difference of 100 V across them (see Figure E2.22). A proton (mass = 1.67×10^{-27} kg and charge = 1.6×10^{-19} C) was placed at the positive plate. To prevent air molecules interfering with the experiments, the volume between the plates was evacuated.

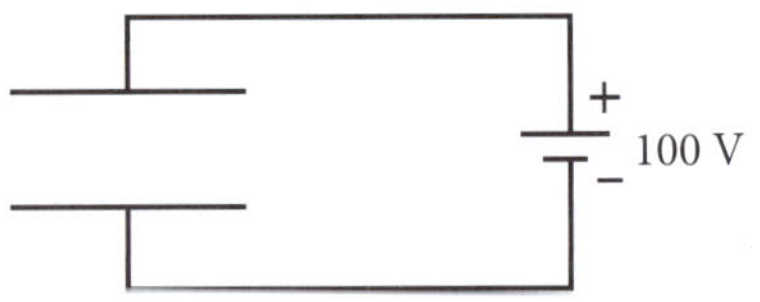

Figure E2.22 Charged parallel plates

a Find the magnitude and direction of the electric field between the plates. (1 mark)

b Find the magnitude of the force on the proton. (1 mark)

c Find the acceleration of the proton and explain why we neglect gravity in this experiment. (2 marks)

d Find the time it would take the proton to reach the negative plate and the final velocity of the charge just before it reaches the negative plate. (2 marks)

e Find the energy gained by the proton moving from one side of the plate to the other. (1 mark)

34 On an ice-skating rink, two skaters push a third skater, as shown in Figure E2.23(a). The skaters were all initially at rest. After the interaction the skaters move apart from one another, as shown in Figure 2.23(b).

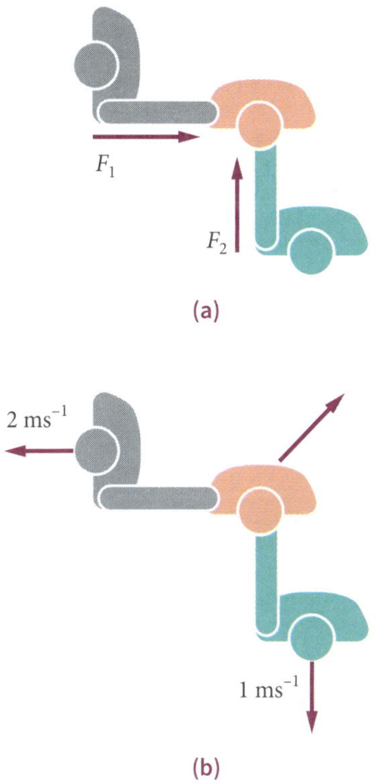

Figure E2.23 Movement of three ice-skaters (viewed from above)

The skater being pushed has a mass of 60 kg, and the two skaters that pushed her each have a mass of 50 kg.

a Find the direction and magnitude of the total momentum of the two equal-mass skaters immediately after they pushed the other skater. (2 marks)

b Determine the velocity of the 60 kg skater after she was pushed. (2 marks)

c Given the skater was pushed for a time of 0.5 s, find the net force applied to the 60 kg skater. (1 mark)

d Explain why this interaction is not an elastic collision. (1 mark)

ANSWERS

SAMPLE YEAR 11 EXAMINATION 1

Objective-response questions

1 **C.** As $v^2 = u^2 + 2as$ and, as $v = 0$, $a = -u^2/2s = \frac{-20^2}{2 \times 40} = -5\ \text{ms}^{-2}$.
Now $F = ma = 2000 \times (-5) = -10^4$ N. **B** and **D** are incorrect as they have the wrong units and **A** is incorrect because a positive force would increase the velocity rather than decrease it.

2 **C.** The reaction to pushing the ground down applies an upwards force to the person. **A** is incorrect as your centre of mass would not go upwards. **B** is incorrect as this would pull you downwards. **D** is incorrect as the acceleration due to gravity is a constant that will not change.

3 **D.** The distance travelled is the absolute values of the area under the graph. In the first 5 s the skateboarder travels 50 m north and in the next 5 s the skateboarder travels 50 m south, giving a total distance of 100 m. **A** and **C** are incorrect because the skateboarder travels north and south. **B** is incorrect because the distance is incorrect.

4 **A.** $a = g \sin\theta$ and hence $g = a/\sin\theta = 2/\sin 20° = 5.8\ \text{ms}^{-2}$ and $W = mg = 10 \times 5.8 = 58$ N. **C** and **D** are incorrect as they have the wrong units for weight and **B** is incorrect as it has the wrong acceleration due to gravity.

5 **B.** The relative velocity is obtained by subtracting the velocity of the observer, as shown in Figure EA1.1. (Note $-40\ \text{ms}^{-1}$ south $= +40\ \text{ms}^{-1}$ north.) **A**, **C** and **D** are all incorrect as they have the wrong direction for the velocity.

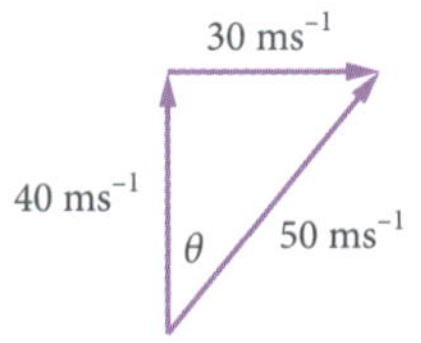

$$\theta = \tan^{-1}\frac{30}{40} = 37°$$

Figure EA1.1 Vector diagram to find relative velocity

6 **D.** Friction is given by $F_{\text{friction}} = \mu F_R$. The net downwards force the sled exerts on the ground would be $W = mg - 60\sin 30°$ and hence $F_{\text{friction}} = \mu(mg - 60\sin 30°)$. **A**, **B** and **C** are incorrect as they all use the wrong reaction force.

7 **A.** The period $T = \frac{60}{12} = 5$ s and $v = f\lambda = \lambda/T = \frac{15}{5} = 3\text{ms}^{-1}$. **B**, **C** and **D** are all incorrect as they have the wrong numerical value.

8 **C.** Momentum must be conserved in the collision. As the change of momentum (impulse) is the same for each player, but in the opposite direction, the sum of the change in momentum for the two-player system is zero. The total momentum of the two-player system therefore remains constant ensuring momentum is conserved. **A**, **B** and **D** are incorrect as each would result in a net change in momentum for the system in the collision, which would mean momentum was not conserved.

9 **D.** The rate of heat loss per unit area of the surface Q is proportional to the thermal conductivity and inversely proportional to the wall thickness. If the wall thickness is halved it will double the rate of heat loss and doubling the thermal conductivity will also double the heat loss. The result would be that heat would be lost four times faster. **A**, **B** and **C** are all incorrect as they predict a lesser or greater heat loss than the expected four times.

10 **B.** A ray must be incident at an angle greater than the critical angle to be totally internally reflected and it must be attempting to pass into a medium where the wave would travel faster (i.e. the second medium must have a lower refractive index than the first medium). **C** and **D** are incorrect because if the angle of incidence was less than the critical angle the ray would move into the second medium. **A** is incorrect because light cannot be totally internally reflected at a boundary where the speed of the wave in the second medium would be lower than the first (i.e. where the refractive index of the second medium is higher than the refractive index in the first medium).

11 **A.** Intensity decreases as the inverse square of the distance from the source. As Sirius is twice as far away as Polaris, we would expect it to be one-quarter as bright as Polaris. **B** is incorrect as it implies intensity decreases as the inverse of distance. **C** is incorrect because it implies distance does not affect brightness. **D** is incorrect because it implies the intensity would increase with distance.

12 **D.** All the blocks will accelerate at the same rate and, as $F = ma$, the force on the block with the greatest mass must be the greatest force. **A** and **B** cannot be correct because the blocks will have the same acceleration and hence must have different net forces applied as each block has a different mass. **C** cannot be correct because the lightest block will require the least force to accelerate it.

13 **B.** The frequency remains constant as the ray moves from one medium to another and, as the ray is refracted towards the normal, the speed of the wave decreases when it enters the water. As $v = f\lambda$, decreasing velocity must result in a decrease in wavelength. **A** is incorrect as the frequency cannot change because the wave must be continuous across the boundary. **C** and **D** are incorrect as both would require light to travel faster in water than air.

14 **A.** The wave would produce beats (a periodic change in loudness). **B** and **D** are incorrect as the wave has a constant period and frequency. **C** is incorrect as the image shows the amplitude of the wave (and hence the loudness) clearly increases and decreases periodically.

15 **D.** Beats are produced by the superposition of two waves with slightly different frequencies. **A** and **C** are incorrect as the two waves being superimposed have the same frequency and hence would not go in and out of phase periodically. **B** is incorrect as a loud and soft sound, even if the tuning forks had slightly different frequencies, would not totally cancel out each other when they were out of phase.

16 **C.** $E = V/d$ and $F = qE$. Hence $F = q(V/d) = 10^{-6} \times \frac{20}{0.02} = 10^{-3}$ N down. Now as $F = ma$, $a = F/m = \frac{10^{-3}}{10^{-6}} = 10^3\ \text{ms}^{-2}$ downwards.
A and **B** are incorrect because it is a negative particle and will be attracted towards the positive plate and repelled from the negative plate, producing a downwards force on the particle. **D** is incorrect because it is the wrong numerical answer.

17 **A.** $I = V/R = \frac{4}{2} = 2$A and, as the resistors are in series, the same current must flow through each resistor. Hence for the 6 Ω resistor,

$V = IR = 2 \times 6 = 12$ V. **A**, **B** and **C** are all incorrect as they would result in a different current flowing through the 6 Ω resistor than that flowing in the 2 Ω resistor.

18 **B.** Only these metals all contain unpaired electrons and hence magnetic domains. **A** is incorrect because copper is not ferromagnetic. **C** is incorrect because aluminium is not ferromagnetic. **D** is incorrect because carbon is not ferromagnetic.

19 **C.** The magnetic field lines are in the right direction, as given by the right-hand grip rule, and the lines get closer together closer to the wire, reflecting the strength of the field. **A** and **D** are incorrect as the direction of the magnetic field lines are incorrect. **B** is incorrect because the magnetic field lines are equally spaced, which would incorrectly imply that the field strength around the wire was uniform (i.e. does not change with distance from the wire).

20 **A.** This represents what happens to the intensity and wavelength of the radiation emitted from an object when it is heated. **B** and **D** are incorrect because the particles would have more kinetic energy in the hotter body, resulting in more radiation being emitted from the body. **C** is incorrect because as the body was heated the particles would oscillate faster and emit higher-frequency radiation, not lower-frequency radiation.

Extended-response questions

21 EM This question tests students' understanding of graphing motion, the concept of inertia, the difference between distance and displacement, and the difference between speed and velocity.

a Total distance = 60 + 60 + 40 + 40 = 200 m ✓

b Average speed = total distance/total time = $\frac{200}{60} = 3.33$ ms^{-1} ✓

As the runner returned to the starting point the total displacement and hence total velocity is zero. ✓

c Velocity = gradient; steepest gradient = $\frac{60}{10} = 6$ m^{-1} ✓ between $t = 0$ and $t = 10$ seconds ✓

d Velocity is equal to the gradient of the position against time graph for each time interval, as shown in Figure EA1.2.

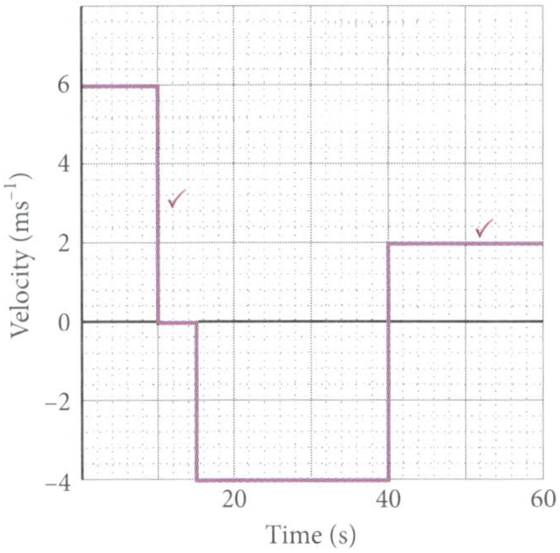

Figure EA1.2 Velocity against time graph

e The runner has mass and therefore it will take time for the runner to change velocity due to his inertia. ✓ The graph shows instantaneous changes in velocity, which would require infinite forces. ✓ The data therefore may not accurately reflect the motion of the runner.

22 EM This question tests students' ability to apply the equations of motion and Newton's second law of motion.

a $F = ma = 2000 \times 5 = 10^4$ N east ✓

b $F_{net} = F_{applied} - f_{friction}$ and hence $10^4 = F_{applied} - 2500$ or $10^4 + 2500 = 12\,500$ N east ✓

$F_{applied} = 10^4 + 2500 = 12\,500$ N east ✓

c $v = at + u = 5 \times 8 = 40$ ms^{-1} east ✓

d Displacement after 8 s can be determined from $s = \frac{1}{2}at^2 + ut$ and, as $u = 0$, $s = \frac{1}{2}(5 \times 8^2) = 160$ m.

For the work done against friction:

$W = (f_{friction})s = 2500 \times 160 = 4 \times 10^5$ J ✓

e Work = $(F_{applied})s = 12\,500 \times 160 = 2 \times 10^6$ J ✓

Power = $E/t = \frac{2 \times 10^6}{8} = 2.5 \times 10^5$ W ✓

f $P = Fv = 2500 \times 40 = 10^5$ W ✓

23 EM This question tests students' understanding of graphing motion, kinetic and potential energy of a falling body, and acceleration due to gravity.

a The area under the graph from the maximum positive velocity (i.e. just after the ball bounces) until the velocity is zero (i.e. at the top of the ball's motion) is $9 \times 1 \times 0.5 = 4.5$ m. ✓

b Acceleration will be equal to the negative gradient of the graph, $g = \frac{9}{1} = -9$ ms^{-2}. ✓

c Kinetic energy is conserved in an elastic collision. ✓ The graph represents elastic collisions as the ball leaves the ground with the same velocity and hence the same kinetic energy as it had just before it hit the ground. ✓

d At $t = 0$ the ball is released from rest and its kinetic energy increases until it hits the ground at $t = 1$ s. ✓ A short time later at $t = 1.25$ s, the ball leaves the ground with the same kinetic energy it had just before it hit the ground. The kinetic energy then decreases until the ball reaches its maximum height, and the velocity and kinetic energy reach zero at $t = 2.25$ s. ✓

e The gravitational potential energy is initially a maximum just before the ball is released. The gravitational potential energy decreases as the ball falls until it reaches zero when the ball comes momentarily to rest on the ground when $t = 1.25$ s. ✓ The gravitational potential energy then increases until it reaches a maximum value when the ball reaches the top of the flight when $t = 2.25$ s. ✓

f The total energy of the ball remains constant. ✓

24 EM This question tests students' understanding of conservation of momentum, impulse, and the difference between elastic and inelastic collisions.

a As the cars are entangled, the mass after the collision is 2000 + 3000 = 5000 kg. The velocity can be deduced from the momentum after the collision.

As $p = mv$, $v = p/m = \frac{10\,000}{5000} = 2$ ms^{-1} north. ✓

b Yes, momentum was conserved ✓ as the sum of the momentum before the collision ($p = 20\,000 - 10\,000 = 10\,000$ Ns north) is equal to the momentum after the collision. ✓

c No, kinetic energy was not conserved ✓ because the sum of the kinetic energy before the collision ($\Sigma\frac{1}{2}mv^2 = 0.5 \times 2000 \times 10^2 + 0.5 \times 3000 \times 3.3^2 = 1.16 \times 10^5$ J) is not equal to the sum of the kinetic energy after the collision ($0.5 \times 5000 \times 2^2 = 1 \times 10^4$ J). ✓

d Calling north positive, $\Delta p = p_{final} - p_{initial} = 4000 - 20\,000 = -16\,000$ Ns or 16 000 Ns ✓ south. ✓

e $F = \Delta p/t = \frac{16\,000}{1} = 16\,000$ N south. ✓

25 **EM** This question tests students' understanding of ray diagrams, wave fronts and Snell's law.

a The answer in Figure EA1.3 should show the ray reflected at the same angle as the incident ray and wavefronts perpendicular to the reflected ray and with the same spacing (i.e. same wavelength) as the incident wave fronts.

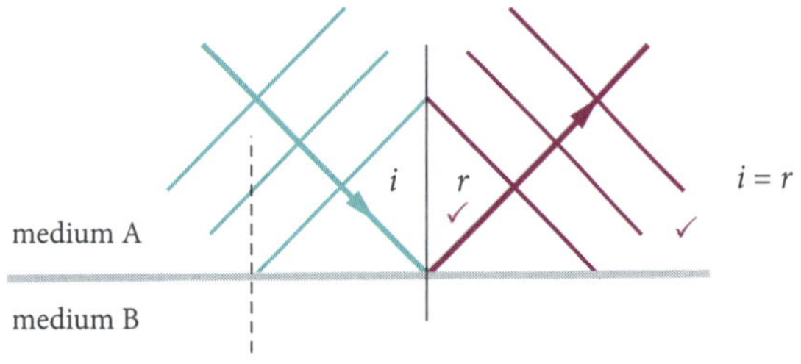

Figure EA1.3 Reflection of a light ray

b Figure EA1.4 shows the angle of refraction is less than the angle of incidence and the wavefronts perpendicular to the refracted ray and closer together (i.e. smaller wavelength) than the wave fronts on the incident ray.

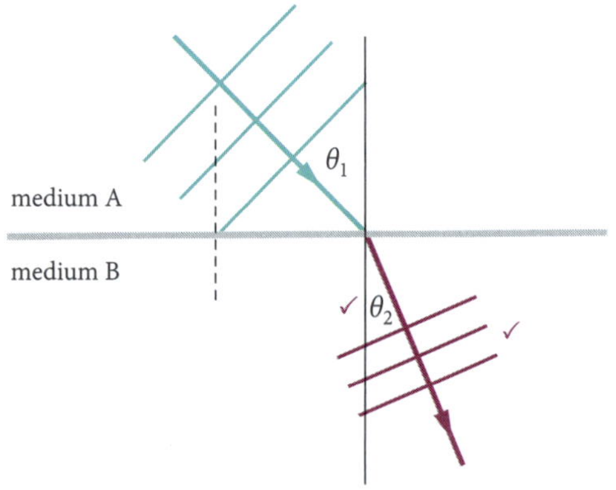

Figure EA1.4 Refracted light ray

c Applying Snell's law, $\sin\theta_1/\sin\theta_2 = v_1/v_2$ and hence:

$\theta_2 = \sin^{-1}((v_2/v_1)\sin\theta_1) = \sin^{-1}\left(\frac{200}{340}\right)\sin 40 = 22.2°$ ✓

26 This question tests students' understanding of the Doppler effect and relative velocity.

a First change the velocity of the car to SI units:

$50\text{ kmh}^{-1} = 50 \times \frac{1000}{60 \times 60} = 13.9\text{ ms}^{-1}$ ✓

Applying the Doppler effect equation:

$$f' = f\frac{(v_{wave} + v_{observer})}{(v_{wave} - v_{source})}$$

This equation is set up for the source moving towards the observer and the observer moving towards the source. In the case at hand, the observer is moving towards the source but the source of the sound is moving away from the observer and we must therefore change the sign of the source velocity in the denominator of the above equation. Hence for this problem:

$$f' = f\frac{(v_{wave} + v_{observer})}{(v_{wave} + v_{source})} \text{ or}$$

$v_{source} = (f/f')(v_{wave} + v_{observer}) - v_{wave}$

$v_{source} = \left(\frac{550}{500}\right)(340 + 13.9) - 340 = 49\text{ ms}^{-1}$ east ✓

b The cars are both moving in the same direction and hence calling east positive:

$v_{relative} = v_{object} - v_{observer} = 13.9 - 49 = -35.1\text{ ms}^{-1}$
$= 35.1\text{ ms}^{-1}$ ✓ west ✓

27 **EM** This question tests students' understanding of inclined planes and the coefficient of friction.

a Down the plane $F = mg\sin\theta = 25 \times 9.8 \times \sin 25° = 103.5$ N ✓

Into the plane $F = mg\cos\theta = 25 \times 9.8 \times \cos 25° = 222$ N ✓

b $f_{friction} = \mu F_R = 0.2 \times 222 = 44.4$ N ✓

c Net force on block down the plane $= 103.5 - 44.4 = 59.1$ N ✓

As $F = ma$, $a = F/m = \frac{59.1}{25} = 2.36\text{ ms}^{-2}$ down the plane ✓

28 **EM** This question tests students' understanding of the changes in potential difference around a circuit, Ohm's law, and series and parallel resistors.

a Sum of parallel resistors, $1/R_T = \frac{1}{6} + \frac{1}{12}$ and hence $R_T = 4\ \Omega$ and therefore the total resistance for the circuit is $4 + 2 = 6\ \Omega$. ✓

b First we must find the total current as this is the current in 2 Ω resistor.

$I_T = V_T/I_T = \frac{24}{6} = 4$ A. Now as there is $V = IR = 4 \times 2 = 8$ V across the 2 Ω, there must be $24 - 8 = 16$ V across the parallel resistors. Hence the potential difference across the 6 Ω resistor is 16 V. ✓ The current $= V/R = \frac{16}{6} = 2.67$ A ✓

c As there is 16 V across this resistor, the power dissipated in the resistor would be given by:

$P = IV = V^2/R = \frac{16^2}{12} = 21.3$ W ✓

d The answer is illustrated in Figure EA1.5.

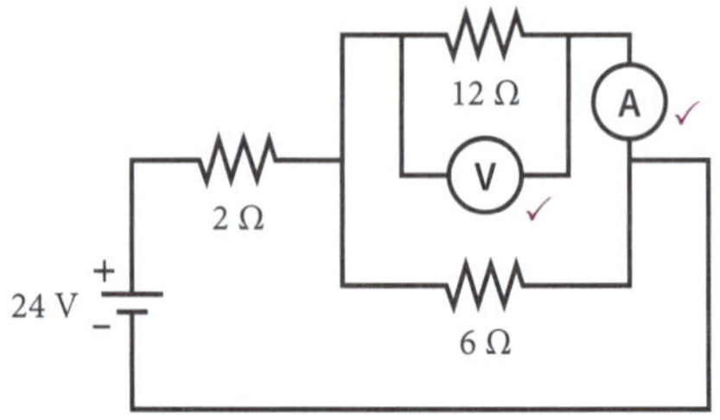

Figure EA1.5 Position of voltmeter and ammeter in circuit

29 **EM** This question tests students' understanding of specific heat capacity, thermal conductivity and electrical power.

a Applying $Q = mC\Delta T = 0.2 \times 4184 \times 70 = 5.86 \times 10^4$ J. ✓

b The power produced by a 25 Ω resistor with a potential difference of 240 V across it would be:

$P = V^2/R = \frac{240^2}{25} = 2.3 \times 10^3$ W ✓

Now power = energy/time, hence $t = E/P = \frac{5.86 \times 10^4}{2.3 \times 10^3} = 25.5$ s ✓

c First convert area to square metres:

$A = \frac{250}{100 \times 100} = 0.025\text{ m}^2$ ✓

The rate at which energy is transmitted through conduction is given by:

$Q/t = kA\Delta T/d = 0.03 \times 0.025 \times \frac{90 - 20}{0.002} = 26$ W ✓

30 **EM** This question tests students' understanding of electric fields and the forces on a charge in an electric field.

a $E = V/d = \frac{200}{0.05} = 4000\text{ Vm}^{-1}$ ✓ downwards ✓

b $W = mg = 5 \times 10^{-5} \times 9.8 = 4.9 \times 10^{-4}$ N ✓ downwards

c The charge must be negative for the field to exert an upwards force on the charge. To levitate the charge, the net force on the charge must be zero and hence:

$F_{electric} = F_{gravity}$ or $qE = mg$ ✓

Therefore $q = mg/E = \frac{4.9 \times 10^{-4}}{4000} = -1.2 \times 10^{-7}$ C ✓

31 EM This question tests students' understanding of ferromagnetic materials, electromagnets and magnetic fields.

a Figure EA1.6 illustrates the answer.

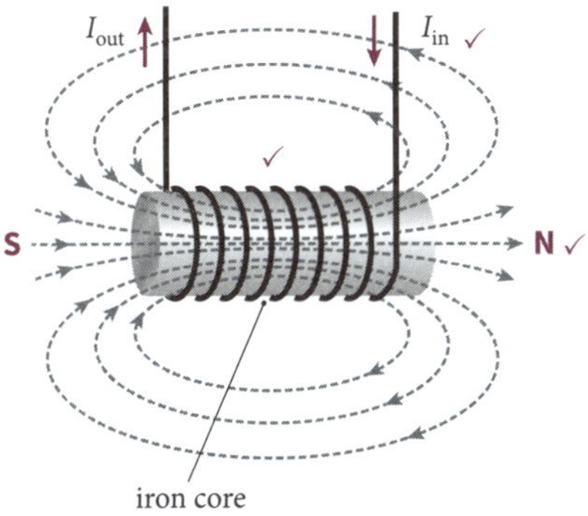

Figure EA1.6 Magnetic field around an electromagnet

b If the iron core was removed, the magnetic field produced would be significantly weakened. ✓ The direction of the field would not change but the strength of the field would decrease. This occurs because the aligned magnetic fields of the magnetic domains ✓ in the iron core would no longer be contributing their magnetic field to the magnetic field of the solenoid. ✓

32 EM This question tests students' understanding of the interaction of two magnetic fields, vectors and Oersted's law.

a In order to deflect the compass to the east, the magnetic field produced by the wire must be directed to the east above the wire. This will only occur if the current flows from south to north in the wire. ✓

b As the compass points at 45°, the two perpendicular magnetic fields acting on the compass must be equal in magnitude. Therefore the field produced by the wire at the compass is $B = 5 \times 10^{-5}$ T east. ✓

c For a long current-carrying wire, $B = \mu_0 I/2\pi r$ ✓ and hence:

$I = 2\pi r B/\mu_0 = \frac{0.04 \times 5 \times 10^{-5}}{2 \times 10^{-7}} = 10$ A north ✓

33 EM This question tests students' understanding of momentum, inclined planes and conservation of energy.

a First we must convert the truck's velocity to SI units:

$\frac{60 \times 1000}{60 \times 60} = 16.67$ ms^{-1}

Now we apply conservation of momentum.

Calling east positive, the momentum before the collision $= 12\,000 \times (16.67) = 2 \times 10^5$ Ns. ✓

Equating the momentum before the collision to the momentum after gives:

$2 \times 10^5 = (2000 + 12\,000)v$ and hence $v = \frac{2 \times 10^5}{14\,000}$

$= 14.29$ ms^{-1} east ✓

b The change in kinetic energy = kinetic energy after – kinetic energy before. ✓

$\Delta K = (\frac{1}{2} \times 14\,000 \times (14.29^2)) - (\frac{1}{2} \times 12\,000 \times (16.67^2))$

$= 2.38 \times 10^5$ J ✓

c The height up the plane can be determined using conservation of energy. As no energy is lost to friction, all the kinetic energy after the collision will be converted into gravitational potential energy.

$\frac{1}{2}mv^2 = mgh$ ✓ and hence:

$h = (\frac{1}{2}mv^2)/mg = \frac{\frac{1}{2} \times 14\,000 \times 14.29^2}{14\,000 \times 9.8} = 10.42$ m ✓

Now as the gradient is 15°, the distance moved up the incline will be:

vertical height/sin θ and hence distance = 10.42/sin 15°

= 40.26 m ✓

SAMPLE YEAR 11 EXAMINATION 2

Objective-response questions

1 **D.** The gradient of the graph is equal to the velocity. The gradient is initially zero (i.e. initial velocity is zero) and then becomes larger in a negative direction (accelerates in the negative direction). **A** and **C** are incorrect as they require a non-zero initial velocity. **B** is incorrect as a positive acceleration would require the gradient to become more positive as time passed.

2 **A.** $w = mg = 50 \times (9.8 - 2) = 390$ N. **C** and **D** are incorrect as they have the wrong units and **B** is incorrect because it assumes the lift is not accelerating.

3 **D.** The mass is at rest so the net force on the mass is zero by Newton's first law. **A, B** and **C** are all incorrect as they imply a net force is applied to the mass. (If this was the case it would be accelerating.)

4 **B.** Here the area under the graph is equal to the change in momentum (i.e. the impulse), and experiment A has an impulse of 12 Ns and experiment B has an impulse of 10 Ns. As the initial velocity is zero, the impulse = $\Delta p = mv$. Hence experiment A will result in a 20% greater final velocity. **A** is incorrect as this would only occur if the area under the graphs was equal. **C** is incorrect as this would require the area under curve B to be greater than the area under curve A. **D** is incorrect as it would require the area under curve A to be twice as great as the area under curve B.

5 **D.** Writing $a = F_{net}/m$ for the system gives $(M_2 g - M_1 g \sin\theta)/(M_1 + M_2)$. **A** is incorrect as it ignores the force of gravity on M_1. **B** is incorrect because the mass of the system should include M_1. **C** is incorrect because it uses the component into the plane rather than down the plane to calculate the force produced by M_1.

6 **B.** Adding this wind direction to the velocity of the helicopter gives the observed velocity. **A, C** and **D** all give the wrong velocity vector when the helicopter's velocity is added to the wind velocity.

7 **C.** The period can be measured from the screen $T = 4 \times 2.5$ ms = 0.01 s and as $f = 1/T = \frac{1}{0.01} = 100$ Hz. **A** and **B** are incorrect as they are the wrong numerical answer, and **D** is incorrect as it is the period not the frequency.

8 **B.** Because the waves are a half-cycle out of phase we have to subtract the amplitude of the small wave from the amplitude of the larger wave. Amplitude of sum = 8 ms – 2 ms = 6 ms. **A, C** and **D** are incorrect as they are the wrong numerical answers.

9 **A.** The sound waves will be diffracted as they pass through the window and the amount of diffraction increases with wavelength. **B** is incorrect as short wavelengths are actually diffracted less

than long wavelengths. **C** is incorrect because refraction refers to waves moving into a different medium and **D** is incorrect because reflection is not related to wavelength.

10 **B.** The force on the charge will be opposite to the direction of the electric field as the charge is negative, but the electrical potential is higher at point B because electrical potential is defined as the point where a positive charge could do more work. **C** and **D** are incorrect as the force on the charge is in the direction of its motion and hence no external force will be required to move it between the points. **A** is incorrect because positive charge is used to define electrical potential difference.

11 **C.** There must be a node at the closed (water) end and an antinode at the open end. The longest wavelength wave (the fundamental or 1st harmonic) that would resonate with the tuning fork would be a quarter of the wavelength produced by the tuning fork. **A** and **B** are wrong because they would require nodes or antinodes at both ends of the tube. **D** is incorrect because we can determine the fundamental mode of oscillation from the information given.

12 **D.** This is one of the ways in which we differentiate between real and virtual images. **C** is incorrect because a virtual image cannot be projected onto a screen as light rays do actually converge to it. **A** and **B** are incorrect because real and virtual images can be created by convex lenses.

13 **A.** The gradient of the graph (V/I) is the resistance and the graph is a straight line for both resistors. A straight-line graph implies the current is proportional to the voltage and hence both resistors are ohmic. **B** and **D** are incorrect as *B* has a lower resistance than *A*. **C** is incorrect because the voltage against current graphs for the resistors are straight-line graphs, which indicates that both resistors are ohmic.

14 **C.** The intensity decreases as the inverse square of distance and hence the intensity on Saturn will be $\frac{1}{10^2}$ = 100th of the intensity on Earth. **B** and **D** are incorrect as they are not the correct numerical answers and **A** is incorrect as it has the wrong units.

15 **A.** By the right-hand grip rule the adjacent-pole faces will be both north and the like poles will repel each other. **B**, **C** and **D** are all incorrect because the magnetic field directions specified are not consistent with the right-hand grip rule.

16 **C.** The ray undergoes refraction and dispersion when it enters and leaves the droplet and is totally internally reflected inside the droplet. **A** is incorrect because the ray is internally reflected rather than scattered. **B** and **D** are incorrect because the ray does not undergo interference in the droplet.

17 **C.** The resistance of metal conductors generally increases as they are heated due to the increased motion of lattice atoms, causing more collisions between the conduction-electron and lattice atoms. **A** is incorrect because an ohmic resistor has a constant resistance and hence we would expect 200 mA to flow in the resistor when 6 V was placed across it. **B** is incorrect as more current would flow at 6 V, not less, if this was the case. **D** is incorrect because these would be expected.

18 **A.** $I = V/R$ and hence adding the current in each of the parallel resistors will give the total current. $I_T = V/R_1 + V/R_2 + V/R_3 = 6/R_1 + 6/R_1 + 6/2R_1 = 15/R_1$. **B**, **C** and **D** are incorrect as they all have the wrong values of current.

19 **D.** This is an example of Newton's third law of motion. The tug boat pulls the barge and the barge exerts an equal-reaction force on the tugboat. The boats accelerate because the tugboat exerts more force on the water than the water exerts on the tugboat and barge. **A** and **B** are incorrect as they violate Newton's third law. **C** is incorrect because there would be no net force on the barge and hence it would not be accelerating.

20 **A.** Assuming the current flows clockwise around the circuit and applying Kirchhoff's loop rule gives $9 - 6I - 3 - 3I = 0$ and hence $I = \frac{2}{3}$ A. As the current is positive it must have been flowing in a clockwise direction. **B** is incorrect as it has the wrong current direction; **C** and **D** are incorrect as they have the wrong numerical answer.

Extended-response questions

21 EM The question tests students' understanding of graphing motion, the difference between distance and displacement, and the meaning of average velocity.

a Gradient = acceleration and the maximum gradient is between $t = 0$ and $t = 10$ s

Hence $a_{max} = \frac{60}{10} = 6\ ms^{-2}$ ✓ north ✓

b Displacement = distance if the velocity is in one direction. Now the area under the graph = displacement. Thus the distance = area under graph = 300 + 300 + 450 = 1050 m. ✓

c Displacement = area under graph between $t = 0$ s and $t = 55$ s, thus $s = 1050 - 500 = 550$ m ✓ north ✓

d Average velocity = total displacement/total time

$= \frac{550}{55} = 10\ ms^{-1}$ ✓ north

22 EM This question tests students' understanding of relative motion in two dimensions and motion in a moving medium.

a $v_{relative} = v_{object} - v_{observer}$. The vector subtraction is shown in Figure EA2.1.

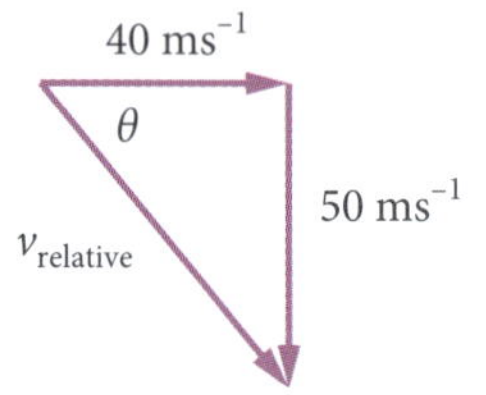

Figure EA2.1 Vector graph

$\theta = \tan^{-1}\left(\frac{50}{40}\right) = 51.3°$; bearing will be given by $90° + 51.3° = 141.3°$

$v_{relative} = \sqrt{40^2 + 50^2} = 64\ ms^{-1}$ ✓ at N141.3° ✓

b In this case we add the velocities, as shown in Figure EA2.2.

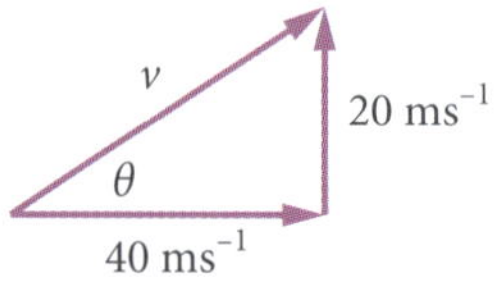

Figure EA2.2 Vector graph

$\theta = \tan^{-1}\left(\frac{20}{40}\right) = 26.6°$ or as a bearing N63.4°

$v = \sqrt{20^2 + 40^2} = 44.7\ ms^{-1}$ ✓ at N63.4° ✓

23 EM This question tests students' understanding of equations of motion, Newton's second law, two-dimensional force diagrams, and friction and power.

a We must first convert the speed to SI units:

$300\ \text{kmh}^{-1} = \dfrac{300 \times 1000}{60 \times 60} = 83.3\ \text{ms}^{-1}$ ✓

Now $v^2 = 2as + u^2$ and as $u = 0$, $a = v^2/2s = \dfrac{83.3^2}{2 \times 400}$

$= 8.68\ \text{ms}^{-2}$ east ✓

b $F = ma = 1500 \times 8.67 = 1.3 \times 10^4$ N east ✓

c Figure EA2.3 illustrates the forces on the car.

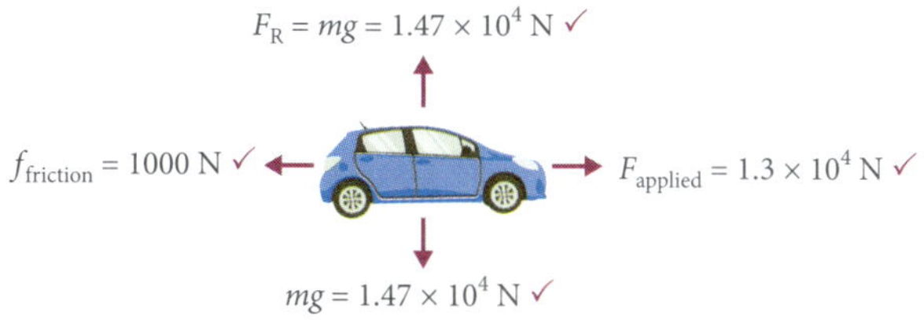

Figure EA2.3 Force on a car

d The applied force will have to be equal to the frictional force to ensure there is no net force on the car. Hence the applied force is 1000 N and the power $P = Fv = 1000 \times 83.3 = 8.33 \times 10^4$ W. ✓

24 EM This question tests students' understanding of wave graphs and the relationship between frequency and wavelength.

a Amplitude = 20 Vm^{-1} ✓

b One complete cycle = λ = 200 m ✓

c $v = f\lambda$ and hence $f = v/\lambda$ ✓ $= \dfrac{3 \times 10^8}{200} = 1.5 \times 10^6$ Hz or 1.5 MHz ✓

25 EM This question tests students' understanding of resonance, standing waves on strings and the wave equation.

a Figure EA2.4 illustrates the waves on the guitar strings.

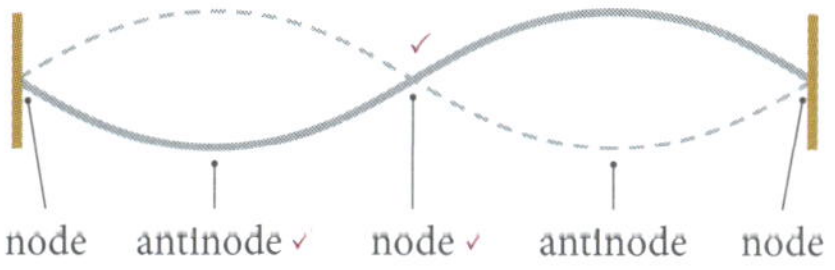

Figure EA2.4 Waves on the guitar string

b The length of the string represents one complete wavelength when it is oscillating in the 2nd harmonic and hence $\lambda = 40$ cm, ✓ and $v = f\lambda = 400 \times 0.4 = 160\ \text{ms}^{-1}$. ✓

26 EM This question tests students' understanding of ray-tracing and image formation in lenses.

a The action through the lens is illustrated in Figure EA2.5.

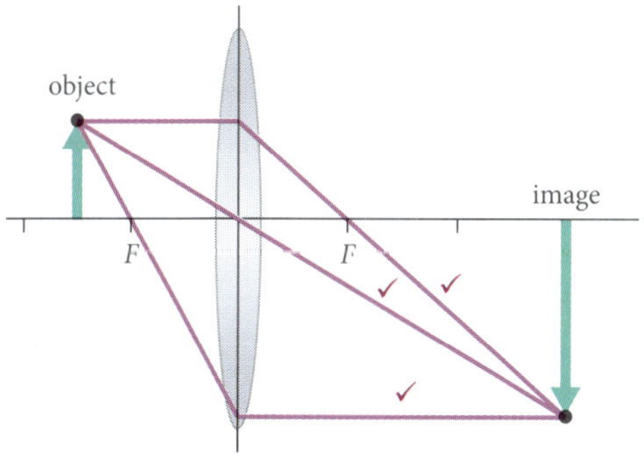

Figure EA2.5 Tracing the rays through the lens

b From Figure A2.5 we see the image is (i) real, ✓ (ii) inverted ✓ and (iii) magnified. ✓

27 EM This question tests students' understanding of specific heat capacity, latent heat capacity and heat.

Energy to heat 200 g of ice from −20 °C to 0 °C =
$Q = mC\Delta T = 0.2 \times 2108 \times 20 = 8432$ J

Energy to melt 200 g of ice = $Q = mL = 0.2 \times 3.34 \times 10^5$
$= 6.68 \times 10^4$ J ✓

Energy to raise the temperature of 200 g of water from 0 °C to 90 °C:
$Q = mC\Delta T = 0.2 \times 90 \times 4186 = 7.53 \times 10^4$ J ✓

Total heat required = $8432 + 6.68 \times 10^4 + 7.53 \times 10^4 = 1.5 \times 10^5$ J ✓

28 EM This question tests students' understanding of Coulomb's law, weight force and vector addition in two dimensions.

a If we call the separation of the charges r, $\sin 10° = (r/2)/(0.2)$ and hence:

$r = (0.2 \sin 10° \times 2) = 6.9 \times 10^{-2}$m = 6.95 cm ✓

b There are three forces on each ball: the force of gravity, the tension in the string and the electrical repulsion. As these forces cause the string holding the ball to make an angle of 10° to the vertical, we can use a vector diagram of the forces to find the unknown force.

From Figure EA2.6:

$F = mg \tan 10° = 0.005 \times 9.8 \times \tan 10°$
$= 8.64 \times 10^{-3}$ N ✓

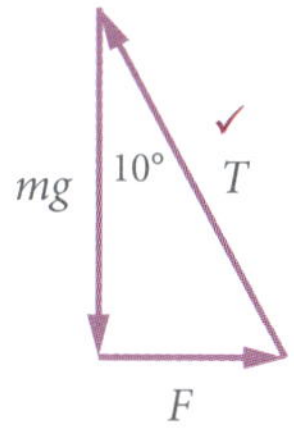

Figure EA2.6 Vector diagram

c As $q_1 = q_2$, $F = (1/4\pi\varepsilon_0)q^2/r^2$ and hence:

$q = \sqrt{Fr^2/(1/4\pi\varepsilon_0)} = \sqrt{\dfrac{(8.64 \times 10^{-3})(6.95 \times 10^{-2})^2}{9 \times 10^9}}$

$q = 6.8 \times 10^{-8}$ C ✓

d The new horizontal force on each ball would be $4 \times 8.64 \times 10^{-3}$ = 0.0346 N and a vector diagram of this force and the weight will enable us to find the new angle (as shown in Figure EA2.7).

From Figure EA2.7:
$\tan\theta = F/mg$ and hence

$\theta = \tan^{-1}\left(\dfrac{0.0346}{0.005 \times 9.8}\right) = 35.2°$ ✓

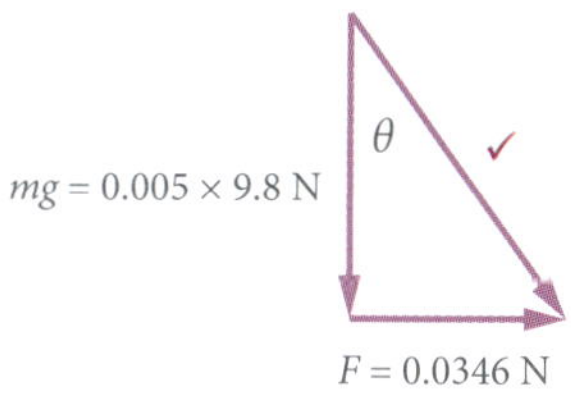

Figure EA2.7 Vector diagram

29 EM This question tests students' understanding of impulse, work, kinetic energy and potential energy, and conservation of energy.

a Gradient = $\dfrac{40}{0.4} = 100\ \text{Nm}^{-1}$ ✓

b The gradient tells us how much force it would take to draw the string of the bow back 1 m. ✓ That is, the gradient tells us how hard it is to draw back the string of this bow.

c Area under the curve = 40 × 0.4 × 0.5 = 8 Nm or 8 J ✓

d If energy is conserved all the stored energy will go into the kinetic energy of the arrow. ✓

Thus work = $W = \frac{1}{2}mv^2$ and hence $v = \sqrt{\dfrac{2W}{m}} = \sqrt{\dfrac{2 \times 8}{0.02}} = 28.3\ \text{ms}^{-1}$ ✓

e Initially all the energy would be potential energy stored in the bow. When the arrow is fired all this elastic potential energy is transformed into kinetic energy in the arrow. ✓ As the arrow rises this kinetic energy will be converted into gravitational potential energy until, at the top of the flight, all the kinetic energy will have been transformed into gravitational potential energy. ✓ As the arrow falls down the gravitational potential energy will be converted into kinetic energy.

At the instant it reaches the firing point, all the potential energy will have been transformed into kinetic energy. ✓

30 **EM** This question tests students' understanding of Snell's law, refractive index, graphing physical relationships and dispersion.

a Gradient = $\frac{1}{0.62} = 1.6$ ✓

As Snell's law states, $\sin i/\sin r = v_1/v_2$, the gradient ($\sin r/\sin i$) is equal to $v_2/v_1 = n$. ✓

b Now as the gradient is equal to v_2/v_1 and $v_2 = 3 \times 10^8$, the speed of light in the glass is:

$v_1 = v_2/\text{gradient} = \frac{3\times10^8}{1.6} = 1.9 \times 10^8$ ms^{-1} ✓

The refractive index of the glass is $n = c/v_1$ thus the gradient of the graph = 1.6. ✓

c Blue light also obeys Snell's law but it travels more slowly than red light in glass. Hence the graph for blue light would be a straight line but with a steeper gradient ✓ (i.e. blue will have a higher refractive index in the glass).

31 **EM** This question tests students' understanding of Ohm's law and resistors in series and in parallel.

a $I = V/R = \frac{10}{80} = 0.125$ A ✓

b First we find the current in the 20 Ω resistor:

$I = V/R = \frac{10}{20} = 0.5$ A

Hence the 10 Ω resistor is sum of the currents in the parallel resistors = 0.5 + 0.125 = 0.625 A. ✓

c The potential difference across each 10 Ω resistor is $V = IR$ = 0.625 × 10 = 6.25 V. ✓ Hence the total potential difference across the circuit is the sum of the potentials around the circuit:

$V = 6.25 + 6.25 + 10 = 22.5$ V ✓

d First calculate the resistance of the parallel resistors:

$1/R = 1/R_1 + 1/R_2 = \frac{1}{20} + \frac{1}{80}$ and hence $R = 16$ Ω ✓

The total resistance for the circuit is therefore:

$R_T = 5 + 16 + 5 = 26$ Ω ✓

32 **EM** This question tests students' understanding of Oersted's law, magnetic fields and the superposition of fields.

a Figure EA2.8 illustrates the magnetic fields.

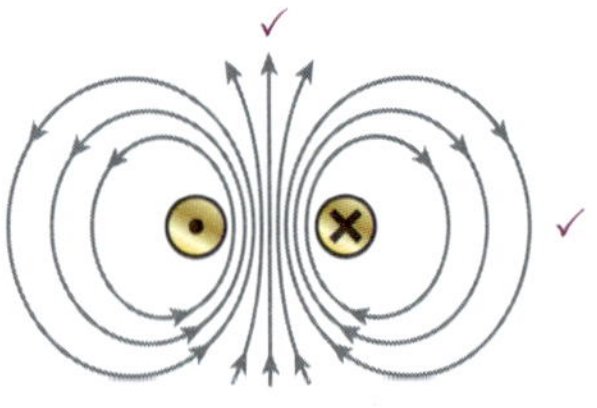

Figure EA2.8 Magnetic fields

b The magnetic field from both wires is in the same direction between the wires and, midway between the wires, the net magnetic field will be twice the field strength from one wire (B). ✓

Hence $B_{net} = 2B = 2(\mu_0/2\pi)(I/r) = 2(2 \times 10^{-7})\left(\frac{5}{0.05}\right)$

$= 4 \times 10^{-5}$ T up the page ✓

c If the current in the wire was doubled, the magnetic field produced by the wire would also double. If we call the initial field produced by this wire B, the new magnetic field would be $2B$. If the current in the wire was reversed, the direction of the magnetic field between the wires produced by the wire would be downwards, which would mean it would be in the opposite direction to the field produced by the other wire. The net magnetic field between the wires would therefore be given by $2B - B = B$. Thus the net field would be:

$B = (2 \times 10^{-7})\left(\frac{5}{0.05}\right) = 2 \times 10^{-5}$ T ✓ down the page ✓

33 **EM** This question tests students' understanding of electric fields, forces on a charge in a magnetic field, Newton's second law, and work and potential difference in an electric field.

a $E = V/d = \frac{100}{0.04} = 2500$ Vm^{-1} downwards ✓

b $F = qE = 1.6 \times 10^{-19} \times 2500 = 4 \times 10^{-16}$ N downwards ✓

c $a = F/m = \frac{4\times10^{-16}}{1.67\times10^{-27}} = 2.4 \times 10^{11}$ ms^{-2} downwards ✓

We can neglect gravity as it is ten orders of magnitude smaller than the acceleration due to the electric field. ✓

d Using the equations of motion with $u = 0$:

$s = \frac{1}{2}at^2 + ut$ and hence $t = \sqrt{\frac{2s}{a}} = \sqrt{\frac{2\times0.04}{2.4\times10^{11}}} = 5.77 \times 10^{-7}$ s ✓

$v = at + u = 2.4 \times 10^{11} \times 5.77 \times 10^{-7} = 1.38 \times 10^5$ ms^{-1} ✓

e We can calculate this several ways: work = Fs, or work = K gained, or work = $qV = 1.6 \times 10^{-19} \times 100$

$= 1.6 \times 10^{-17}$ J ✓

34 **EM** This question tests students' understanding of conservation of momentum in two dimensions, elastic collisions and impulse.

a Adding the momentum of the two skaters:

$p_1 = mv = 50 \times 2 = 100$ Ns west and

$p_2 = mv = 50 \times 1 = 50$ Ns south

The addition of these momentum vectors is shown in Figure EA2.9.

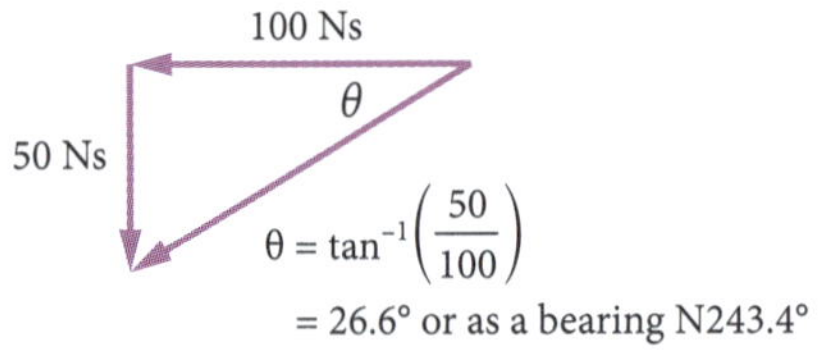

Figure EA2.9 Vector diagram

From Figure A2.9, the sum of momentum

$= \sqrt{50^2 + 100^2} = 112$ Ns ✓ at N243.4° ✓

b Here we can use conservation of momentum. Because the initial momentum of the skaters was zero, the pushed skater must have a final momentum equal in magnitude and opposite in direction to the momentum of the other skaters after they had pushed her. This will ensure that the sum of the momentum of the skaters after the interaction is zero.

Hence the momentum of the 60 kg skater is:

$p = 112$ Ns N63.4° ✓

Now as $p = mv$, $v = p/m = \frac{112}{60} = 1.9$ ms^{-1} at N63.4° ✓

c $F = \Delta p/t = \frac{112}{0.5} = 224$ N at N63.4° ✓

d Kinetic energy must be conserved in an elastic collision. The interaction of the skaters is therefore not an elastic collision because the skaters were originally at rest with no kinetic energy. However, after the interaction, all the skaters are moving and hence have kinetic energy. ✓

INDEX

FORMULAE SHEET

Motion, forces and gravity

$$s = ut + \frac{1}{2}at^2$$

$$v^2 = u^2 + 2as$$

$$\Delta U = mg\Delta h$$

$$P = \frac{\Delta E}{\Delta t}$$

$$\sum \frac{1}{2}mv^2_{\text{before}} = \sum \frac{1}{2}mv^2_{\text{after}}$$

$$\Delta \vec{p} = \vec{F}_{\text{net}}\Delta t$$

$$\omega = \frac{\Delta\theta}{t}$$

$$\tau = r_{\perp}F = rF\sin\theta$$

$$v = \frac{2\pi r}{T}$$

$$U = -\frac{GMm}{r}$$

$$v = u + at$$

$$\vec{F}_{\text{net}} = m\vec{a}$$

$$W = F_{||}s = Fs\cos\theta$$

$$K = \frac{1}{2}mv^2$$

$$P = F_{||}v = Fv\cos\theta$$

$$\sum m\vec{v}_{\text{before}} = \sum m\vec{v}_{\text{after}}$$

$$a_{\text{c}} = \frac{v^2}{r}$$

$$F_{\text{c}} = \frac{mv^2}{r}$$

$$F = \frac{GMm}{r^2}$$

$$\frac{r^3}{T^2} = \frac{GM}{4\pi^2}$$

Waves and thermodynamics

$$v = f\lambda$$

$$f = \frac{1}{T}$$

$$d\sin\theta = m\lambda$$

$$n_{\text{x}} = \frac{c}{v_{\text{x}}}$$

$$I = I_{\text{max}}\cos^2\theta$$

$$Q = mc\Delta T$$

$$f_{\text{beat}} = \left|f_2 - f_1\right|$$

$$f' = f\frac{(v_{\text{wave}} + v_{\text{observer}})}{(v_{\text{wave}} - v_{\text{source}})}$$

$$n_1\sin\theta_1 = n_2\sin\theta_2$$

$$\sin\theta_{\text{c}} = \frac{n_2}{n_1}$$

$$I_1r_1^2 = I_2r_2^2$$

$$\frac{Q}{t} = \frac{kA\Delta T}{d}$$

FORMULAE SHEET *(continued)*

Electricity and magnetism

$$E = \frac{V}{d}$$

$$V = \frac{\Delta U}{q}$$

$$W = qV$$

$$W = qEd$$

$$B = \frac{\mu_0 I}{2\pi r}$$

$$B = \frac{\mu_0 N I}{L}$$

$$\Phi = B_{||} A = BA\cos\theta$$

$$\varepsilon = -N\frac{\Delta\Phi}{\Delta t}$$

$$\frac{V_\text{p}}{V_\text{s}} = \frac{N_\text{p}}{N_\text{s}}$$

$$\vec{F} = q\vec{E}$$

$$F = \frac{1}{4\pi\varepsilon_0}\frac{q_1 q_2}{r^2}$$

$$I = \frac{q}{t}$$

$$V = IR$$

$$P = VI$$

$$F = qv_\perp B = qvB\sin\theta$$

$$F = lI_\perp B = lIB\sin\theta$$

$$\frac{F}{l} = \frac{\mu_0}{2\pi}\frac{I_1 I_2}{r}$$

$$\tau = nIA_\perp B = nIAB\sin\theta$$

$$V_\text{p} I_\text{p} = V_\text{s} I_\text{s}$$

Quantum, special relativity and nuclear

$$\lambda = \frac{h}{mv}$$

$$K_\text{max} = hf - \phi$$

$$\lambda_\text{max} = \frac{b}{T}$$

$$E = mc^2$$

$$E = hf$$

$$\frac{1}{\lambda} = R\left(\frac{1}{n_\text{f}^2} - \frac{1}{n_\text{i}^2}\right)$$

$$t = \frac{t_0}{\sqrt{\left(1-\frac{v^2}{c^2}\right)}}$$

$$l = l_0\sqrt{\left(1-\frac{v^2}{c^2}\right)}$$

$$p_\text{v} = \frac{m_0 v}{\sqrt{\left(1-\frac{v^2}{c^2}\right)}}$$

$$N_\text{t} = N_0 e^{-\lambda t}$$

$$\lambda = \frac{\ln 2}{t_{\frac{1}{2}}}$$

Reprinted 2019, 2022, 2024, 2026

ISBN 978 1 74125 677 2

Pascal Press
PO Box 250
Glebe NSW 2037
www.pascalpress.com.au

Publisher: Vivienne Joannou
Project editor: Mark Dixon
Edited by Karen Enkelaar
Answers checked by Adam Sloan
Indexed by Puddingburn Publishing Services
Cover, page design and typesetting by Sonia Woo
Printed by Vivar Printing/Green Giant Press

Students
All care has been taken in the preparation of this study guide, but please check with your teacher or the NSW Education Standards Authority about the exact requirements of the course you are studying as these can change from year to year.

The validity and appropriateness of the internet addresses (URLs) in this book were checked at the time of publication. Due to the dynamic nature of the internet, the publisher cannot accept responsibility for the continued validity or content of these web addresses.